Healthcare Systems Engineering

Healthcare Systems Engineering

Paul M. Griffin
Harriet B. Nembhard
Christopher J. DeFlitch
Nathaniel D. Bastian
Hyojung Kang
David A. Muñoz

WILEY

This book is printed on acid-free paper. ♾

Published by John Wiley & Sons, Inc., Hoboken, New Jersey.
Published simultaneously in Canada.

For general information about our other products and services, please contact our Customer Care Department within the United States at (800) 762-2974, outside the United States at (317) 572-3993 or fax (317) 572-4002.

Wiley publishes in a variety of print and electronic formats and by print-on-demand. Some material included with standard print versions of this book may not be included in e-books or in print-on-demand. If this book refers to media such as a CD or DVD that is not included in the version you purchased, you may download this material at http://booksupport.wiley.com. For more information about Wiley products, visit www.wiley.com.

Library of Congress Cataloging-in-Publication Data is available:

ISBN 9781118971086 (Hardcover)
ISBN 9781118971109 (ePDF)
ISBN 9781118971093 (ePub)

Cover design: Wiley
Cover image: Blue tech background © Godruma/iStockphoto

Printed in the United States of America

SKY10028090_071221

Contents

Acknowledgments xi

Chapter 1 The Healthcare Delivery System 1

Overview 1
1.1 Healthcare Delivery Components 2
1.2 Major Stakeholders 6
1.3 Global Issues in Health 6
1.4 Drivers for Healthcare Systems 13
Questions and Learning Activities 22
References 23

Chapter 2 Complexity and Systems in Healthcare 25

Overview 25
2.1 Taking a Systems Approach to Healthcare 25
2.2 Complex Adaptive Systems 28
Case Study 2A: Complexity in Chronic Kidney Disease 31
2.3 Systems Thinking and System Dynamics 34
Case Study 2B: Systems Thinking and Causal Loop Diagrams in CKD 40
Case Study 2C: System Dynamics and Stock Flow Diagrams in CKD 45

Questions and Learning Activities 48
References 49

Chapter 3 Patient Flow 53

Overview 53
3.1 Healthcare Settings and Clinical Workflows 53
3.2 Patient Flow through a Hospital 64
3.3 Care Transitions 71
3.4 Process Mapping 75
3.5 Queuing 78
Case Study 3: ED Crowding – A Patient Flow Solution 80
Questions and Learning Activities 83
References 83

Chapter 4 Healthcare Financing 91

Overview 91
4.1 Financing Models for Health Services 91
4.2 Compensation Models for Providers 95
4.3 Cost Allocation and Charges 97
4.4 Capital Budgeting 103
Questions and Learning Activities 107
References 108

Chapter 5 Health Data and Informatics 111

Overview 111
5.1 Healthcare Data 116
5.2 Electronic Health Records 121
5.3 Health Information Exchange 130
5.4 Publicly Reported Healthcare Data 132
Case Study 5: Health Informatics at a Hospital—A 10-Year Journey 136
Questions and Learning Activities 139
References 140

Chapter 6 Lean 141

Overview 141
6.1 Lean Philosophy and Methods 141
6.2 Drivers for Lean Healthcare Systems 143
6.3 A Toolset for Eliminating Wastes 146
6.4 Value Stream Mapping 148
6.5 A3 150
6.6 5S 157
6.7 Kanban 160
6.8 Lean Implementations 164
6.9 Lean Thinking 165
Questions and Learning Activities 166
References 166

Chapter 7 Six Sigma 169

Overview 169
7.1 Six Sigma Philosophy 169
7.2 Six Sigma Quality 170
Case Study 7A: Quantifying Complexity in Translational Research 201
Case Study 7B: Evaluating Collaboration in Translational Research 205
Case Study 7C: Resource Allocation in Translational Research 208
Questions and Learning Activities 210
References 212

Chapter 8 Reliability and Patient Safety 217

Overview 217
8.1 Human Reliability 217
8.2 Errors in Healthcare 218
8.3 Medication Errors 220
8.4 Patient Falls 222

8.5 Human Factors and Ergonomics for Patient Safety 227
Questions and Learning Activities 236
References 237

Chapter 9 Health Analytics 245

Overview 245
9.1 Data Mining 245
Case Study 9A: Predicting Parkinson's Disease Using Data Mining 251
9.2 Data Visualization 255
Case Study 9B: Data Visualization in Obesity Counseling 269
9.3 Social Network Analysis 271
Case Study 9C: SNA in a Pediatric Intensive Care Unit (PICU) 277
9.4 Data Envelopment Analysis 280
Case Study 9D: Using DEA for Finding Best-in-Class Hospitals in a Network 285
9.5 Multicriteria Decision Making 287
Case Study 9E: Use of GP to Optimize Helicopter Emplacement at Medical Treatment Facilities 291
Questions and Learning Activities 292
References 293

Chapter 10 Capacity Management 297

Overview 297
10.1 Capacity Management Challenges 297
10.2 Managing Nursing Units 299
10.3 Managing Operating Rooms 310
10.4 Managing Diagnostic Units 316
10.5 Nurse Staffing and Scheduling 317
Questions and Learning Activities 320
References 321

Chapter 11 Healthcare Logistics 323

Overview 323
11.1 Facility Location 323
Case Study 11: Location of Federally Qualified Healthcare Centers 337
11.2 Home Healthcare Routing and Scheduling 341
Questions and Learning Activities 347
References 349

Chapter 12 Health Supply Chains 351

Overview 351
12.1 Forecasting Demand 351
12.2 Inventory Control 357
12.3 Healthcare Distribution 363
12.4 Coordinating Activities in the Supply Chain 369
Questions and Learning Activities 375
References 377

Chapter 13 Infection Control 379

Overview 379
13.1 Historical Perspective 379
13.2 Infection Control Classification 381
13.3 Checklists for Infection Control 383
13.4 The Case of Sepsis 385
13.5 Mathematical Modeling of Hospital Infection Control 388
Case Study 13: Impact of Interventions on TB Infection in a Clinic 398
Questions and Learning Activities 402
References 403

Index 405

Acknowledgments

This book evolved after a series of annual workshops by the Penn State Center for Integrated Healthcare Delivery Systems, projects funded by the National Science Foundation Center for Health Organization Transformation (ICURC 1067885), and course development for IE 568 Healthcare Systems Engineering. These outlets allowed us to explore and develop several aspects of this emerging domain.

We would like to acknowledge and thank all of the colleagues, faculty, staff, and students who have been a part of these endeavors, especially Michael Beck, Diane Brannon, Beth Colledge, Yining Chen, Cheng Chi, William Curry, Nasr Ghahramani, Xuemei Huang, Marija Jankovic, Mehmet Kilinc, Min-Jung Kim, Lisa Korman, Jennifer Kraschnewski, Hyunji Lee, Deirdre McCaughey, Colleen Rafferty, Madhu Reddy, Jaideep Sood, Conrad Tucker, Monifa Vaughn-Cooke, Steven Wagman, Beatrice Winkler, Renfei (Iris) Yan, and Sai Zhang.

We are deeply grateful to Liz Welker for her help in editing this book and developing some of the artwork.

The encouragement and funding support provided by Charles Schneider through the Service Enterprise Engineering 360 initiative is very much appreciated, as is the support of Virginia and Joseph Mello.

The people at Wiley made this endeavor a reality. We want to acknowledge and thank our editor, Amanda Shettleton, and also Margaret Cummins, Nanda Gopal, and Michael New.

Healthcare Systems Engineering

Chapter **1**

The Healthcare Delivery System

"In nothing do men more nearly approach the gods than in giving health to men."

—Cicero

Overview

Health care (or **healthcare**) is the maintenance or restoration of the human body by the treatment and prevention of disease, injury, illness and other physical and mental impairments. Healthcare is delivered by trained and licensed professionals in medicine, nursing, dentistry, pharmacy, and other allied health providers. The quality and accessibility of healthcare varies across countries and is heavily influenced by the *health policies* in place. It is also and dependent on demographics, social and economic conditions.

A health system (healthcare system or health care system) is organized to facilitate the delivery of care. The World Health Organization (WHO) defines health systems as follows:

> A health system consists of all organizations, people and actions whose primary intent is to promote, restore or maintain health. This includes efforts to influence determinants of health as well as more direct health-improving activities. A health system is therefore more than the pyramid of publicly owned facilities that deliver personal health services. It includes, for example, a mother caring for a sick child at home; private providers; behavior change programs; vector-control campaigns; health insurance organizations; occupational health and safety legislation. It includes inter-sectoral action by health staff, for example, encouraging the ministry of education to promote female education, a well-known determinant of better health. (*Everybody's Business: Strengthening Health Systems to Improve Health Outcomes. WHO's Framework for Action*, 2007)

WHO goes on to say that:

> A good health system delivers quality services to all people, when and where they need them. The exact configuration of services varies from country to country, but in all cases requires a robust financing mechanism; a well-trained and adequately paid workforce; reliable information on which to base decisions and policies; well-maintained facilities and logistics to deliver quality medicines and technologies. ("World Health Organization. Health Systems," n.d.)

1.1 HEALTHCARE DELIVERY COMPONENTS

The delivery of healthcare to a patient population depends on the systematic provision of services. WHO suggests that "People-centered and integrated health services are critical for reaching universal health coverage. *People-centered care* is care that is focused and organized around the health needs and expectations of people and communities, rather than on diseases. Whereas *patient-centered care* is commonly understood as focusing on the individual seeking care (the patient), people-centered care encompasses these clinical encounters and also includes attention to the health of people in their communities and their crucial role in shaping health policy and health services. Integrated health services encompass the management and delivery of quality and safe health services so that people receive a continuum of health promotion, disease prevention, diagnosis, treatment, disease-management, rehabilitation and palliative care services, through the different levels and sites of care within the health system, and according to their needs throughout the life course."

Table 1.1 summarizes the major types of levels and sites of care components and gives some examples of providers and the conditions they address. While there is no universal definition of each type, there is some consensus in usage (except where specifically noted). Improvement of the healthcare system will depend on the provider professionals performing as a team that can act and influence patients as they may transition from one care delivery mode to another.

While Table 1.1 shows delivery types as distinct, in practice there is often overlap and intersection. Primary care can be delivered in urgent care settings (e.g., walk-in clinics). Emergency rooms may often be the *de facto* provider of primary care. Similarly, quaternary care may be an extension of tertiary care.

The *International Classification of Primary Care*, Second Edition (ICPC-2), is a reference (accepted by WHO) that allows classification of

Table 1.1 Delivery of Healthcare Services

Type	Delivery Focus	Providers	Conditions/Needs
Primary care	• Day-to-day healthcare • Often the first point of consultation for patients	• Primary care physician, general practitioner, or family or internal medicine physician • Pediatrician • Dentist • Physician assistant • Nurse practitioner • Physiotherapist • Registered nurse • Clinical officer • Ayurvedic	• Routine check-ups • Immunizations • Preventive care • Health education • Asthma • Chronic obstructive pulmonary disease • Diabetes • Arthritis • Thyroid dysfunction • Hypertension • Vaccinations • Oral health • Basic maternal and child care
Urgent care	• Treatment of acute and chronic illness and injury provided in a dedicated walk-in clinic • For injuries or illnesses requiring immediate or urgent care but not serious enough to warrant an ER visit • Typically do not offer surgical services	• Family medicine physician • Emergency medicine physician • Physician assistant • Registered nurse • Nurse practitioner	• Broken bones • Back pain • Heat exhaustion • Insect bites and stings • Burns • Sunburns • Ear infection • Physicals

(continued)

Table 1.1 Delivery of Healthcare Services (*Continued*)

Type	Delivery Focus	Providers	Conditions/Needs
Ambulatory or outpatient care	• Consultation, treatment, or intervention on an outpatient basis (medical office, outpatient surgery center, or ambulance) • Typically does not require an overnight stay	• Internal medicine physician • Endoscopy nurse • Medical technician • Paramedic	• Urinary tract infection • Colonoscopy • Carpal tunnel syndrome • Stabilize patient for transport
Secondary or acute care	• Medical specialties typically needed for advanced or acute conditions including hospital emergency room visits • Typically not the first contact with patients; usually referred by primary care physicians	• Emergency medicine physician • Cardiologist • Urologist • Dermatologist • Psychiatrist • Clinical psychologist • Gynecologist and obstetrician • Rehabilitative therapist (physical, occupational, and speech)	• Emergency medical care • Acute coronary syndrome • Cardiomyopathy • Bladder stones • Prostate cancer • Women's health
Tertiary care	• Specialized highly technical healthcare usually for inpatients • Usually patients are referred to this level of care from primary or secondary care personnel	• Surgeon (cardiac, orthopedic, brain, plastic, transplant, etc.) • Anesthesiologist • Neonatal nurse practitioner • Ventricular assist device coordinator	• Cancer management • Cardiac surgery • Orthopedic surgery • Neurosurgery • Plastic surgery • Transplant surgery • Premature birth • Palliative care • Severe burn treatment

Table 1.1 Delivery of Healthcare Services (*Continued*)

Type	Delivery Focus	Providers	Conditions/Needs
Quaternary care	• Advanced levels of medicine that are highly specialized and not widely accessed • Experimental medicine • Typically available only in a limited number of academic health centers	• Neurologist • Ophthalmologist • Hematologist • Immunologist • Oncologist • Virologist	• Multi-drug-resistant tuberculosis • Liver cirrhosis • Psoriasis • Lupus • Myocarditis • Gastric cancer • Multiple myeloma • Ulcerative colitis
Home and community care	• Professional care in residential and community settings • End-of-life care (hospice and palliative)	• Medical director (physician) • Registered nurse • Licensed practical nurse • Certified nursing assistant • Social worker • Dietitian or nutritionist • Physical, occupational, and speech therapists	• Post-acute care • Disease management teaching • Long-term care • Skilled nursing facility/assisted living • Behavioral and/or substance use disorders • Rehabilitation using prosthesis, orthotics, or wheelchairs

the patient's reason for encounter (RFE) with primary care or general care ICPC-2). The classification structure addresses the problems or symptoms/complaints, infection, injuries, diagnosis managed, and interventions. It also codes processes such as medical exams, laboratory tests, and how the encounter was initiated (e.g., by a provider or other person), referrals to physician/specialist, referrals to a clinic/hospital. A simplified two-page version is available that makes it conducive for use by a range of medical providers. A systematic review of the literature on ICPC showed that it has been used with the greatest frequency in the Netherlands, Australia, United States, Norway, United Kingdom, and France (Mariana et al., 2009). As the tool becomes more widespread, it may also become a source of data on the reason for healthcare delivery consultation from the perspective of the patient.

1.2 MAJOR STAKEHOLDERS

There are many stakeholders in the healthcare system, including patients, caregivers, healthcare providers, insurers, and institutions, as well as employers and regulators. Major stakeholders are outlined in the Table 1.2 which is from the Agency for Healthcare Research & Quality (AHRQ).

As illustrated in Table 1.2, different stakeholders play different roles and have different needs and desires from the healthcare system. Often, these perspectives may be in conflict; e.g., some pharmaceutical companies may want to pursue a profit-maximizing strategy while some policy makers may want to increase access. Further, there are asymmetries in information between the parties, for example, in the provider-patient relationship. At the end of the day, however, developing approaches that can build partnership and collaboration as well as improving communication between the various stakeholders will be essential to fully realize value-based healthcare. This is clearly demonstrated in the Institute for Healthcare Improvement's access-quality-cost triangle.

1.3 GLOBAL ISSUES IN HEALTH

Healthcare varies significantly by country. This includes how healthcare is financed, who is covered, what services are delivered, and the corresponding health outcomes from the system. We discuss each of these below.

Global Spending

As will be discussed in Chapter 4, healthcare is financed in many different ways, ranging from private insurance to universal coverage. Further, the amount of spending is quite different by country. Figure 1.1 provides data on some of the Organization for Economic Cooperation and Development (OECD) countries. In 2011, the United States spent $8,508 per capita (in U.S. dollars) while New Zealand spent $3,182 (in U.S. dollars, accounting for purchasing power parity). According to the World Bank

Table 1.2 Stakeholder Groups

Stakeholders	Stakeholders' Perspective
Consumers, patients, caregivers, and patient advocacy organizations	It is vital that research answer the questions of greatest importance to those experiencing the situation that the research addresses. Which aspects of an illness are of most concern? Which features of a treatment make the most difference? Which kinds of presentation of research results are easiest to understand and act upon?
Clinicians and their professional associations	Clinicians are at the heart of medical decision making. Where is lack of good data about diagnostic or treatment choices causing the most harm to patients? What information is needed to make better recommendations to patients? What evidence is required to support guidelines or practice pathways that would improve the quality of care?
Healthcare institutions, such as hospital systems and medical clinics, and their associations	Many healthcare decisions are structured by the choices of institutional healthcare providers, and institutional healthcare providers often have a broad view of what is causing problems. What information would support better decisions at an institutional level to improve health outcomes?
Purchasers and payers, such as employers and public and private insurers	Coverage by public or private purchasers of healthcare plays a large role in shaping individual decisions about diagnostic and treatment choices. Where does unclear or conflicting evidence cause difficulty in making the decision of what to pay for? Where is new technology or new uses of technology raising questions about what constitutes a standard of care? What research is or could be funded?
Healthcare industry and industry associations	The manufacturers of treatments and devices often have unique information about their products.
Healthcare policymakers at the federal, state, and local levels	Policymakers at all levels want to make healthcare decisions based on the best available evidence about what works well and what does not. Comparative effectiveness research/patient-centered outcomes research can help decision makers plan public health programs, design health insurance coverage, and initiate wellness or advocacy programs that provide people with the best possible information about different healthcare treatment options.
Healthcare researchers and research institutions	Researchers gather and analyze the evidence from multiple sources on currently available treatment options.

Source: Agency for Healthcare Research and Quality, U.S. Department of Health and Human Services, The effective health care program stakeholder guide. http://www.ahrq.gov/research/findings/evidence-based-reports/stakeholderguide/chapter3.html

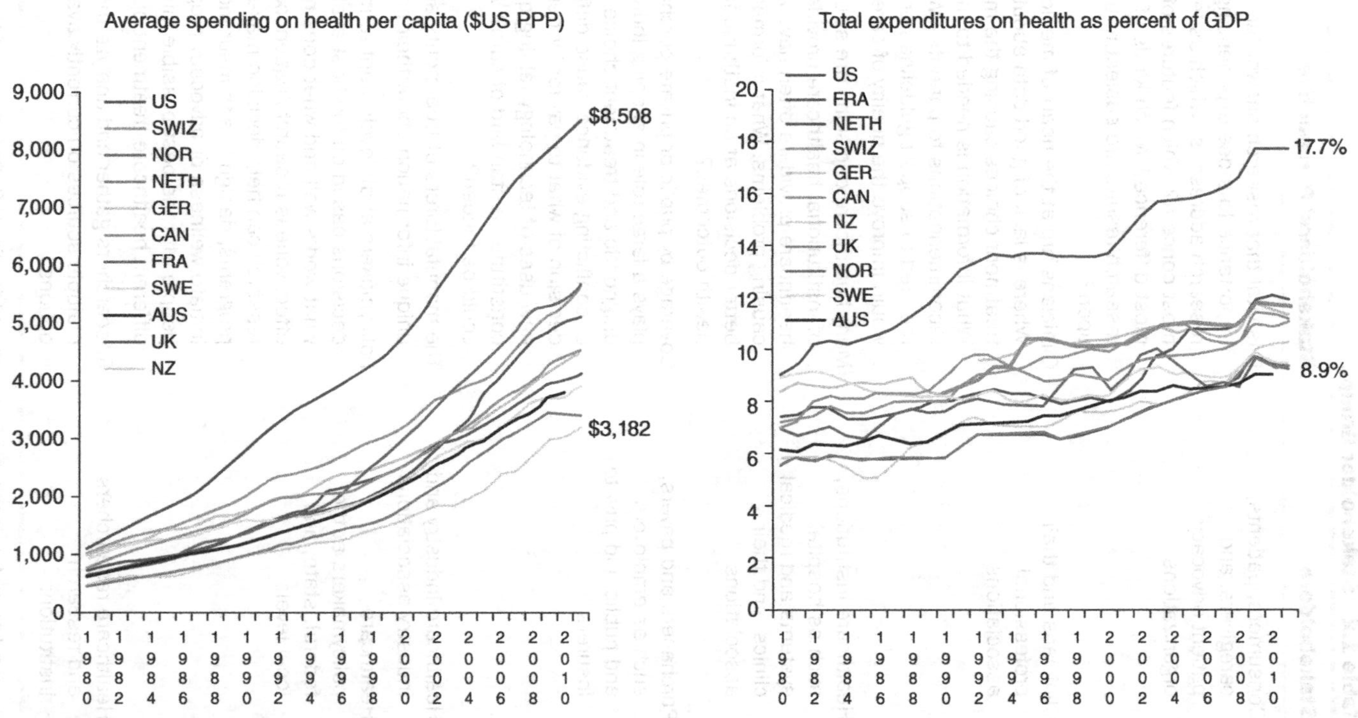

Note: $US PPP - purchasing power party.

Figure 1.1 Comparison of Healthcare Spending for OECD Countries, 1980–2011 ***Source:*** **Commonwealth Fund (2014)**

(2015), the country with the lowest healthcare expenditures in 2011 as a percentage of gross domestic product (GDP) was Timor-Lest (0.7%), while the highest was Tuvalu (18.5%), with the United States coming in second place (17.7%). Further, in Tuvalu 99.9% of the total was public spending. This value was 47.1% for the United States, and the global average was 59.6%.

Spending in and of itself is not the best measure of healthcare for a country. What is important is the value that is received as a result of the spending, that is, the resulting health outcomes.

Global Outcomes

There are several outcomes that are commonly used as a measure of health, including life expectancy at birth by gender, malnutrition prevalence, and infant mortality rate. Although healthcare spending per person in the United States was more than double that in New Zealand, New Zealand performed better on all three outcomes (infant mortality rate of 5% compared to 6%, life expectancy at birth for females of 83 versus 81, and malnutrition prevalence of 0% compared to 0.5%). Among the higher income countries, the United States performed poorly on most measures compared to its peers.

There is little agreement, however, on what the best outcome measures are, and thus it proves difficult to directly compare healthcare systems. For example, in the United States, many have argued that the ability to choose healthcare providers is highly valued. Further, the United States pays much higher prices for prescription drugs compared to other countries due to government laws that protect the special interests of the pharmaceutical industry. These kinds of issues are not necessarily a reflection of inefficiency in the healthcare system.

A report that compares OECD countries was released by the Commonwealth Fund (2014). In this comparison, five classes of outcomes were used: quality care, access to care, efficiency, equity, and healthy lives (details of the measures are found in the report). The results of the study are shown in Figure 1.2. The United Kingdom ranked first in eight of the measures, and had the lowest cost per capita in the group; it was rated overall as the best healthcare system. The United States ranked worst in the comparison in spite of the much higher rate of spending. The authors of the study argue that a key reason for the poor performance by the United States is the lack of universal health insurance. The lack of insurance coverage is a primary driver of lack of *access* and lack of *equity*. Another key reason stated

COUNTRY RANKINGS

Top 2*
Middle
Bottom 2*

	AUS	CAN	FRA	GER	NETH	NZ	NOR	SWE	SWIZ	UK	US
OVERALL RANKING (2013)	4	10	9	5	5	7	7	3	2	1	11
Quality Care	2	9	8	7	5	4	11	10	3	1	5
Effective Care	4	7	9	6	5	2	11	10	8	1	3
Safe Care	3	10	2	6	7	9	11	5	4	1	7
Coordinated Care	4	8	9	10	5	2	7	11	3	1	6
Patient-Centered Care	5	8	10	7	3	6	11	9	2	1	4
Access	8	9	11	2	4	7	6	4	2	1	9
Cost-Related Problem	9	5	10	4	8	6	3	1	7	1	11
Timeliness of Care	6	11	10	4	2	7	8	9	1	3	5
Efficiency	4	10	8	9	7	3	4	2	6	1	11
Equity	5	9	7	4	8	10	6	1	2	2	11
Healthy Lives	4	8	1	7	5	9	6	2	3	10	11
Health Expenditures/Capita, 2011**	**$3,800**	**$4,522**	**$4,118**	**$4,495**	**$5,099**	**$3,182**	**$5,669**	**$3,925**	**$5,643**	**$3,405**	**$8,508**

Notes: * Includes ties. ** Expenditures shown in $US PPP (purchasing power parity): Australian $ data are from 2010.

Figure 1.2 OECD County Health Rankings *Source:* Commonwealth Fund (2014)

is the United States is lagging behind other countries in the sophistication of the health information system, which makes coordinated care difficult to achieve. The United States also has high levels of chronic conditions including diabetes, obesity, and congestive heart failure and hence scores low in *health lives.*

The Economist (2014) performed a 166-country health outcome report. Figure 1.3 shows a plot of ranking based on health outcomes versus ranking on healthcare spending. The outcome measure was a function of life expectancy at age 60, adult mortality in 2012, disability-adjusted life years (a measure of years of life lost due to poor health), and health-adjusted life expectancy. They found that health outcomes (and hence ranking) were correlated with health spending. Further, they found several regional differences. For example, Asia, Europe, and North America make up the top tier; Latin America, the Middle East, and former Soviet countries make up the middle tier; and the lower tier was made up almost exclusively of African countries. Japan, Singapore, and South Korea

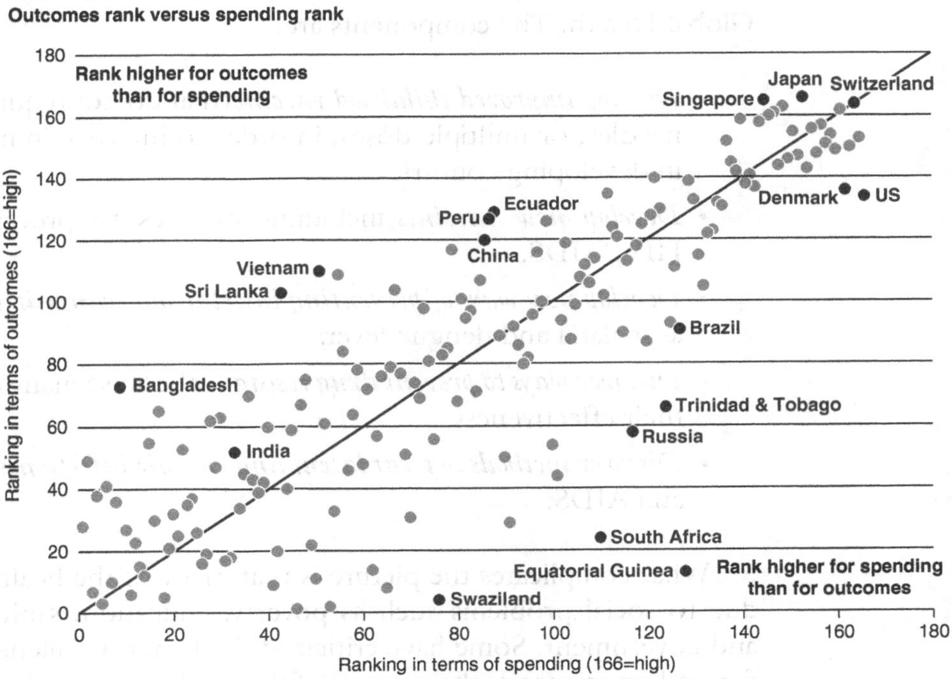

Figure 1.3 Health Outcomes Rank versus Spending Rank by Country
***Source: The Economist* (2014)**

performed well in outcomes per spending, and the United States was a poor-value healthcare system (33rd on outcomes index).

Unique Challenges

One of the more troubling aspects of global health is the growing gaps in health outcomes. For example, the WHO World Health Report (2013) states that 35% of African children were at higher risk of death in 2013 compared to 2003. African adults above 30 have a higher death rate than they had 30 years ago. HIV/AIDS is killing 5,000 persons daily in the 15- to 59-year-old age group (and 1,000 children daily below the age of 15) in sub-Saharan Africa. In fact, HIV/AIDs is responsible for 60% of all child deaths in Africa. Life expectancy increased globally by roughly four months per year from 1955 to 2002, but the gap between developed and developing countries also grew over this range. Further, in 2002, over 10 million children (5 years or younger) died; 98% of these deaths occurred in developing countries.

As a response, the Gates Foundation launched Grand Challenges in Global Health. The components are:

- *Develop improved childhood vaccines* that do not require refrigeration, needles, or multiple doses, in order to improve immunization rates in developing countries.
- *Develop new vaccines,* including vaccines to prevent malaria and HIV/AIDS.
- *Develop new ways of preventing insects from transmitting diseases* such as malaria and dengue fever.
- *Discover ways to prevent drug resistance* because many drugs are losing their effectiveness.
- *Discover methods to treat latent and chronic infections* such as hepatitis and AIDS.

What complicates the picture is that many of the health outcomes are due to social problems such as poverty, education, sanitation, housing, and government. Some have criticized the Grand Challenges as being too focused on science at the expense of these other issues, as well as being too narrowly focused on HIV, malaria, and tuberculosis. It also ignores the delivery and resource allocation issues. In response, the Grand Challenges are updated regularly (e.g., a current focus on women and girls).

The Centers for Disease Control and Prevention (CDC, 2011) released a Healthy People 2020 Report that discusses approaches to improve global health outcomes. They emphasize the importance of global disease detection, response, prevention, and control strategies. They also stress the importance of quickly responding to infections disease threats (e.g., severe acute respiratory syndrome [SARS], Ebola) as well as real-time infectious disease surveillance. Specific chronic conditions called out in the report are diabetes and obesity, mental illness, substance abuse (including tobacco use), and injuries.

It is clear that global health presents many unique challenges. Much of it involves improving access to care and reducing the cost of care. However, it is also important for these changes to be considered in concert with the social issues of primary education, extreme poverty, effective governments, shelter, and clean water and sanitation.

1.4 DRIVERS FOR HEALTHCARE SYSTEMS

There are several important drivers needed to improve healthcare delivery. These include appropriate financing mechanisms, improving access to a primary source of care, and continued advances in technology. Although not an exhaustive list, in this section we discuss the most important of them.

Financial

High costs are one of the most frequently cited barriers for effective health-care delivery. Several factors contribute to these costs including advances in technology, population aging, incentives, the price of prescription drugs, and the wealth of the country. The health industry is somewhat unique in that prices tend to increase with technological advances. In comparison, advances in manufacturing technology bring the costs of production down, which are then passed on to the consumer. In healthcare, technological advances can help to increase life expectancy (which bring a corresponding demand), but they can also simply be more expensive, with little or no additional efficacy. Proton therapy for prostate cancer is one such example. It costs over twice the amount of standard radiation therapy, although there has not been shown to be an increase in efficacy. In spite of this, there was a 67% increase in the number of cases paid by Medicare between 2006 and 2009 (Jarosek et al., 2012).

Much of healthcare spending occurs at the end of life. In 2006 in the United States, for example, Medicare spent on average $38,975 per descendant compared to $5,993 per survivor. The Centers for Medicare & Medicaid Services (CMS) estimates that 27% to 30% of total Medicare spending goes to the 5% of beneficiaries who die each year. Elderly patients are also more likely to have serious chronic conditions. Part of the challenge is helping patients and their families to make the most appropriate choices of care. This includes better ways to explain risks and outcomes of medical procedures. In addition, there is currently little internalization of the costs by the patient or family in many cases. Both of these issues can lead to unnecessary, ineffective, or unwanted treatments.

Drug prices differ significantly by country and for some can be a significant burden. The United States pays the highest drug prices in the world, which have doubled in the past decade. In 2012, 11 of the 12 drugs approved by the Food and Drug Administration (FDA) had a cost of over $100,000 per year (Experts in Chronic Myeloid Leukemia, 2013). Some of the high price is due to the cost of bringing a new drug to market, which includes research and development and extensive clinical trials. However, much of the reason for high drug prices in the United States is simply due to government policy. According to Alpern, Stauffer, and Kesselheim (2014), many firms are taking advantage of laws that require insurers to include expensive drugs in their coverage. Further, they can buy the rights to inexpensive generics and block out competitors. One example is a drug for parasite infection (albendazole), which sold for $5.92 per day in 1996 when it was developed. Currently, the price is $119.58 per day.

Several other reasons may also contribute to high costs, including the overuse of specialty care, rising administrative costs, physician fees, and malpractice costs. Government policy, consumer demand, and market incentives all play a strong and interconnected role in defining costs. Developing a sustainable financing model that provides value-based medicine is of utmost importance; this may be unique for each country. We discuss different financing models in Chapter 4.

The Dartmouth Atlas for Healthcare has documented significant geographic differences in healthcare costs, with no significant differences in health outcomes. The conclusion is that there can be significant healthcare operational inefficiencies that lead to these high costs. Focusing on identifying and removing these inefficiencies may also be of importance in reducing costs.

Population Health and Wellness

Historically, people have not incurred a significant component of the cost of their behaviors, including smoking, excessive drinking, or eating unhealthy foods. Many have argued that this has led to an increase in chronic conditions. Perhaps the most commonly mentioned condition is obesity. Roughly 10% of all medical spending in the United States is due to obesity (Finkelstein et al., 2009). It is estimated that by 2018, 43% of Americans will be obese and the resulting healthcare costs will quadruple.

Of course, obesity is not the only chronic condition from behavioral choices. There are over 6 million deaths annually attributable to smoking. The CDC estimates that in the United States, over $300 billion of annual medical costs (including productivity loss) is due to smoking. They also estimate that the cost of excessive drinking in the United States costs over $220 billion each year.

In order to encourage people to be more involved in their health, several types of *wellness programs* have been developed. The most common is when an employer or insurance provider gives rewards, typically financial, for weight loss, smoking cessation, or diabetes management. This can come in the form of subsidized gym memberships, time off during the day to work out, or cash. Alternatively, there can be a penalty for behavior. For example, if you are smoker, then a "penalty" is assessed by the provider. For example, a smoker may need to pay a $300/year penalty each year to obtain coverage. The support of penalty is typically not only for the employee, but also for the employee's family.

A study done by Berry, Mirabito, and Baun (2010) showed a return of $2.71 to the employer for each $1 invested in the program. RAND (2013) also found significant improvements among participants in smoking cessation and weight reduction/control, but not in cholesterol control. Further, the number of wellness programs is growing, and it is generally believed that properly constructed wellness programs in general have a positive impact. Over half of U.S. employers currently offer some type of wellness plan. Some of the stated keys to success stated by RAND are clear messaging, easy access to wellness activities, and making it a strategic priority.

Equity

WHO defines *equity* as

> the absence of avoidable or remediable differences among groups of people, whether those groups are defined socially, economically, demographically, or geographically. *Health inequities* therefore involve more than inequality with respect to health determinants, access to the resources needed to improve and maintain health or health outcomes. They also entail a failure to avoid or overcome inequalities that infringe on fairness and human rights norms.

Similar to the case of health outcomes, there is no agreed-upon method for measuring equity in health. This is an extremely important issue. Limited resource allocation decisions are made based on these measures, and it is essential that they be given to the appropriate need.

One approach developed by Reidpath and Allotay (2009) used disability-adjusted life years (DALYs) as the key health outcome measure. Gross national product (GNP) was used as the measure of population wealth. Equity is defined as DALYs per capita weighted by the per capita GNP. The key conclusion is that it isn't enough to look for health inequalities. Economic factors also need to be considered, since wealthier countries tend to have much better health infrastructure compared to their poor counterparts.

By any measure, there are large health inequities across the globe. For example, WHO estimates that life expectancy in Malawi is 47 years compared to 83 years in Japan. Further, Norway has 40 physicians per 10,000 persons, while Myanmar has 4 physicians per 10,000. Inequities tend to be larger in cities and are highly related to education, employment, and income. They also vary significantly by gender and race/ethnicity.

So why is equity in health so important? An excellent report by Margaret Whitehead (2000) summarizes this as well as any. She argues that:

- There is consistent evidence that disadvantaged groups have a poorer survival chance.
- Large gaps in mortality can also be seen between urban and rural populations and between different regions in the same country.
- There are great differences in the experience of illness. Disadvantaged groups not only suffer a heavier burden of illness than others but also experience the onset of chronic illness and disability at younger ages.
- Other dimensions of health and well-being show a similar pattern of blighted quality of life.

It is worth mentioning that although some inequalities in healthcare may be unavoidable (someone living in a warmer region is more likely to get malaria than someone living in a very cold region), the notion of equity implies that the differences that exist can be changed, and that there is a moral and ethical responsibility to do so.

Quality—First, Do No Harm

Although not a part of the Hippocratic Oath, a phrase taught to almost every medical student is "first, do no harm." In other words, no matter what we do in healthcare delivery, our primary concern is that none of our actions should harm the patient. The term *harm*, however, is a controversial one. For example, extending a person's life may be considered a harm if procedures are given that the patient didn't want.

As an example of patient harms, let's consider the condition of sepsis. Septic patients take up approximately 25% of intensive care unit (ICU) bed capacity, making up over a million hospitalizations annually in the United States. Early recognition, treatment, and management of sepsis can significantly improve outcomes. For example, survival rates decrease by 7.6% for each hour of delay in antimicrobial administration at the onset of septic shock. The efficient and effective transfer of sepsis patients into and out of the ICU is a key component of reducing patient harms. The slow transfer of patients into the ICU has been shown to lead to increased morbidity and mortality. Each hour of delay into the ICU increases ICU mortality by 8%, and patients with certain high-risk vital signs (e.g., critical cardiac arrest risk triage score [CART]) delayed by 18 to 24 hours were found to have a 52% mortality rate in the ICU, significantly higher than their nondelayed counterparts. Unexpected events during ICU transfers are common, occurring 67% of the time. These include equipment errors (39%), patient/staff management issues (61%), and serious adverse outcomes (31%), including major physiological derangement (15%) and death (2%). Communication lapses are also common during patient handoff and over shift changes due in large part to increased memory load at those transitions. These lapses include medication errors, omission of pending tests, and lack of responsibility handoff.

Quality programs have been developed in almost all hospitals with the goal of improving patient safety and reducing patient harms. The Institute of Medicine (IOM) defines quality this way: "Quality is the extent to which health services for individuals and populations increase the likelihood of desired health outcomes and are consistent with current professional knowledge" (Institute of Medicine, 2008). Many other valid definitions of quality relate and build on this one.

The quality and accessibility of healthcare varies across countries and is heavily influenced by the health policies in place. It is also dependent on demographics, social, and economic conditions. Several factors have placed increased importance on quality programs. These include the increase in many parts of the world of hospital-acquired infections, the increase in country interconnectedness that leads to faster spread of infectious disease, and the increase in obesity and aging and the corresponding increase in hospital falls.

Technological advances such as tracking systems and information dashboards can provide information in rapid fashion to aid in a more timely response that helps to reduce harms. However, simple but well-defined processes where everyone knows their role can also be extremely helpful. Examples include the use of hand washing programs, increasing visibility of patients from nursing stations, and checklists. One of the most famous examples is the intensive care checklist protocol developed by Pronovost (2006). It was estimated that over 18 months, this simple intervention saved the state of Michigan 1,500 lives and $100 million.

Electronic Health Records

The *electronic health record* (EHR), also called the electronic medical record [EMR] is in its simplest form a digital version of the paper charts in the clinician's office. However, EHRs now include a broad range of information that covers the total health of the patient in real time and securely. In the United States, the passage of the Health Information Technology for Economic and Clinical Health Act (HITECH Act) in 2009 helped to initiate the adoption of the EHR and supporting technology. Prior to 2009, only 20% of physicians were utilizing electronic patient records.

The IOM (2008) defined 12 key attributes of an EHR:

1. Provides active and inactive problem lists for each encounter that link to orders and results; meets documentation and coding standards.
2. Incorporates accepted measures to support health status and functional levels.
3. Ability to document clinical decision information; automates, tracks, and shares clinical decision process/rationale with other caregivers.

4. Provides longitudinal and timely linkages with other pertinent records.
5. Guarantees confidentiality, privacy, and audit trails.
6. Provides continuous authorized user access.
7. Supports simultaneous user views.
8. Access to local and remote information.
9. Facilitates clinical problem solving.
10. Supports direct entry by physicians.
11. Cost measuring/quality assurance.
12. Supports existing/evolving clinical specialty needs.

Software related to the EHR is the *practice management system* (PMS), which manages administrative and financial information. This includes patient insurance information, patient scheduling, and billing. In addition, there can be a patient portal (PP), which provides online services to the patient. This may include online scheduling, prescription refills, and clinical information on patient visits to the provider. In order to encourage EHR adoption, a Meaningful Use program was put in place that authorizes CMS to provide incentive payments to hospitals that implement or upgrade EHR and can demonstrate how it is used in a significant (or meaningful) way.

According to the Healthcare Information and Management Systems Society (HIMSS, 2010), England has been the biggest investor in EHR. Further, the Asia Pacific region is the largest growth region, but the major barrier to global adoption is cost.

There are several potential benefits from EHR adoption. These include reductions in human and medical errors, a more streamlined workflow for the clinician, better patient tracking over time, and easier information access. However, in addition to cost, there are other important challenges for adoption. First and foremost is interoperability, that is, the ability of information technology systems and software to communicate and exchange data. Key to addressing interoperability is the establishment of standards. Other important issues are security of the data and privacy concerns.

Successful implementation of EHR has the potential to transform healthcare delivery by increasing the connectivity between components that allows for coordinated care. It can also help improve patient participation in their healthcare through records access. It is clear that global adoption, however, will take significant time and effort.

Point of Care

In many countries, there is limited capacity for healthcare resources such as the emergency department (ED) and ICU. Overcrowding of these resources can lead to poor health outcomes for patients, increased length of stay, and increased costs. In many cases, the overcrowding may be due to overuse. EDs provide a full range of services, regardless of a patient's ability to pay. There is interest, therefore, in moving the point of care for the patient to an appropriate source.

At one extreme is to make the patient the point of care through the use of devices, sensors, applications, and information technology. Consider the following hypothetical case: Luka and her parents were alerted that she had asthma through a balloon inflation game at school. Her air quality is monitored through a wearable patch in her shirt, and she is assisted in taking her medicine with reminders from her phone and reports to her physician. Dosing is personalized based on patch results and a sensor built into her respirator that measures lung capacity and compares results to his historical baselines. Although realization of this scenario would require significant advances, particularly on the information technology component, it would greatly reduce ED visits by Luka (note that asthma is one of the greater reasons for ED visits among children) and provide her with better outcomes through tailored and coordinated care.

Telehealth (or telemedicine) is another enabling technology for patient-centered point of care. It allows for diagnosis and management of conditions, and can effectively support patient education. Telehealth can use a variety of technologies, including video, remote monitoring, and smartphone. Telehealth has been shown to be effective in several different studies. For example, telehealth interventions were found to be effective for individuals' self-care of heart failure (Radhakrishnan & Jacelon, 2012).

Medical tourism occurs when a patient seeks care in another country. This can occur when patients in less-developed countries seek services from a more developed country that they don't have access to in their home country. More recently, however, tourism has occurred when patients in developed countries seek services at a lower price. An industry of health tourism has developed in order to serve as the intermediary. In some cases, geographic regions have developed around a particular industry.

For example, the border town of Los Algodones in Mexico has focused on dental tourism. In a population of 5,500, there are 350 dentists. Several supporting dental labs have also located there. The result is prices that are less than one-third of the corresponding American prices.

Personalized Medicine

Advances in technology have allowed for customization of care to the individual. This is known as *personalized medicine* (also known as *precision medicine*). The FDA has defined personalized medicine as providing "the right patient with the right drug at the right dose at the right time." However, it can be more broadly defined as tailoring all stages of care (prevention, diagnosis, treatment, and follow-up) to the individual.

An illustration of personalized medicine is in the use of baseline comparisons. In traditional medicine, population statistics from clinical trials and other studies are used to establish baseline conditions (blood pressure, A1c levels, body mass index, low-density lipoprotein (LDL) cholesterol, etc.). If a patient has a test of his LDL cholesterol, for example, he may be categorized as having a low, medium, or high level. Patients with a high level may be prescribed a drug to help bring the level down since there has been an association found between LDL cholesterol and cardiovascular disease. Whether a patient is classified as "at risk" is based on population studies. However, these statistics are based on averages and are typically not stratified by specific characteristics of the patient. The prescription, therefore, may not actually help the patient. A recent paper in *Nature* (Schork, 2015) looked at the top 10 highest-grossing drugs in the United States, and found that they help only between 1 in 4 and 1 in 25 of the patients who take them. Crestor, for example, which is the most commonly prescribed drug for cholesterol, helps only 1 in 20 patients who take it.

Advances in information technology, including big and wide data, along with new devices have allowed for the inclusion of data that are specific to the individual, including their genetics, the environment in which they live, and real-time sensing of patient data. These allow for the move from general clinical trials to individual trials (called *N-of-1 trials*). Advances in genetic testing and genome sequencing have greatly helped to move the field.

The following case illustrates the promise of personalized medicine (McMullan, 2015):

> In 2005 Stephanie Haney, now 45, had a pain on her right side that wouldn't go away. It hurt when she coughed or sneezed. She was pregnant, so she didn't investigate the cause, assuming perhaps she'd broken a rib. Two years later, she was diagnosed with stage 4 lung cancer. After undergoing chemotherapy, Haney began taking Tarceva (erlotinib) in 2008. But three years later, the drug was no longer keeping the tumors at bay. Prompted by friends and an insistent doctor, she had genetic testing on her tumors, which showed they were ALK (anaplastic lymphoma kinase) positive. This gave her doctor a major clue as to which drugs were most likely to work (or not). Haney was able to start taking Xalkori (crizotinib), designed specifically for ALK-positive lung cancer tumors. She joined a clinical trial for Xalkori in Philadelphia, two and a half hours away. Three years later, her tumors were barely visible.

In order for personalized medicine to be fully successful, considerable advances need to be made in the EHR, since there will be massive amounts of data that will need to be managed and analyzed. Further, there are still many issues to be worked through, including privacy and data ownership. Finally, it will require the coordination of efforts across providers to collect and share data.

QUESTIONS AND LEARNING ACTIVITIES

1. Briefly review the state of healthcare in any country or region in the 1800s and trace its history to present day. Consider, for example, what has happened with medical schools, hospitals, health insurance, pharmaceuticals, and medical equipment over the past 200+ years.
2. What are some of the most common reasons for accessing a physician in any country or region?
3. What are some of the most common reasons for accessing a physician in another country (i.e., what is referred to as "medical tourism")?
4. Compare healthcare delivery systems between two countries considering factors such as healthcare quality, access, efficiency, and equity.
5. Investigate the congruence between different healthcare ranking systems, such as those used by the World Health Organization, the Commonwealth Fund, or others.
6. Map relationships between stakeholders in the healthcare system and identify points of conflict.

REFERENCES

Ackoff, R. L. *Re-creating the Corporation: A Design of Organizations for the 21st Century*. Oxford University Press, 1999.

Alpern, J. D., Stauffer, W. M., & Kesselheim, A. S. (2014). High-cost generic drugs: Implications for patients and policymakers. *New England Journal of Medicine, 371*(20), 1859–1862.

Berry, L., Mirabito, A. M., & Baun, W. B. (2010). What's the hard return on employee wellness programs? *Harvard Business Review, 88*(12), 104–112.

Centers for Disease Control and Prevention (2011). Healthy people 2020. Retrieved from http://www.healthypeople.gov/2020/topics-objectives/topic/global-health

Experts in Chronic Myeloid Leukemia. (2013). The price of drugs for chronic myeloid leukemia (CML) is a reflection of the unsustainable prices of cancer drugs: From the perspective of a large group of CML experts. *Blood, 121*(22), 4439–4442.

Everybody's Business: Strengthening Health Systems to Improve Health Outcomes. *WHO's Framework for Action*. (2007). Retrieved from http://www.who.int/healthsystems/strategy/everybodys_business.pdf

Finkelstein, E. A., Trogdon, J. G., Cohen, J. W., Dietz, W. (2009). Annual Medical Spending Attributable to Obesity: Payer- and Service-specific Estimates. *Health Affairs, 28*(5): w822–831.

HIMSS (2010). Electronic Health Records—A Global Perspective. Retrieved from: http://www.himss.org/files/HIMSSorg/content/files/Globalpt1-edited%20final.pdf

Institute of Medicine (2008). Committee on Data Standards for Patient Safety: Board of Health Care Services. Key Capabilities of an Electronic Health Record System: Letter Report. Washington, DC: The National Academies Press.

International Classification of Primary Care, Second edition (ICPC-2). (n.d.). Retrieved from http://www.who.int/classifications/icd/adaptations/icpc2/en/

Jarosek, S., Elliot, S., Virnig, B. A. (2012). Proton beam radiotherapy in the U.S. Medicare population: growth in use between 2006 and 2009. Data Point Publication Series, Data Points #10. Retrieved from http://www.ncbi.nlm.nih.gov/books/NBK97147/pdf/Bookshelf_NBK97147.pdf

Maani, K. E., and Cavana, R. Y. "Systems Thinking." *System Dynamics: Managing Change and Complexity (New Zealand: Pearson Education, 2007)* 7 (2007).

Mariana, A., Autran, M., Teresa, M., Almeida, C. G. N. De, Coeli, C. M., Moreno, B., … Brazil, F. (2009). International Classification of Primary Care: A systematic review, 409348.

McMullan, D. (2015). What is personalized medicine? Genome, Retrieved from http://genomemag.com/what-is-personalized-medicine/#.VXSKzs9VhBc

Pronovost, P., Needham, D., Berenholtz, S., et al. (2006). "An intervention to decrease catheter-related bloodstream infections in the ICU." *N. Engl. J. Med. 355*(26): 2725–3.

Radhakrishnan, K., & Jacelon, C. (2012). Impact of telehealth on patient self-management of heart failure: a review of literature. *Journal of Cardiovascular Nursing*, *27*(1), 33–43.

RAND (2013). Workplace Wellness Programs Study. Retrieved from: http://www.rand.org/content/dam/rand/pubs/research_reports/RR200/RR254/RAND_RR254.pdf

Schork, N. J. (2015). Personalized medicine: Time for one-person trials. *Nature 520*(7549): 609–611.

The Commonwealth Fund (2014). Mirror Mirror On the Wall—How the US Healthcare System Compares Internationally. Retrieved from file:///C:/Users/Griffin/Desktop/1755_Davis_mirror_mirror_2014.pdf

The Economist (2015). Health Outcomes and Cost: A 166 Country Comparison. Retrieved from www.eiu.com/healthcare

The World Bank Data (2015). Retrieved from http://data.worldbank.org/

World Health Organization. Health Systems. (n.d.). Retrieved from http://www.who.int/topics/health_systems/en/

World Health Organization (2013). The World Health Report. Retrieved from http://www.who.int/whr/en/

Chapter 2

Complexity and Systems in Healthcare

"The systems that fail are those that rely on the permanency of human nature, and not on its growth and development."

—**Oscar Wilde**

Overview

The word *system* tends to evoke ideas of interrelated parts or components that cooperate in some way. Examples of systems include natural systems such as the ocean currents, the solar system and ecosystems, and designed systems such as mechanical systems, software systems, and social-economic systems. **Complex adaptive systems** is an approach to addressing structures that have to change based on their environment in order to survive. **Systems thinking** is a scientific framework for understanding the change and complexity of a system as an interconnected whole rather than components in isolation. **System dynamics** is mathematical, computer-based modeling for framing, understanding, and discussing the complexity of systems; other approaches include knowledge management, soft systems methodology, and strategic options development and analysis. In this chapter, we focus on complex adaptive systems, systems thinking, and systems dynamics as approaches to studying the dynamic cause and effect in healthcare.

2.1 TAKING A SYSTEMS APPROACH TO HEALTHCARE

A National Academy of Engineering (NAE) and Institute of Medicine (IOM) report entitled *Building a Better Delivery System* (2005) suggested that the main cause of the crises in safety, quality, cost, and access of the American healthcare system can be explained by complexity.

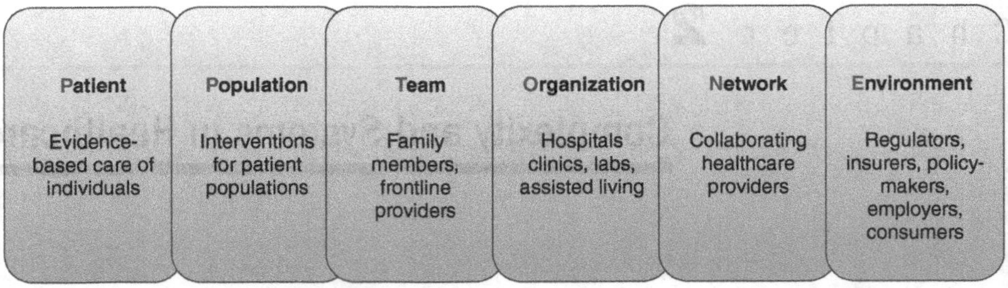

Figure 2.1 Six Levels of the Healthcare System

The complexity of the healthcare system stems from six interconnected levels—**the patient, the population, the team, the organization, the network, and the political and economic environment (PPTONE)** as shown in Figure 2.1. These levels are composed of several components, such as multiple clinicians within a clinic, or multiple clinics in a hospital. In addition to the complex structure of the healthcare system, the dynamic associations between and within levels and their corresponding components yield high degrees of complexity (Rardin, 2007).

"A system is *not* the sum of its parts—it is the product of their *interactions* (Ackoff, 1999). [Furthermore], a system subsumes its parts and can itself be part of a larger system" (Maani & Cavana, 2007). Plsek suggests that a definition of a system is "the coming together or parts, interconnections and purpose" (Plsek, 1996). He also asserts that while systems can be broken down into parts which are interesting in and of themselves, the real power lies in the way the parts come together and are interconnected to fulfill some purpose. As described in Figure 2.1, the healthcare delivery system consists of various levels, for example, the organization level has hospitals, clinics, and laboratories that are interconnected through the flows of patients and information to fulfill the purpose, which is maintaining and improving health. Plsek (1996) explains it this way:

> The intuitive notion of various system "levels," such as the microsystem and macrosystem, has to do with the number and strength of interconnections between the elements of the systems. For example, a doctor's office or clinic can be described as a microsystem. It is small and self-contained, with relatively few interconnections. Patients, physicians, nurses, and office staff interact to produce diagnoses, treatments, and information. In contrast, the health care system in a community is a macro-system. It consists of numerous microsystems (doctor's offices, hospitals, long-term care facilities, pharmacies, Internet websites, and so on) that are linked to provide

continuity and comprehensiveness of care. Similarly, a thermostat and fan comprise a relatively simple microsystem. Combine many of these, along with various boiler, refrigerant, and computer-control micro-systems, and one has a macro-system that can maintain an office building environment.

By taking a systems approach to healthcare, we mean that we can rigorously explore and explain the interactions of the healthcare system using established principles. The *Building a Better Delivery System: A New Engineering/Health Care Partnership* report identified many ways in which systems-engineering tools can be successfully applied to address healthcare-related issues. In follow-up, an Institute of Medicine (IOM)/NAE discussion paper observed:

> ... because health care alone does not necessarily translate to improvement in health, there is a need to integrate all the systems and subsystems that influence health; and separately optimizing each component does not optimize the overall system results. (Kaplan et al., 2013)

At the core, this chapter is concerned with the development and delivery of a better healthcare *system*. The sense-making framework in Figure 2.2

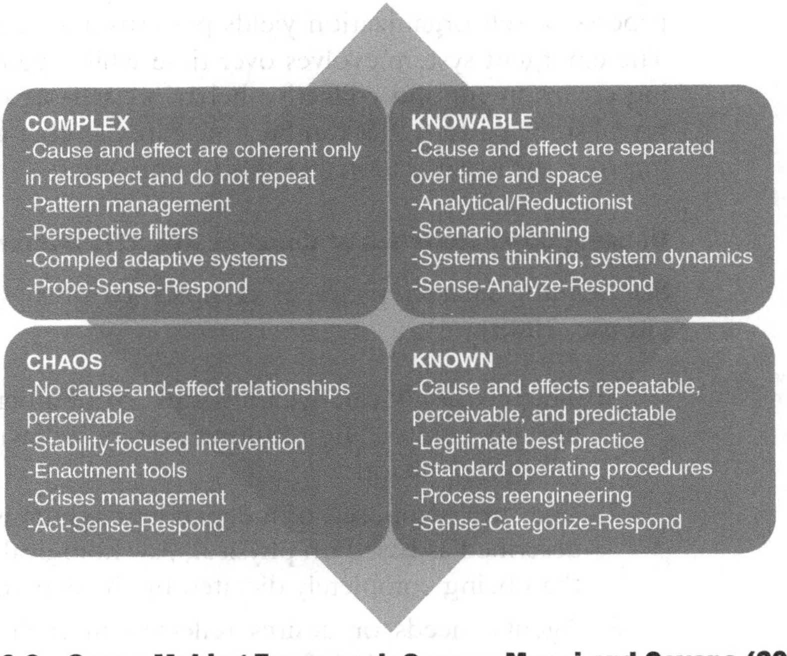
COMPLEX
-Cause and effect are coherent only in retrospect and do not repeat
-Pattern management
-Perspective filters
-Compled adaptive systems
-Probe-Sense-Respond
KNOWABLE
-Cause and effect are separated over time and space
-Analytical/Reductionist
-Scenario planning
-Systems thinking, system dynamics
-Sense-Analyze-Respond
CHAOS
-No cause-and-effect relationships perceivable
-Stability-focused intervention
-Enactment tools
-Crises management
-Act-Sense-Respond
KNOWN
-Cause and effects repeatable, perceivable, and predictable
-Legitimate best practice
-Standard operating procedures
-Process reengineering
-Sense-Categorize-Respond

Figure 2.2 Sense-Making Framework *Source:* Maani and Cavana (2007)

suggests a starting point for consideration about what type of methods and tools should be used for a particular problem or aspect of a problem. Complex adaptive systems are used for complex problems. Systems thinking and system dynamics are used for knowable problems.

2.2 COMPLEX ADAPTIVE SYSTEMS

Complex adaptive systems are a collection of micro-systems that adapt to the environment in order to improve or survive. Typical examples of such systems include ecosystems, social systems, cyberspace, the stock market, and so on. In such systems, the cause and effect are coherent only in retrospect and do not have a fixed one-to-one occurrence. The two key characteristics of complex adaptive systems include self-organization and emergence. There is no single centralized control structure that controls system behaviors. The "agents" in the system have the freedom to respond to inducements in many different and fundamentally unpredictable ways. Therefore, the process where the overall system behavior is formed is governed by any agent inside or outside of the system, arising as a result of the interactions between the elements of the system (self-organization). This process of self-organization yields patterns and regularities (emergence). The emergent system evolves over time while agents adapt to the changing environment and collective behaviors self-organize. Close study may reveal some pattern that can be anticipated so that actions can be taken to improve the organizational or global outcomes.

Defining Characteristics of Complex Adaptive Systems

Six defining characteristics of complex adaptive systems are as follows (Rouse, 2008):

1. They are nonlinear, dynamic, and do not inherently reach fixed equilibrium points. The resulting system behaviors may appear to be random or chaotic.
2. They are composed of independent agents whose behavior can be described as based on physical, psychological, or social rules, rather than being completely dictated by the dynamics of the system.
3. Agents' needs or desires reflected in their rules are not homogeneous, and therefore their goals and behaviors are likely to

conflict—those conflicts or competitions tend to lead agents to adapt to each other's behaviors.

4. Agents are intelligent, learn as they experiment and gain experience, and change behaviors accordingly. Thus, overall system behavior inherently changes over time.
5. Adaptation and learning tends to result in self-organizing and patterns of behavior that emerge rather than being designed into the system. The nature of such emergent behaviors may range from valuable innovations to unfortunate accidents.
6. There is no single point(s) of control—system behaviors are often unpredictable and uncontrollable, and no one is "in charge." Consequently, the behaviors of complex adaptive systems usually can be influenced more than they can be controlled.

The last point bears some amplification. Given that no one is in charge of a complex adaptive system, the best way to approach the management of such systems is with organizational behaviors that embody a collaborative design rather than an authoritarian design. Table 2.1 suggests the perspective of constructive organizational behaviors in complex adaptive systems as compared to other traditional systems (Rouse, 2008).

Complexity in Healthcare

A healthcare system comprising many different elements is inherently complex. In healthcare, the "agents" include human beings, making unexpected behaviors an eventuality. Healthcare systems include diverse

Table 2.1 Comparison of Organizational Behaviors

	Traditional System	Complex Adaptive System
Roles	Management	Leadership
Methods	Command and control	Incentives and inhibitions
Measurement	Activities	Outcomes
Focus	Efficiency	Agility
Relationships	Contractual	Personal commitments
Network	Hierarchy	Heterarchy
Design	Organizational design	Self-organization

Source: Rouse (2008)

agents such as patients, care providers, payers, and other stakeholders. The individual and collective behavior of the diverse agents can manifest itself at all levels. For instance, between the levels of the system, such behavior can lead to positive consequences, such as innovation and breakthroughs, or negative consequences, such as error or waste. Trust can evolve between the providers at the team level as well as between those providers and their patient. Policy created at the environment level can lead to collaborations among units at the organizational level. In other words, *each* level of the healthcare system can be viewed as a complex adaptive system, and *all* levels of the healthcare system can be viewed as a complex adaptive system where the phenomena of interest unfolds in dynamic ways. In order to enhance this complex system, we need an approach that decomposes the complexity while analyzing nonlinear relationships among agents from an integrated perspective. Without an understanding of the system complexity, improvement efforts can lead to unintended consequences.

McDaniel, Lanham, and Anderson (2009) suggest some important studies of complexity in healthcare:

> Health care organizations have been well studied as CAS (Anderson, Issel, & McDaniel, 2003; McDaniel & Driebe, 2001; Miller et al., 1998; Zimmerman et al., 1998). Although no real consensus exists on the set of characteristics that define a CAS, the following set of five key characteristics captures the major concepts from the literature (Beinhocker, 2006; Cilliers, 1998; Maguire, McKelvey, Mirabeau, & Oztas, 2006; Waldrop, 1992): (a) diverse agents that learn, (b) nonlinear interdependencies, (c) self-organization, (d) emergence, and (e) coevolution. We are not attempting in this section to give a deep review of complexity science or CASs theory. Maguire et al. (2006) have done a recent and well-regarded survey of complexity science and organizational studies. McDaniel and Driebe (2001) applied CAS theory to the analysis of HCOs. These works may be consulted by those interested in a more comprehensive treatment of these topics than the one provided here. Health care organizations have diverse agents that learn (Cilliers, 1998) including providers, patients, and other stakeholders. Diversity is often a source of creativity and problem-solving ability (McDaniel & Walls, 1997) but can also be a source of communication difficulties. Learning is not one-dimensional, focusing on uncertainty reduction, but it also incorporates learning aimed at uncertainty absorption (Boisot & Child, 1999). Relationships among agents are usually nonlinear (Capra, 2002; Kauffman, 1995). Outputs may be disproportional to inputs; small inputs can produce large outcomes; and large inputs can produce small outcomes.

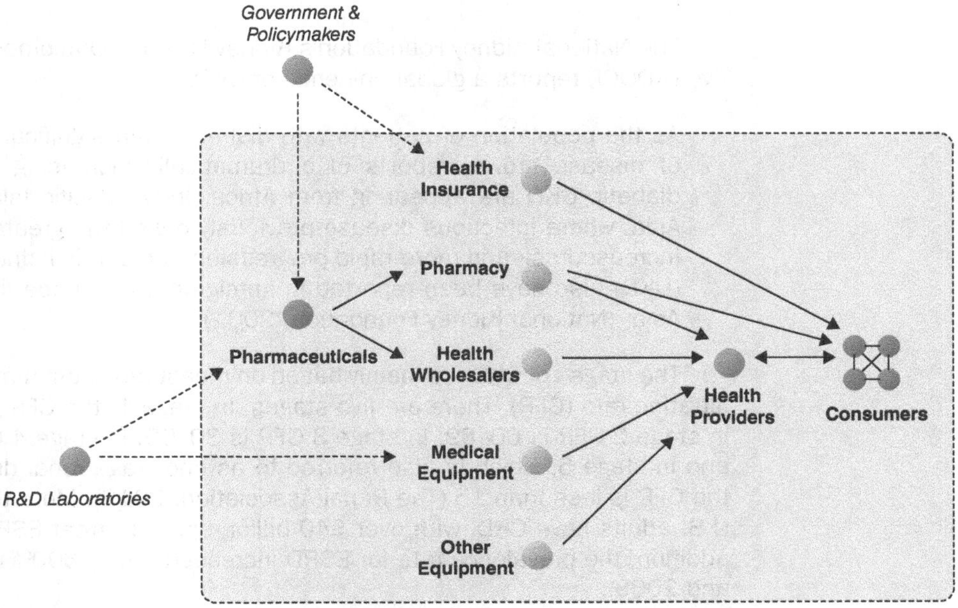

Figure 2.3 Summary of the Complexity of Five Markets in the Healthcare Delivery Network *Source:* Rouse (2008)

Figure 2.3 shows a high-level view of the overall healthcare delivery network based on service value (Basole & Rouse, 2008, Rouse, 2008). The references mentioned here have not been added to the reference list of this chapter. The reader may refer to the original paper for the citations. Each node in the network includes many companies and other types of enterprises. Determining which nodes of the network are involved in any one particular healthcare transaction would involve on the order of 1 billion binary calculations.

CASE STUDY 2A: COMPLEXITY IN CHRONIC KIDNEY DISEASE

A major function of the kidneys is to remove waste products and excess fluid from the body. Chronic kidney disease (CKD) is a condition characterized by a gradual loss of kidney function over time. The most common causes of CKD are diabetes and high blood pressure, and incidences of CKD tend to increase as a result of factors such as obesity and age.

The National Kidney Foundation's Kidney Disease Outcomes Quality Initiative (KDOQI) reports a global epidemic of CKD:

> As the population of patients with diabetes with significant duration of disease grows, reports of a dramatically increasing burden of diabetic CKD are appearing from Africa, India, Pacific Islands, and Asia, where infectious disease previously posed the greatest threat. Increased risk and more rapid progression of diabetic kidney disease (DKD) also have been reported in immigrants to Europe from South Asia. (National Kidney Foundation, 2002)

The stages of CKD are mainly based on measured or estimated glomerular filtration rate (GFR). There are five stages. In stage 1, the GFR is 90 or more; in stage 2 GFR is 60–89; in stage 3 GFR is 30–59; in stage 4 GFR is 15–29; and in stage 5, which is also referred to as end-stage renal disease (ESRD) the GFR is less than 15 (The Renal Association, 2013). More than 20 million U.S. adults have CKD, with over $40 billion spent to treat ESRD in 2009. In addition, the prevalence rate for ESRD increased nearly 600% between 1980 and 2009.

One of the significant problems in combating CKD is that most people do not know they have it. Figure 2.4 shows the percentage of people who reported having CKD by stage among those who participated in the National Health and Nutrition Examination Survey (NHANES).

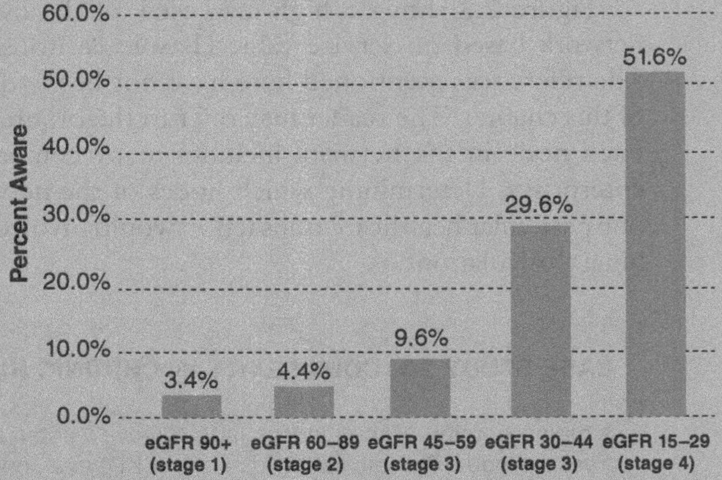

Figure 2.4 People Who Were Aware of Their Disease by eGFR 1999–2012 *Source:* National Health and Nutrition Examination Survey

CKD detection and control can be viewed as a complex adaptive system. Figure 2.5 shows an overall framework for this problem. It emphasizes that screening effectiveness can be increased by medical research, but that is far from the deciding factor on disease detection and control. In the case of CKD, for example, the KDOQI guidelines illuminate the need for primary care physicians to devote greater attention to the care of patients with CKD. Figure 2.6 presents a high-level flowchart of the KDOQI quick guide to evidence-based CKD care for the primary care physician (Fox, Voleti, Khan, Murray, & Vassalotti, 2013).

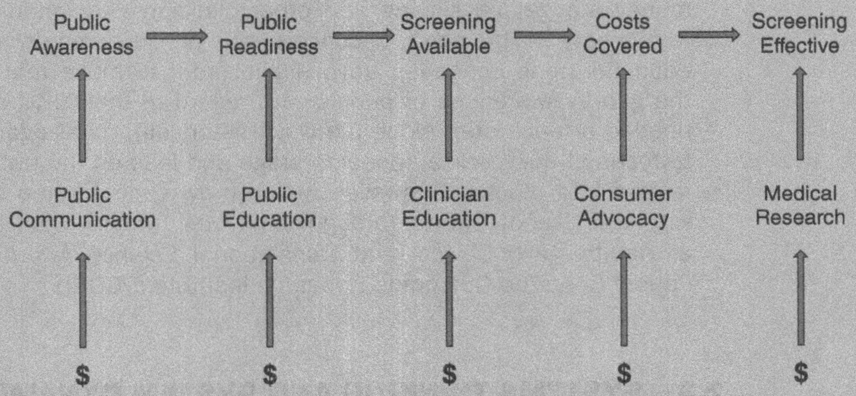

Figure 2.5 A Framework for Disease Detection in a Complex Adaptive System *Source:* Adapted from Rouse (2000)

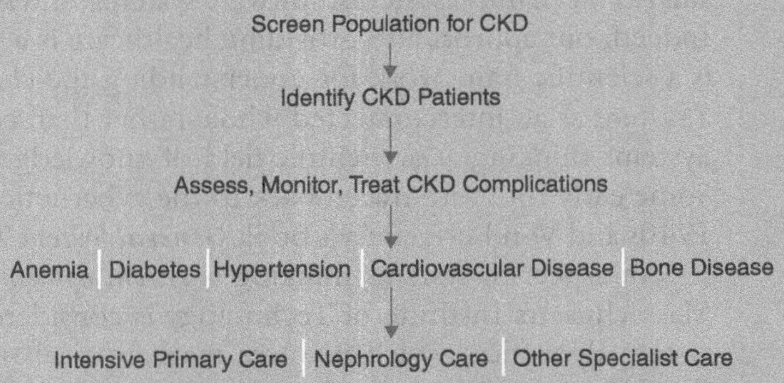

Figure 2.6 High-Level Flowchart of CKD Care for the Primary Care Physician

Interviews with a group of physicians suggested that many family practice physicians are not fully familiar with the KDOQI guidelines. Poor education as well as inconsistent treatment practices have aggravated the problem. Limited understanding—of both patients and their healthcare providers—about the severity of renal dysfunction contributes to the progression of the disease and hampers the provision of optimal care for CKD. Additionally, variations in practice for CKD treatment worsen health outcomes of patients with CKD, decrease satisfaction of patients and care providers, and increase the burden of complications associated with CKD.

Healthcare systems would have to bear the cost—in both time and money—to get physicians and other clinicians educated about CKD. This would have to be done in concert with public communication, publication education, and consumer advocacy in order to make real progress against the gap in awareness of people with regard to their CKD status. Funding is needed at *each* step along the way. Historically, most agency funding tends to focus at the medical research stage and less on the translation of clinical results into practice. However, we are now seeing some shift in this trend with the rise of efforts through initiatives through the National Institutes of Health (NIH) Clinical and Translational Science Award (CTSA), and the Patient-Centered Outcomes Research Institute (PCORI).

2.3 SYSTEMS THINKING AND SYSTEM DYNAMICS

The word *system* tends to evoke ideas of interrelated parts or components that cooperate in some way. Examples of systems include natural systems such as the ocean currents, the solar system and ecosystems, and designed systems such as mechanical systems, software systems, and social-economic systems. Indeed, our approach to providing healthcare is a system. *Systems thinking* is a scientific framework for understanding the change and complexity of a system as an interconnected whole rather than components in isolation. Systems thinking as a scientific field of knowledge has evolved over time. Some early roots are traced back to the cybernetics work of Weiner in the 1940s and Von Bertalanffy's book *General System Theory* in 1954.

An aspect of systems thinking is *system dynamics*. Jay Forrester of the Massachusetts Institute of Technology is considered to be the founder of system dynamics, which involves mathematical modeling and computer simulation modeling for framing, understanding, and discussing the complexity of systems. His books include *Industrial Dynamics* (Pegasus Communications, 1961); *Principles of Systems*, 2nd ed. (Pegasus

Communications, 1968); *Urban Dynamics* (Pegasus Communications, 1969); and *World Dynamics* (Wright-Allen Press, 1971). Although the field of systems thinking has evolved in some overlapping ways with complex adaptive systems discussed in the previous section, one of the tenets that distinguishes systems is that cause and effect are separated over time and space versus being coherent only in retrospect (see Figure 2.2).

The definition of the term *systems thinking* is quite varied and the role and relationship of system dynamics to systems thinking has also been debated. Forrester held the view that systems thinking is a small part of the system dynamics approach. By contrast, systems investigator Barry Richmond held that systems thinking had a narrow impact on making inferences about behavior based on the underlying structure. Instead, his view was that system dynamics modeling formed the bulk of systems thinking. Systems thinking has been mainly used to gain insight into improving a system by understanding how the structure of a complex system is organized and how the components of the system influence one another. System dynamics that also have these functions of understanding have been widely applied toward identifying and testing high-leverage policies due to their quantitative modeling capabilities. Nowadays, many people see system dynamics modeling as one of a broad range of tools and methods encompassed by systems thinking. Figure 2.7 summarizes these viewpoints. A fuller history of the origins of systems thinking, system dynamics, and management thought is given in Maani & Cavana (2007).

Dimensions of Systems Thinking

Systems thinking has three distinct but related dimensions: paradigm, language, and methodology. Figure 2.8 summarizes these dimensions.

> **Paradigm** can be explained by the type of thinking: forest thinking—seeing the big picture; dynamic thinking—recognition that things change; operational thinking—an understanding of the "physics of operations"; and closed-loop thinking—recognition that the end (effect) can influence the means (cause/s).
>
> The **language** of systems thinking has specific attributes that follow from its tool kit. Foremost, it is visual, with representations done in a precise way according to certain rules. As a language, it has to translate perceptions. In the end, the visual communication emphasizes interdependencies of the elements.

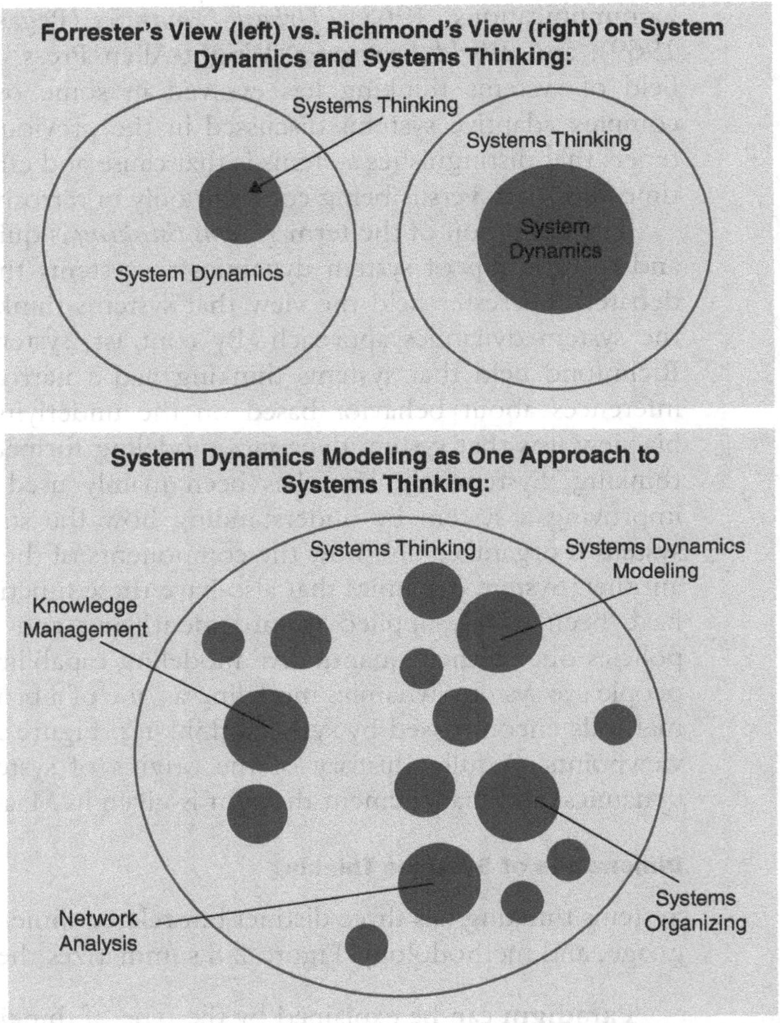

Figure 2.7 Three Views on the Definition and Role of Systems Thinking
***Source:* Initiative on the Study and Implementation of Systems (ISIS), Chapter 5, p. 112)**

The **methodology** of systems thinking incorporates learning technologies and tools to understand complex structure. These tools include causal loop maps and stock and flow models, which are closely associated with system dynamics, as well as computer simulation, learning laboratories, and group model building.

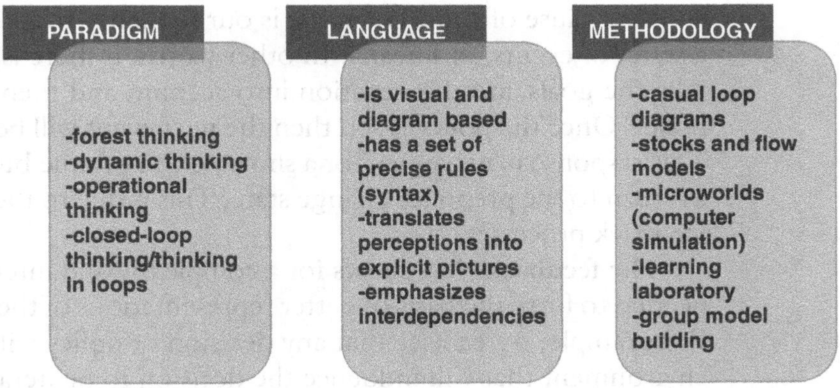

Figure 2.8 Dimensions of Systems Thinking ***Source:*** **Maani and Cavana (2007, pp. 7–8)**

Linear vs. Feedback Perspectives

The *linear* perspective has an event-oriented view of the world. Figure 2.9 shows an example of the linear view. The (nonlinear) *feedback* perspective has a closed-loop thinking or thinking-in-loops view of the world. Figure 2.10 shows an example of the feedback view.

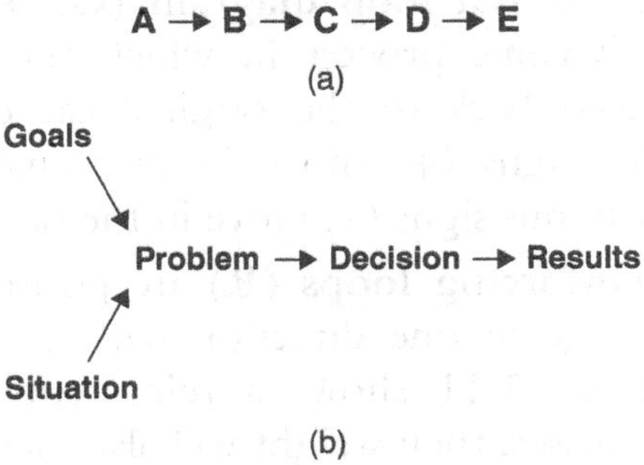

Figure 2.9 Linear View Example

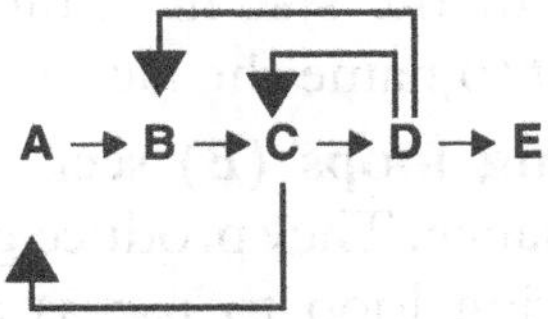

Figure 2.10 Feedback View Example

One cause of ***policy resistance*** is our tendency to interpret experience as a series of events, or linearly. In other words, if there is a problem, we may take the goals and the situation into account and then make a decision or policy. Once the policy is set, then the next event will be the desired results. This response may happen for a short period of time but then there may be a return to the prepolicy change state. This is due to the system's nonlinear feedback processes.

The feedback view allows for a complexity and interconnectivity that is needed to form the basis of better representations of the underlying system. For example, we can see that any decision or policy will have to exist in an environment that will influence the decision to be iterated upon, and each will have side effects or unintended consequences.

Methodologies

Causal Loop Diagrams

All systemic behavior can be described through two basic processes:

- Reinforcing
- Balancing

The **causal loop diagram (CLD)** is a conceptual tool that reveals a dynamic process in which the chain effect(s) of a cause is/are traced back to the original cause (effect). Variables connected by plus signs (+) move in the same direction, while variables linked by minus signs (–) move in the opposite direction.

Reinforcing loops (R) are positive feedback loops that compound change in one direction with even more change in that direction. Figure 2.11 shows a reinforcing loop for eating. If food intake increases, then weight will also increase. If food intake decreases, then weight will also decrease. Note that in constructing the causal loop diagram, the name of the factors must be neutral and the outcomes a result of the system reinforcing nature. For example, it would be *incorrect* to name the factors as "greater weight" and "more food."

Balancing loops (B) seek an equilibrium or some desired level of performance. They produce goal-seeking behaviors. Figure 2.12 shows a balancing loop to use as a self-control intervention for a balance in weight. Included is the acceptable or desirable weight. As weight

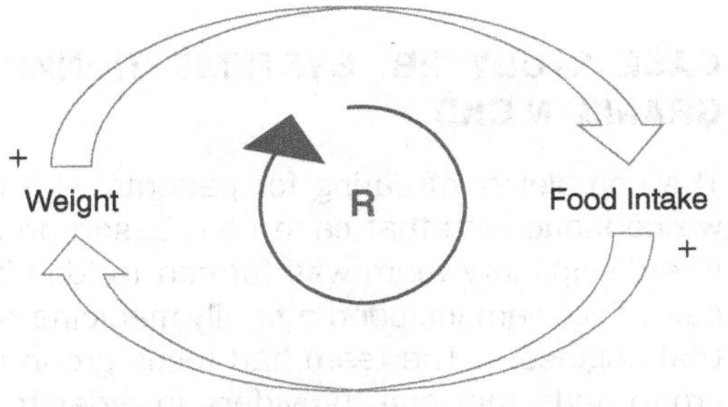

Figure 2.11 Reinforcing Loop on Eating and Weight *Source:* Sterman (2000)

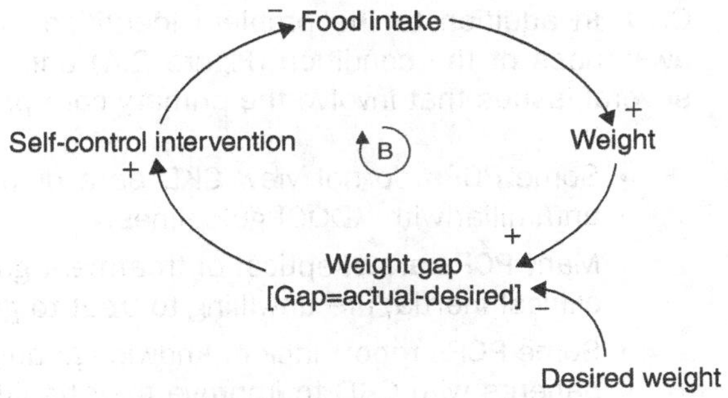

Figure 2.12 Balancing Loop to Use a Self-Control Intervention

increases, the gap between the actual and desired weight also increases. As that gap increases, the use of a self-control intervention may increase, which counterbalances or reduces food intake. As weight gets low, the weight gap is also low and the use of intervention decreases. The process continues until the gap is stabilized.

Every link in a system contains a delay. Important delays may be labeled as needed. For example, in Figure 2.11 it may be significant to add a delay between food intake and weight changes and between the use of a self-control intervention and changes in food intake.

The purpose of using causal loop diagram models is typically to answer key questions about the system, such as:

- Which gaps are driving our system when and by how much?
- How accurately do we know what each of the gaps is?
- How are we monitoring the gaps?
- What are the different ways in which we can close the gaps?

CASE STUDY 2B: SYSTEMS THINKING AND CAUSAL LOOP DIAGRAMS IN CKD

The complexity of caring for patients with CKD was highlighted in Case 2a; we continue with that case here based on a study in Kang et al. (2016). An interdisciplinary team was formed to identify effective interventions for CKD care. The team included a family medicine physician, nephrologist, and industrial engineers. The team had focus group interviews with a patient advisory group and other care providers in order to understand the gaps in care for CKD. These discussions led to determining the scope of the case and the designing of interventions.

We noted multiple barriers that hamper the provision of optimal care for CKD. In addition to the problem identified previously with patients lacking awareness of the condition (Figure 2.4) and how it is diagnosed, there are several issues that involve the primary care physicians (PCPs):

- Some PCPs do not view CKD as a distinct medical condition and are unfamiliar with KDOQI guidelines.
- Many PCPs are skeptical of treatment goals for hypertension or, due to clinical inertia, are unwilling to treat to goals.
- Some PCPs report lack of knowledge and skill to educate and motivate patients with CKD to improve their health, and fear overwhelming their patients with too much information if CKD is discussed as an entity separate from diabetes or hypertension.
- Engaging patients in self-management has been difficult because of time and reimbursement constraints in the current healthcare system, as well as a perceived lack of educational materials for patients.

To improve CKD outcomes in this environment, we considered three interventions that relate to the KDOQI guidelines:

1. Education and implementation of the adapted KDOQI guidelines for PCPs.
2. Education and implementation of the adapted KDOQI guidelines for care managers.
3. Care coordination between PCPs and nephrologists, including early referrals to nephrology care.

A causal loop diagram was developed to support prospective planning and analysis of the system-level interventions. This model is particularly effective because it can demonstrate the interrelationships among patients, providers, and policies; identify feedback loops established from the dynamic relationships of variables; and predict the effects of the interventions. We address the first of these interventions in this case report.

Educational interventions targeting family practice physicians have been shown to improve compliance with KDOQI guidelines (Wentworth, Fox, Kahn, Glaser, & Cadzow, 2011). The causal loop diagram in Figure 2.13 shows how physician education about KDOQI guidelines may impact patients with stage 3 CKD in the long run. The double bar on the arrow represents delays. The letters *R* and *B* in a circle correspond to the reinforcing and the balancing loops. Our study showed that the physician education intervention will lead to a delay in disease progression of the patients. To see this, Figure 2.13 can be read and interpreted as follows:

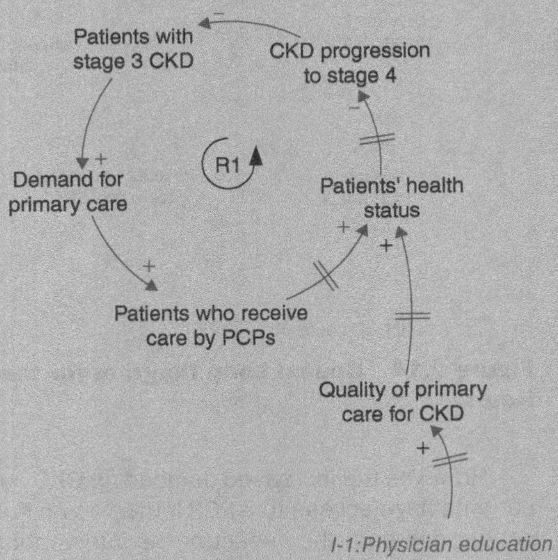

Figure 2.13 Causal Loop Diagram for the Long-Run Impact of Educating PCPs about the KDOQI Guideline

Educating PCPs about the KDOQI guidelines (marked in red italic) with regular assessment of their knowledge will improve the *quality of primary care for CKD* by increasing their adherence to the guidelines for patients with stage 3 CKD, such as controlling risk factors causing complications and using different screening measures for comorbidities. In the long term, enhanced CKD care management in primary care settings improves *patients' health statuses*, which prevents some patients in stage 3 from progressing to stage 4. That is, *patients' health statuses* and *disease progression* move in the opposite (–) direction. As disease progression slows down for this patient population and more patients stay longer in stage 3, *the number of patients with stage 3 CKD* increases above what it otherwise would have been. The increased patient population leads to an increase in *demand for primary care* (+). As more *patients received timely, managed care by PCPs*, patients' health statuses become better.

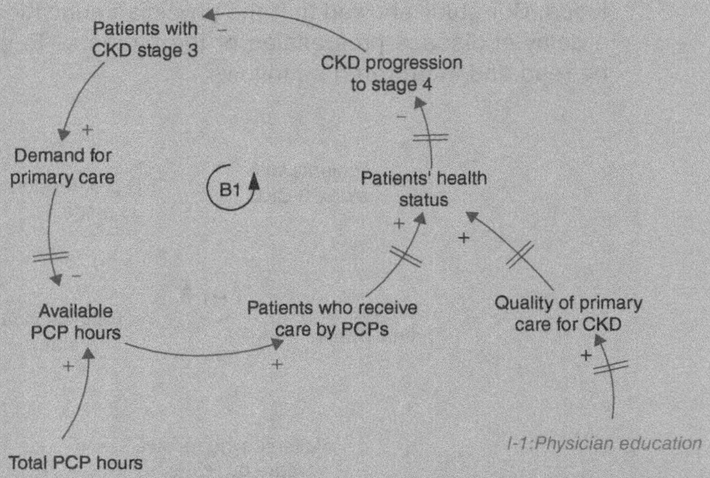

Figure 2.14 Causal Loop Diagram for the Impact of the Limited PCP Hours

However, the increased demand for CKD care can be satisfied only when all patients have access to a PCP. If there is no sufficient number of PCPs to meet patient demand, the impact of the intervention is limited. Figure 2.14 shows this result through a balancing feedback loop. The *total PCP hours* reflect the total amount of time that PCPs can provide services in a given practice. The *available PCP hours* reflect a portion of the total time that PCPs have available. This time availability is impacted by the *patient demand for primary care*. These factors move in the opposite (–) direction; for example, if patient demand for primary care rises, PCPs' available hours decrease as physicians fulfill the growing demand. As their available time becomes limited, fewer *patients who need regular care* would be able to receive timely primary care. The limited access to PCPs leads to poor CKD care management over time and, as a result, there would be a faster *progression of the disease to stage 4*.

The development of the causal loop diagrams enabled us to identify key feedback loops and understand the complexity of a system for CKD care. Due to complexities in a healthcare system, many studies adopted approaches that isolate one part of a healthcare delivery system rather than looking at the entire system. This tendency has been identified as a core impediment to improvement (Institute of Medicine [U.S.] Roundtable on Value & Science-Driven Health Care, 2011; Kaplan et al., 2013; Reid, Compton, Grossman, & Fanjiang, 2005). By employing systems thinking, this study overcame the limitations of the perspectives that look at relationships between variables in a linear fashion and that focus on only one part of a system. The in-depth understanding of a complex system involving delays also suggested caution against overoptimistic expectations about the immediate impact of the interventions on the system.

Stock Flow Model

The structure and dynamics of complex systems can be represented by stocks and flows, along with feedback. Stocks are accumulated quantities within the system, such as cash, inventory, population, or level of knowledge. Stocks describe the condition of the system and would continue to exist even if all the flows in the system were brought to a halt. The state of stocks changes only through their inflow and outflow rates.

Flows are the changes to the stocks that occur during a period of time, such as revenue earned during the month, people who died during the year, and new knowledge gained over the year. The flows in the system are usually the outcomes of decisions by management or external (exogenous) forces outside management's control. Flows are changed by stocks and by auxiliary variables.

A systems dynamics model often refers to a stock flow computer simulation model. Stock flow models are used to develop effective policies and to test their long-term impacts at a system-level. Figure 2.15 shows a basic structure of a stock flow model. In system dynamics, stocks are denoted by a rectangle, which represents a repository. Flows are denoted by an arrow, along with a valve that controls the amount of inflow and outflow.

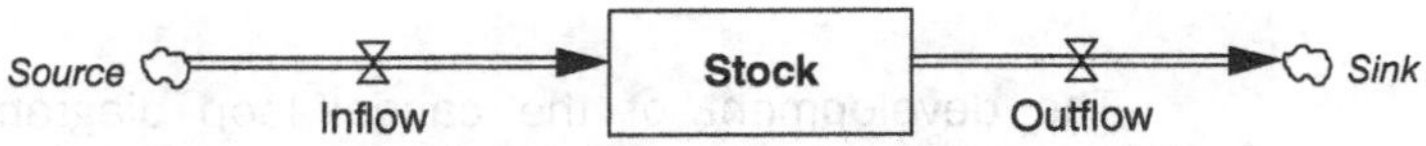

Figure 2.15 Basic Structure of a Stock Flow Model

Mathematically, stocks can be computed by the following integral equation:

$$\text{Stock(t)} = \int_{t_0}^{t} [\text{Inflow}(x) - \text{outflow}(x)]dx + \text{Stock}(t_0)$$

where inflow(x) and outflow(x) represent the level of the inflow and the outflow at time x between the initial time t_0 and the current time t. That is, the value of the stock at time t is the net difference between the inflow and the outflow between time t_0 and t plus the initial value of the stock at time t_0. The net change in a stock can be represented by the following differential equation:

$$\frac{d(Stock)}{dt} = \text{Inflow}(t) - \text{outflow}(t)$$

On the other hand, flow can be determined in various ways including constants, mathematical functions of stocks and auxiliary variables, and graphical representations.

The following are steps to develop a stock flow model:

1. Develop a systems map (e.g., causal loop diagram) to understand the overall structure of the system.
2. Define key variables and determine variable types (stock, flow, endogenous factors, exogenous factors, etc.).
3. Build the basic structure of a stock flow diagram.
4. Collect data and define relationships of the variables.
5. Examine steady-state conditions.
6. Calibrate a simulation model and adjust parameters.
7. Perform sensitivity analysis.
8. Design and test policies.
9. Determine the optimal policies.

Stock flow modeling has been widely applied in many areas in healthcare systems to evaluate interventions and support decision making. The application ranges from improving patient flow in a healthcare delivery system (Behr & Diaz, 2010; Brailsford, Lattimer, Tarnaras, & Turnbull, 2004; Lattimer et al., 2004) to preventing chronic diseases (Homer, Hirsch, Minniti, & Pierson, 2004; Homer et al., 2008; Homer, Jones, et al., 2004; Jones et al., 2006) to improving public health (Atun, Lebcir, Drobniewski, & Coker, 2005; Dangerfield, Fang, & Roberts, 2001; Royston, Dost, Townshend, & Turner, 1999).

CASE STUDY 2C: SYSTEM DYNAMICS AND STOCK FLOW DIAGRAMS IN CKD

Case 2b underscored the feedback loops arising from the healthcare system for CKD care and the role of a causal loop diagram to understand the potential impact of interventions; we continue with that case here based on a study in Kang (2015). As a next step, we are interested in exploring the dynamics of how patients progress from CKD stage 3 to CKD stage 5 (refer to p. 32 for definitions of the stages) under care management. We also seek to test various scenarios and determine the optimal policy.

Figure 2.16 shows the basic structure of the stock flow diagram where the stocks are patients with CKD and the flows are rates of change to the stocks. As the CKD incident rate and the death rate change, the stock of undiagnosed patients with CKD changes. They will be diagnosed as having CKD stage 3, 4, or 5 without care management (i.e., no care management (NCM). Patients in stages 3–5 (NCM) move to a stock of CKD stages 3–5 with care management (CM) as they engage in care management. Patients with CKD stage 5 would either be on dialysis or have a kidney transplant in order to survive. Both patients without and with care management may advance to the next stages as their disease progresses or they may die at the current stage. It was assumed that patients do not go back to earlier stages. The stock flow model in this case was developed in Vensim® software (Ventana Systems, Inc.).

A simulation model was built based on both quantitative and qualitative data. The model was improved with respect to a set of key parameters by conducting model calibration. We used an optimization function to determine the parameters that provide the best fit between a past time-series data set and the variables of interest. Four scenarios were then devised to measure the individual effect of the four interventions. Figure 2.17 shows how the four interventions affected the number of patients with

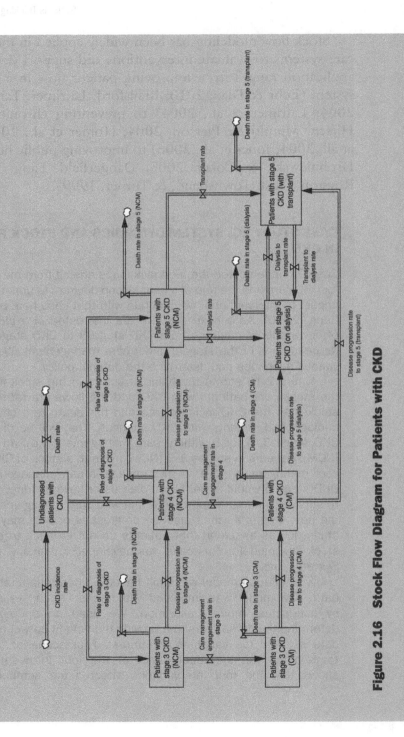

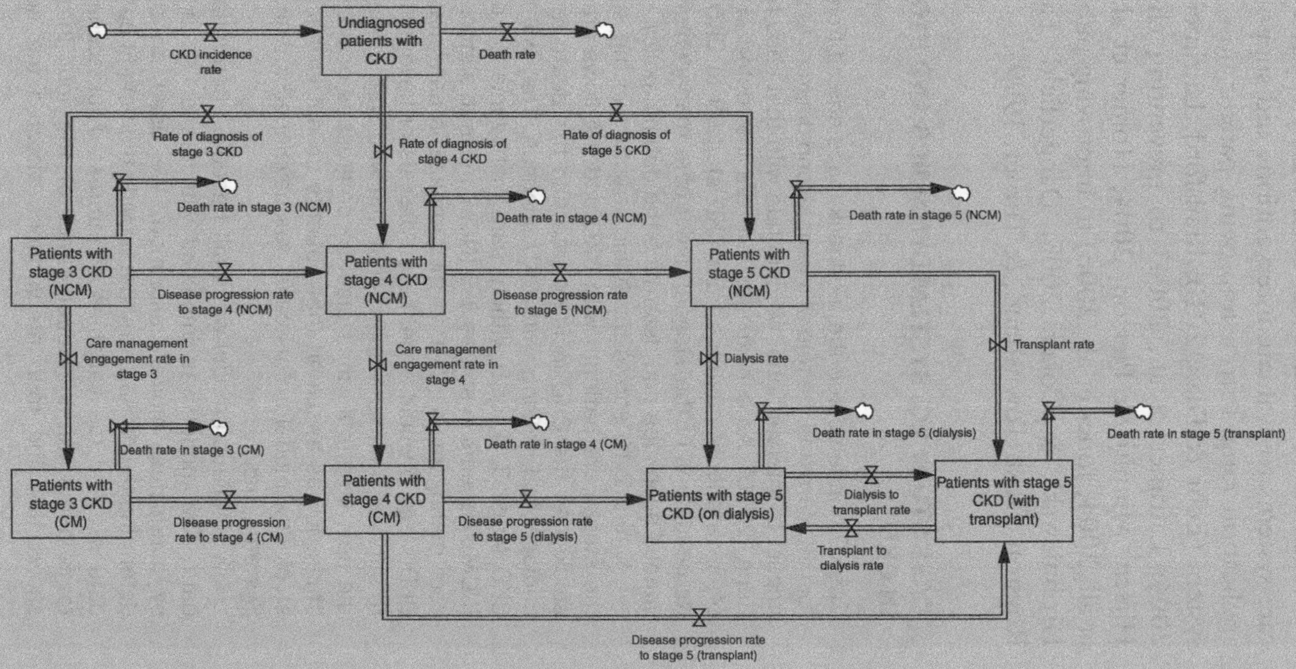

Figure 2.16 Stock Flow Diagram for Patients with CKD

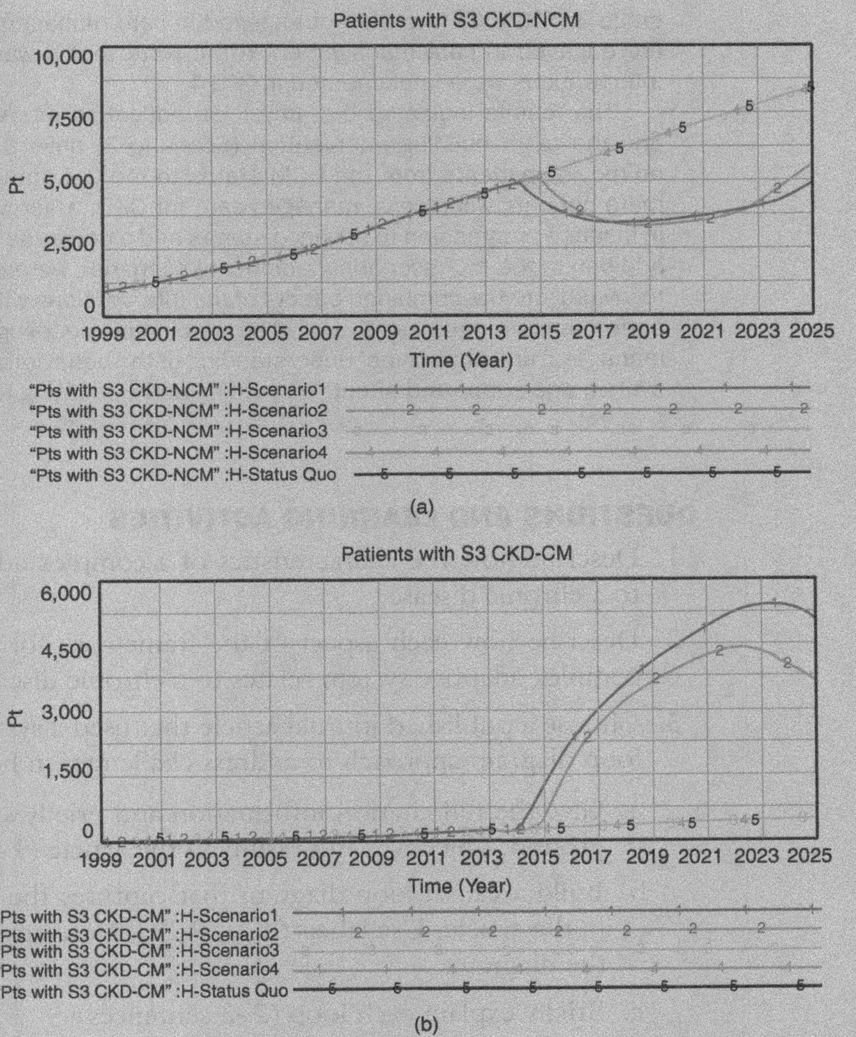

Figure 2.17 Chart Showing How the Four Interventions Affected the Number of Patients with Stage 3 (S3) CKD Who Are Not Engaged in Care Management (NCM) and Who Are Engaged in Care Management (CM)

stage 3 (S3) CKD who are not engaged in care management (NCM) and who are engaged in care management (CM). Note that it was assumed that the interventions were implemented in 2014.

The results indicated that physician education intervention (Scenario 1) and the team building intervention (Scenario 2) have the greatest impacts on moving patients from the NCM status to the CM status. This implies that more patients will receive managed care for CKD, which would lead to delays in disease progression to advance stages and a decrease in mortality rates. In addition to the four scenarios, additional scenarios were designed and tested to investigate the combined effects of the interventions with respect to various performance measures. Overall, this stock flow modeling enhanced hospital managers' and physicians' understanding of the behaviors of the complex care system and supported effective planning and evaluating for the interventions.

QUESTIONS AND LEARNING ACTIVITIES

1. Describe how the characteristics of a complex adaptive system relate to a chronic disease.
2. Describe how each aspect of the framework for disease control as a complex adaptive system relates to a chronic disease.
3. Choose a published journal article that used a systems thinking causal loop diagram approach to address challenges in healthcare.
    a. Give the full citation information and briefly summarize the background, objective, and results in the article (1–2 paragraphs).
    b. Build a causal loop diagram that captures the dynamics described in the article. Use labels (+, –, R, B). Note any important delays in the diagram.
    c. Briefly explain each loop (2–3 sentences).
    d. Explain the limitations of the causal loop diagram model in addressing the topic(s) presented in the article.
4. Review literature that proposed or implemented interventions to address one of the following issues:

    Asthma
    Cardiomyopathy
    Chronic kidney disease (CKD)

Crohn's disease
Diabetes
Glaucoma
Hypertension
Parkinson's disease

a. For each article, give the full citation information and briefly summarize the background, objective, methods, findings, and limitations of the study (2–3 paragraphs).
b. Based on the literature review, develop a causal loop diagram capturing key dynamics arising from the problem. Name your loops with proper labels (positive, negative loops). Note any important delays in your diagram.
c. Explain in a paragraph how each loop captures the dynamics described.
d. Propose two interventions to improve the current system. Provide rationales behind the interventions. Suppose/presume possible consequences of the interventions.
e. Expand the causal loop diagram capturing the changes caused by the interventions.
f. Evaluate impacts of the interventions on the system. What are the expected/unexpected impacts? How does causal loop diagram modeling help estimate potential consequences?

REFERENCES

Atun, R. A., Lebcir, R., Drobniewski, F., & Coker, R. J. (2005). Impact of an effective multidrug-resistant tuberculosis control programme in the setting of an immature HIV epidemic: system dynamics simulation model. *International Journal of STD & AIDS*, *16*(8), 560–70. doi:10.1258/0956462054679124

Behr, J. G., & Diaz, R. (2010). A System Dynamics Approach to Modeling the Sensitivity of Inappropriate Emergency Department Utilization. In S.-K. Chai, J. Salerno, & P. Mabry (Eds.), *Advances in Social Computing SE - 9* (Vol. *6007*, pp. 52–61). Springer Berlin Heidelberg. doi:10.1007/978-3-642-12079-4_9

Brailsford, S. C., Lattimer, V. A., Tarnaras, P., & Turnbull, J. C. (2004). Emergency and on-demand health care: modelling a large complex system. *Journal of the Operational Research Society*, *55*(1), 34–42. doi:10.1057/palgrave.jors.2601667

Dangerfield, B. C., Fang, Y., & Roberts, C. A. (2001). Model-based scenarios for the epidemiology of HIV/AIDS: the consequences of highly active antiretroviral therapy. *System Dynamics Review*, *17*(2), 119–150. doi:10.1002/sdr.211

Fox, C. H., Voleti, V., Khan, L., Murray, B., & Vassalotti, J. (2013). A Quick Guide to Evidence-Based Chronic Kidney Disease Care for the Primary Care Physician. Implement Science.

Homer, J., Hirsch, G., Minniti, M., & Pierson, M. (2004). Models for collaboration: how system dynamics helped a community organize cost-effective care for chronic illness. *System Dynamics Review*, *20*(3), 199–222. doi:10.1002/sdr.295

Homer, J., Jones, A., Seville, D., Essien, J., Milstein, B., & Murphy., D. (2004). The CDC's Diabetes Systems Modeling Project: Developing a New Tool for Chronic Disease Prevention and Control. In *22nd International Conference of the System Dynamics Society* (Vol. *2004*, pp. 25–29). Retrieved from http://climateinteractive.org/about/writing/Diabetes Systems Modeling Project-Jones.pdf

Homer, J., Milstein, B., Wile, K., Pratibhu, P., Farris, R., & Orenstein, D. R. (2008). Modeling the local dynamics of cardiovascular health: risk factors, context, and capacity. *Preventing Chronic Disease*, *5*(2), A63. Retrieved from http://www.pubmedcentral.nih.gov/articlerender.fcgi?artid=2396963&tool=pmcentrez&rendertype=abstract

Institute of Medicine (U.S.). Roundtable on Value & Science-Driven Health Care and National Academy of Engineering. (2011). *Engineering a learning healthcare system: a look at the future*.

Jones, A. P., Homer, J. B., Murphy, D. L., Essien, J. D. K., Milstein, B., & Seville, D. A. (2006). Understanding diabetes population dynamics through simulation modeling and experimentation. *American Journal of Public Health*, *96*(3), 488–94. doi:10.2105/AJPH.2005.063529

Kang, H., Nembhard, H. B., Curry, W. J., Ghahramani, N. and Hwang, W. (2016). "A Systems Thinking Approach to Prospective Planning of Interventions for Chronic Kidney Disease," *Health Systems*, To appear.

Kaplan, G., Bo-Linn, G., Carayon, P., Pronovost, P., Rouse, W., Reid, P., & Saunders, R. (2013). Bringing a Systems Approach to Health.

Discussion Paper, Institute of Medicine and National Academy of Engineering, Washington, DC. http://www.iom.edu/systemsapproaches.

Lattimer, V., Brailsford, S. C., Turnbull, J., Tarnaras, P., Smith, H., George, S., ... Maslin-Prothero, S. (2004). Reviewing emergency care systems I: insights from system dynamics modelling. *Emergency Medicine Journal*, *21*(6), 685–91. doi:10.1136/emj.2002.003673

Maani, K., & Cavana, R. (2007). *Systems thinking, system dynamics: managing change and complexity* (2nd ed.). North Shore, N.Z: Pearson Education New Zealand.

McDaniel Jr,, R. R., Lanham, H. J., & Anderson, R. A. (2009). Implications of complex adaptive systems theory for the design of research on health care organizations. *Health Care Manage Rev*, *34*(2), 191–199. doi:10.1097/HMR.0b013e31819c8b38.Implications

National Kidney Foundation. (2002). *KDOQI Clinical Practice Guidelines for Chronic Kidney Disease: Evaluation, Classification, and Stratification.* Retrieved from https://www.kidney.org/professionals/kdoqi/guidelines_ckd/p4_class_g3.htm

Plsek, P. (1996). Redesigning Health Care with Insights from the Science of Complex Adaptive Systems. In *Crossing the Quality Chasm* (pp. 309–322).

Rardin, R. (2007). *Research Agenda for Healthcare Systems Engineering*. Final report. Arlington, VA: National Science Foundation, Grant No. 0613037.

Reid, P., Compton, W., Grossman, J., & Fanjiang, G. (2005). *Building a Better Delivery System: A New Engineering/Health Care Partnership*. Washington, D.C. Retrieved from http://www.nap.edu/openbook.php?record_id=11378

Rouse, W. B. (2000). Managing Complexity: Disease Control as a Complex Adaptive System. *Information, Knowledge, Systems Management*, *2*(2), 145–165.

Rouse, W. B. (2008). Health Care as a Complex Adaptive System : Implications for Design and Management. *The Bridge*, *Spring*.

Royston, G., Dost, A., Townshend, J., & Turner, H. (1999). Using system dynamics to help develop and implement policies and programmes in health care in England. *System Dynamics Review*, *15*(3), 293–313. doi:10.1002/(SICI)1099-1727(199923)15:3<293::AID-SDR169>3.0.CO;2-1

Sterman, J. (2000). *Business Dynamics: Systems thinking and modeling for a complex world* (1st ed.). The McGraw-Hill Companies.

The Renal Association. (2013). CKD stages. Retrieved from http://www.renal.org/information-resources/the-uk-eckd-guide/ckd-stages#sthash.Q6Y5eKuU.dpbs

Wentworth, A.L., Fox, C.H., Kahn, L.S.,, Glaser, K., & Cadzow, R. (2008) Two Years After A Quality Improvement Intervention For Chronic Kidney Disease Care In A Primary Care Office. *American Journal of Medical Quality*

Chapter 3

Patient Flow

"Consistency in quality means not allowing the ordinary rush of business and even extraordinary events slow or suspend the process."

—Juran Institute, Inc.

Overview

Patient flow is the process by which patients move through a hospital system and transition between healthcare settings. Improving patient flow can lead to fewer unnecessary process steps; better provider-to-patient ratios; and fewer waits, delays, and cancellations (Institute for Healthcare Improvement, 2003). As a result, this improvement in patient flow often leads to improved safety and quality of care, as well as better access for those patients without it.

This chapter provides the fundamentals of patient flow in healthcare settings. It explains different types of healthcare settings in the United States, and discusses major patient flow and care transition issues within and between those settings. This chapter also details the use of process mapping for understanding and improving patient flow and clinical workflow processes.

3.1 HEALTHCARE SETTINGS AND CLINICAL WORKFLOWS

It is important to understand the main types of healthcare settings since patient flow is different in each setting. Table 3.1 provides a summary of the clinical workflow elements that are involved in these settings. The top level of the table shows the input, process, output, and duration components. The middle level shows the access and whether patient arrivals are mostly

Table 3.1 Clinical Workflows in Healthcare Settings

Clinical Workflow Elements	Primary Ambulatory Care	Specialty Ambulatory Care	Emergency Care	Outpatient Surgery Center	Operating Room Care	Inpatient Care	Long-Term Care
Input: arrival for requested service encounter	Arrival Registration Chief complaint	Arrival Registration Chief complaint	Arrival Triage Registration Chief complaint	Arrival Registration Preop prep	Surgical decision Preadmit Preop prep	ED OR Direct admit	Home referral Hospital discharge
Throughput process: service provisions	Vital signs Room Provider Labs/tests Medications Documentation Referral	Vital signs Room Provider Labs/tests Medications Documentation	Provider Labs/tests Consults Documentation	Anesthesia Incision Procedure Close	Anesthesia Incision Procedure Close	Nursing care providers Room Procedures Labs/tests Care coordination	Nursing care, therapies; convales-cence
Output: departing encounter	Instruction review Patient education Check-out Schedule	Instruction review Patient education Check-out Schedule	Observation Admit policy Instruction review Patient education Discharge Death	Recovery Instruction review Patient education Check-out Schedule	Recovery Input to hospital	Instruction review Patient education Discharge Check-out Schedule Home health Input to nursing home, rehab, long-term acute care hospital, or skilled nursing facility	Home Return to hospital Death

Duration: targeted durations of care	15–30 minutes	15–30 minutes	1–6 hours or longer	1–6 hours	1–6 hours or longer	1–6 days or longer	1–6 weeks or longer
Access: ability to get care	In person Scheduled Urgent Virtual	Scheduled	Emergent Urgent Convenient No options (care safety net)	Elective scheduled Block time	Emergent Scheduled Block time	Admissions policy based	Admissions policy based
Random: not knowing when or that patients are arriving for care	Same day acute visits	Same day acute visits	All unscheduled	Scheduled	Scheduled but acute add-on or emergencies	Unscheduled and scheduled	Scheduled and controlled by policy
Predicted: having an idea of the service need pattern	Scheduled acute visits Preventive care visits	Scheduled	Trend per hour Scheduled	Scheduled	Schedule (block vs. resources)	Trend per hour with discharges	Scheduled and controlled by policy

(*continued*)

Table 3.1 (Continued)

Clinical Workflow Elements	Primary Ambulatory Care	Specialty Ambulatory Care	Emergency Care	Outpatient Surgery Center	Operating Room Care	Inpatient Care	Long-Term Care
Demand: request or need for services	Increasing volume with decreased time per visit	Increasing requests but access limited by schedule capitation	Increasing volumes especially complex diagnostics and social needs	Managed by schedule Frequently weekdays	Daily variation based on block schedule or day of the week	Increasing but more difficult to meet insurance criteria for payment	Increasing with aging population and push to decrease hospital length of stay and volume
Capacity: room or ability to provide service	Rooms Number of providers Staff Laboratories Virtual space	Rooms Providers Staff Laboratories	Space Number of providers	OR room Surgery Anesthesia Equipment	OR room Surgery Anesthesia Equipment	ICU rooms Stepdown rooms Floor beds Observation space Advance testing capabilities Nursing staff	Beds Nursing and support staff
Waits/Queue: non-value-added time/buildup of patients in the system	Waiting room/lobby In room Tests (increasing)	Waiting room/lobby In room Tests	Waiting room/lobby In room Tests Admits (boarders)	Waiting room/lobby	Waiting room/lobby Wait for block time Anesthesia Turnaround time	Transfers ED boarders PACU backup	Awaiting bed availability at home, or hospital

random or predicted. The bottom level shows the waits/queue, demand, and capacity; waiting is the chasm between demand and capacity. Details are discussed in the subsequent sections.

Primary Ambulatory Care

Primary care is a patient's main source for routine and preventative medical care, ideally providing continuity and integration of healthcare services. All family physicians, and many pediatricians and internists, practice primary care. The aims of primary care are to provide the patient with a broad spectrum of care. These services may be acute or chronic, preventive, diagnostic, or curative. The intent of the relationship between a patient and *primary care physician (PCP)* is to be able to care over a period of time and to coordinate all the care that the patient receives. Primary care settings can include group practices, primary care clinics, and hospitals. Historically, this is the nidus of the patient-physician relationship. The patient knew the doctor as a person, and the doctor knew the patient and his family and home environment. That concept of the "small-town doctor" who cared for one from birth to death was the idealistic view. In some areas of the United States and abroad, PCPs also provide procedural-based care (colonoscopy, appendectomy, etc.), but this type of care is most frequently referred to a specialist in those areas (gastroenterologist, surgeon, etc.).

With the decrease in number of medical students entering primary care, movement to group practices and limitations on time reimbursed in care, the single-practitioner PCP who cares for the same patient from birth to death, outpatient or inpatient, for procedures and preventative care is a rarity. In most places, that solo practice is only of historical significance. Today, to replace the person-to-person relationship, this has led to the concept of *medical homes*, where a team of care providers provides comprehensive, patient-centered, and coordinate care, as defined and measured by medical home certification. Instead of the individual person remembering everything about the individual patient, a group of providers use electronic medical records, care coordinators, and outcomes data to measure and care for individuals and groups of patients.

Specialty Ambulatory Care

Ambulatory care is medical care provided on an outpatient basis, including evaluations, procedures, diagnostics, treatment, and ongoing care. The service can extend beyond the physician visit to include therapy or procedural-based services. Primary ambulatory care is described in the preceding section. Specialty ambulatory care has unique features compared to PCP services. While the PCP may see patients for a variety of issues and concerns, acute or chronic, the specialty ambulatory practice is frequently referral based and focuses on one organ system, set of disease processes, or specific procedures. For example, cardiologists deal primarily with issues related to the heart. A subspecialist in cardiology is the "invasive cardiologist" who primarily performs cardiac catheterizations and other procedures, or the electrophysiologist, who specifically deals with the electrical conduction system of the heart. A patient would not likely set up an appointment to see an electrophysiologist for an acute skin infection. Therefore, in most clinical settings, the specialist ambulatory visit is driven by "referrals." This process includes a "previsit" evaluation, frequently with the PCP, who would then send the patient to the specialist. Some specialists accept "self-referrals" without screening, while others require a previsit data set or questionnaire before agreeing to establish a provider-patient relationship. For example, when arranging to see a neurosurgeon, many would require a computed tomography (CT) scan or magnetic resonance imaging (MRI) of the brain to evaluate need for the visit. Increasingly, specialists are employing advanced practice clinicians (APCs), who are either physician assistants (PAs) or nurse practitioners (NPs) to see patients in the ambulatory setting. The reason for this may be efficiency, clinician preference, economics, or (likely) a combination of all three. The economic reality is that specialists, especially those who primarily perform procedures, have a high demand by hospitals and patients to perform those procedures. Concomitantly, these procedures generate revenue for the health system and the provider. The equivalent financial reward for time spent doing a procedure is exponentially larger than spending the same amount of time in the clinic seeing patients. Additionally, many providers who gravitate to specialty procedural-based services enjoy that aspect of care. In increasing demand for service and priorities, the role of an APC to see the patient in the clinic while the specialist is performing the procedure has been increasingly deployed.

Emergency Care

Emergency care is a pattern of healthcare in which a patient is treated for a brief but severe episode of illness, or sequel of an accident or other trauma. Designed initially for emergencies only, such as heart attacks, car accidents, strokes, and cardiac arrests, the provision of emergency care has extended to urgent care and generally unplanned emergency care. In some areas of the healthcare system, emergency care had also been used as access to healthcare for those who are uninsured or underinsured, independent of acuity. In the United States, emergency medicine has been described as the healthcare safety net. As the patient population gets older, and the access to a PCP is less available, emergency care is also used as a complex diagnostic center. Emergency care is usually provided in a hospital by specialized personnel using complex and sophisticated technical equipment and materials, and it may involve intensive or emergency care. This pattern of care is often necessary for only a short time, unlike chronic care. Recently, and based on local regulations, free-standing emergency departments have been designed to provide emergency care in a free-standing facility. Emergency care is the source of the majority of inpatient hospital admissions, ranging from well over 90% in community hospitals to slightly less than 50% in some tertiary/quaternary care academic facilities. As hospitals begin to fill, the capacity (beds) is limited, and the continued demand of the patients (admits, many from the ED) begin to create a mismatch. The wait (queue) that develops results in *boarding* of admitted patients in the ED. As a domino effect, the admitted patients occupy bed capacity in the ED, and therefore emergency care patients begin to queue (wait) because there is not space. This highlights the interdependency of the health system, and also leads to needed innovation for process/care. These are discussed in detail later in the chapter.

Outpatient Surgical Center

Outpatient surgical care is a hybrid design of ambulatory care and operating room care. These free-standing centers provide a venue for procedures, frequently under anesthesia, where the patient does not require an overnight stay. The arrival and initiation of these scheduled procedures occur in the ambulatory care. While arrival and departure for the surgical center is similar to ambulatory workflow, the preparation, preop, intraprocedure, and recovery are similar to operating room care. This venue and system of care was borne out of the desire for hospitals to maintain operating rooms for more complex cases.

Operating Room Care

Operating room care is a type of acute care focused on major procedures and occurs mostly in the hospital setting. These procedures can be elective or emergent, and require a surgeon, anesthesia provider, and support staff in a sterile room with requisite equipment for the procedure. These procedures range in complexity from a few minutes to perform an exam (e.g., exam under anesthesia [EUA]) that would require minimal resources (room, table, surgeon/examiner, anesthesia provider, support staff) to a complex multiple-hour procedure. A complex example might include a liver transplant with a living related donor. This would require multiple surgical teams working on multiple patients in multiple rooms, and specific equipment over a period of nearly a day. This high-intensity environment is a source of great care, but also great risk and high cost. As such, there has been much focus on the standardization of care, costs, and outcomes of surgical procedures in the recent decade.

All operating room care starts with the decision to operate. In an emergent situation, this may occur in seconds, but in most cases, this is a collective decision between the surgeon and the patient, resulting in "informed consent" to perform a procedure. This initiates the process of scheduling the room, equipment, time, preparation, anesthesia plan, and postprocedural recovery plan. Given that most surgical procedures are scheduled, this allows for process mapping and planned resource utilization. The culture of surgical care, however, is based on "block time" for the surgeon, rather than a resource allocation model. Simply, the surgeon is allocated a specific number of rooms and times per week to schedule his/her procedures. Savvy systems have considered intraoperative and downstream (hospital resources) when allocating surgical time, but most continue on the traditional block model.

Once scheduled, the anesthetic plan is created and coordinated with the surgeon. Anesthesiologists have a number of options to provide services for the procedure. Anesthesiologists are also frequently allocated rooms by which they will provide anesthesia each day of surgery, sometimes allocated to specific surgical types (vascular, transplant, pediatrics, etc.) when specific skill sets are available. As you can see, this complex preprocedure scheduled plan has a number of steps requires prior to the actual procedure itself.

Upon arrival at the OR, even when emergent, a surgical checklist is reviewed. Figure 3.1 shows the World Health Organization (WHO) surgical safety checklist, which is employed in one shape or form in many ORs. The checklist reviews the different preprocedural preparation at sign-in (arrival). This occurs in the preprocedural area, prior to anesthesia administration. When the patient is taken into the actual operating room, a series of steps is reviewed prior to surgical incision. Once completed, the actual incision, procedure, and closure of that procedure are completed. The duration, complexity, and details of the procedure are dependent on the surgical type. Again, an EUA will be much simpler than a liver transplant. Upon surgical completion, prior to leaving the operating room, the surgical team will complete the checklist in order to "sign out" the case. Patients are then transported for anesthesia recovery, and subsequently placed in an inpatient setting. For patients able to go home, the process of discharge is similar to an outpatient visit, and those procedures are typically performed in an ambulatory surgical center. Given the local cultures of operating rooms and

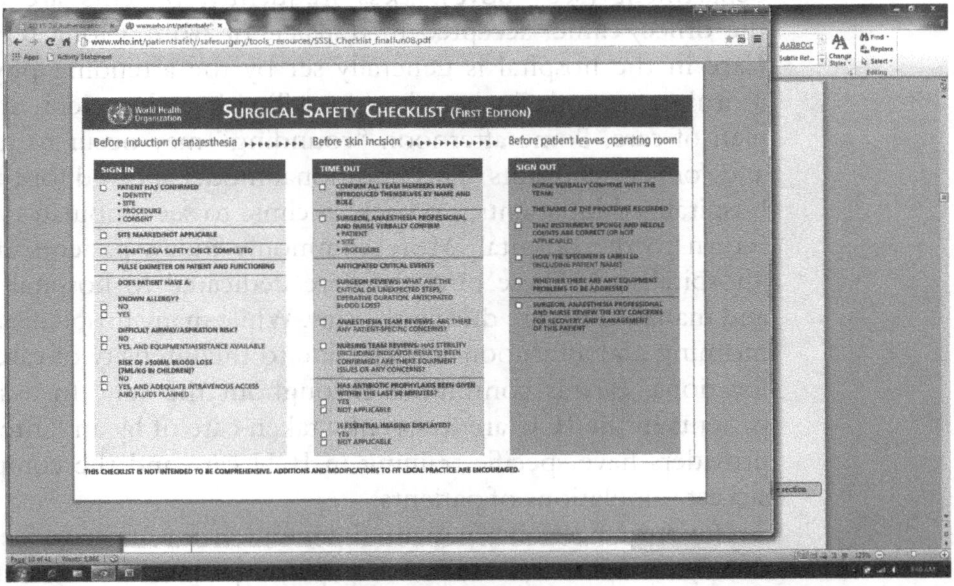

Figure 3.1 Surgical Safety Checklist (2008)

the fact that most are scheduled procedures, there is opportunity for flow science for improvement, at a standard, but with localization.

Inpatient Care

Inpatient care is acute care in the hospital setting. This may be on "the floor," which is a unit that is frequently service or disease specific, or the "step-down" unit, which is a location that has higher-intensity resources and equipment, or the intensive care unit (ICU). The ICU is the highest resource area of a hospital. Care is provided by nurses, therapists, and support staff, based on the orders of the physician directing the care. The "attending physician of record" is the authoritative decision maker, while the members of his/her team and the staff at the hospital form the interdisciplinary group that cares for the patient. The resource allocation may be one nurse for every four patients on the floor, and expand to two nurses per patient in some intensive care settings for specific patients. In the hospital, the interdisciplinary team may expand to "consultants," where the attending physician of record may request the opinion of another physician (specialist) for the care of that patient's disease process. This consultant physician will provide recommendations in the care of the patient.

Admission to inpatient care frequently starts in the emergency department (see above). Less frequently, patients are admitted from the office/clinic, accepted in transfer, or from a long-term care facility. Care in the hospital is generally set by the attending physicians during morning "rounds," where he/she will visit each patient, and set the daily plan of care. Some afternoon "rounding" may occur to update the plan and/or review results. This traditional model allowed for physicians to see hospitalized inpatients, go back to clinic to see ambulatory patients, then return to the hospital. More commonly today, patients are admitted to "hospitalists." These physicians are dedicated to hospital inpatient care, and manage the care during the stay. While many hospitalists still adhere to morning and afternoon rounds, due to the intensity of care in all hospital locations, care is continuous throughout the stay. In a similar fashion, patients in the ICU are frequently taken care of by an "intensivist." These providers have specific training in ICU care and the complexities of the sickest populations of patients.

Communications, patient flow, and handoffs have become tremendously complex issues. For example, a patient arrives at the ED and is taken care of by the emergency physician. The severity of illness is so high that the patient is admitted to the ICU, where the intensivist manages the patient's care. After a few days, the patient improves to where he/she is transferred to the step-down unit and subsequently "floor level" under the care of the hospitalist. The complexity of care and the advancement of modern medicine require providers to have very detailed and specialized knowledge at each stage of care. However, during the course of three to four days, there may have been three attending physicians of record, more than 10 nurses, and many more support staff who have cared for the patient, none of whom knew that person five days ago. Unlike years past, where the PCP would care for the patient at each stage of the hospitalization as well as ambulatory setting, the care transitions require transfer of information (handoff) to ensure continuity of care. Each transition of care and step in the process has the potential for missed information, care adjustments, and simple changes in options-based decision making. Additionally, with pressures to measure "quality" in hospital care and the drive to decrease length of stay in the hospital, these increasingly rapid transitions between providers and locations are even more important.

Observation care is a unique process of care that flows similarly to an inpatient care provision but is considered an ambulatory service by healthcare insurance. Sometimes referred to as "23hr obs," patients with specific conditions, such as chest pain, asthma, and congestive heart failure, may have an acute episode that requires more than a few hours of emergency care or a clinic visit, but *may* not require a multiple-day inpatient hospitalization. Therefore, insurers and providers designated specific diseases and processes for caring for this patient population in a location that would have access to most inpatient hospital services (diagnostic testing, consulting specialists, etc.) but was designated as outpatient care within 24 hours. From a flow perspective, it functions similar to inpatient care, but in a condensed 24-hour period.

Long-Term Care

Long-term care refers to health, mental health, social, and residential services provided to a temporarily or chronically disabled person over an extended period of time with a goal of enabling the person to function as independently as possible. The primary consumers of long-term care are people who have chronic and/or complex health problems accompanied by functional disabilities (Williams & Torrens, 2002). Long-term care settings can include nursing homes, assisted living facilities, and home care.

Various long-term care facilities or types have come into play. Rehabilitation facilities have gained acceptance in the United States over the past decade, providing extended inpatient and outpatient care to recover and strengthen patients after a hospitalization or procedure. Rehabilitation services are based on the care of the physiatrist, who has specialty training in rehabilitation medicine, and frequently include physical, occupational, speech, and language therapists.

Long-term acute care hospitals (LTACHs) are facilities in which patients who no longer need immediate traditional hospital care, but still require care at a hospital level in a subacute or longer-term facility, may be accessed. The best example to illustrate the role of the LTACH is for patients who have suffered a devastating illness or injury and are on a ventilator. The inpatient hospital care provides services for recovery, but the patient may be dependent on a ventilator to breath. In transferring to an LTACH, experts in ventilator "weaning" may work to provide nutrition, rehabilitation services, and medical care over a period of weeks to months to get that individual patient strong enough to be removed from a ventilator.

Skilled nursing facilities (SNFs, pronounced "sniffs") are subacute nursing facilities that provide extensive, higher-intensity, and additional nursing care that cannot be provided in a traditional nursing home for the post-discharge hospital patient. They may not require the intense medical care provided by LTACHs but more nursing-based care.

Nursing homes are another long-term care facility, and service levels provided are variable based on individual homes. In general, care can range from independent living to full-time nursing care. Independent living areas are where patients live independently in apartments or homes and are visited occasionally by on-site care providers. They can be "upgraded" in service to the nursing units, where they live in single rooms and are provided nursing care with frequency as dictated by the need. From a flow perspective, as the population ages and the desire to decrease inpatient hospitalization rises, this area of care will be increasingly important.

3.2 PATIENT FLOW THROUGH A HOSPITAL

Many hospitals in the United States have been plagued by operational inefficiencies and capacity problems, while the pressure for patient flow improvement has increased. Poorly managed patient flow in a hospital not only compromises patient outcomes but also increases financial burdens on hospitals. This is often the case because support staff are unable to keep up and the providers have less time to focus on delivering personalized, patient-centered care. Thus, effective management and improvement of both patient flow and clinical workflow processes in all healthcare settings is paramount.

The Institute for Healthcare Improvement (IHI) indicated that the major sources of inefficient patient flow stem from the variability throughout and between the healthcare settings (IHI, 2003). According to the IHI, by "developing the ability to shape, predict, and manage variability and to allocate resources appropriately at the front line of care can improve patient outcomes, increase staff morale and retention, reduce costs, and improve quality of life for both patients and caregivers" (IHI, 2003). Further, the IHI suggests that smoothing the flow of patients in and out institutions can help to reduce wide fluctuations in occupancy rates and prevent surges in patient visits that lead to overcrowding, poor handoffs, and delays in care. We consider some of the main contributors to patient flow, including admission and discharge, ED capacity and crowding, ED boarding, OR scheduling, and variability.

Improving patient flow through a hospital, however, requires a systematic approach to understand the system as a whole and manage the variability effectively. These efforts should be extended to pre– and post–acute care systems to smooth patient flow between care settings and continuum of care. Efforts to make more efficient use of existing resources by smoothing patient census and improving admission, handoffs, and discharge processes can have significant benefits not only in terms of improved safety and more continuous care, but also in terms of increased efficiency and revenue.

Admission and Discharge

As previously noted, transitions and continuity of care between providers and venues has become essential to quality of care. Improving patient flow during the key patient transitions, admission and discharge, remains a focus. In the process of hospitalization, admission is the starting point and discharge the ending point. From a healthcare perspective, admission is a transition to more intensity of care, and discharge is the transition to a less care-intensive environment. During the hospitalization, there are frequently new diagnoses, complex care provisions, and medication additions and removals in a very short period of a few days. Therefore, because there will be different providers caring for the patient in the hospital, clarifying the care plan at arrival (transition from PCP to hospitalist) is essential in the diagnostic and therapeutic success of the hospitalization. Transfer of data is required, and robust electronic medical record (EMR) systems can facilitate such care. Additionally, standard operating procedures or checklists may also provide guidelines in onboarding hospitalized patients. In addition to quality of transition, the actual bed acquisition to initiate the transition can be challenging. When multiple sources (ED, OR, clinic, nursing home) vie for the same resources (hospital beds), the patients must wait (queue) in those locations or be transferred to other facilities. One way to eliminate that queue is to manage it, providing resources available (testing, consultation, etc.) while the bed resource is being mobilized. Obviously, in a limited capacity system, the bed will not become available until another patient is discharged.

Another way hospital systems attempt to eliminate congestion and improve patient is by discharging patients early in the day. Similar to any other service that provides beds (hotels) or seats (restaurants), that space cannot be occupied by another person until it is unoccupied, cleaned, and prepared for the next person. This is no different in the hospital, in concept. In order to get a patient discharged, the transition of care must be clear, and the information provided to the patient, family, and next caregiver must be transitioned. Similarly, the patient must no longer physically occupy the room, and the room must be cleaned and prepared. To have more beds available, hotels have a checkout time, frequently 11 A.M., a transition time for cleaning and turnover (11 A.M. to 3 P.M.), and then a check-in time (after 4 P.M.). Hospitals had attempted to copy this approach by setting discharge times as early in the day as possible. However, clinical variables, such as changes in care plan, emergencies for other patients diverting limited resources away from discharging, and lack of motivation to leave a high-intensity care area, have limited success.

ED Capacity and Crowding

Many existing efforts to improve patient flow in acute care settings focus on emergency departments, which are experiencing record volumes as more and more physicians direct their patients to the ED rather than admitting them themselves. The increase in volume tends to increase wait times in the ED. According to a survey by the American Hospital Association (2012), half of hospitals indicated that their ED is at or over capacity. The nationwide average waiting time in EDs increased 45 minutes in 1998–2000 to 55 minutes in 2008–2010 (Hing & Bhuiya, 2012). The delays often cause some sick or injured patients, who grow tired of waiting, to leave without being seen (LWOS).

Many studies have investigated the causes of ED crowding; they can be classified into input, throughput, and output factors in the ED process (Asplin et al., 2003). The common input factors are increased ED visits, including unnecessary visits and high patient acuity. Inadequate staffing is one of the significant throughput factors. Hospital bed shortage and inpatient boarding are frequent output factors and they are not normally controlled by the ED (Derlet & Richards, 2000; Derlet, Richards, & Kravitz, 2001; Hoot & Aronsky, 2008; Niska, Bhuiya, & Xu, 2010).

To mitigate these problems, many potential solutions have been suggested. Improved access and availability of primary care can help nonurgent patients avoid ED visits (Billings, Parikh, & Mijanovic, 2000; Grumbach, Keane, & Bindman, 1993). Separate care for minor injuries such as a fast-track area has contributed to reducing waiting times in the ED (Cooke, Wilson, & Pearson, 2002; Fernandes, Christenson, & Price, 1996). Several studies have implemented physician-initiated triage, where ED residents (Svirsky et al., 2013), physician assistants, and attending physicians (Rowe et al., 2011) see patients and order tests for patients in triage. A comprehensive approach to reinventing emergency department flow was demonstrated in the physician-directed queuing (PDQ) methologolgy (DeFlitch, Geeting, & Paz, 2015) Team-based care also affects patient flow (Muntlin Athlin, von Thiele Schwarz, & Farrohknia, 2013). The integration of various technologies—such as EMRs, computerized provider order entry (CPOE), and real-time locating system (RTLS)—has helped hospitals enhance patient flow in the ED (Amini, Otondo, Janz, & Pitts, 2007; Coleman, Hammerschmith, & Duvall, 2013; Georgiou et al., 2013; Stahl, Drew, Leone, & Crowley, 2011).

Optimizing patient flow in the ED is imperative in many hospitals. However, the ED crowding problem affects not only ED processes and people involved in the process, but also the entire hospital system because EDs are interrelated to the other parts of the system through which patients move for care. For example, patient flow in the ED affects patient flow in the OR, which is a major revenue generator and an expensive resource at hospitals across the United States.

Boarding

Limits in ED capacity may also result in "boarding," the practice by which patients in need of admission wait in the ED for a bed to open up on an inpatient unit (IHI, 2003). The American College of Emergency Physicians (ACEP) defines a boarding patient as "a patient who remains in the emergency department after the patient has been admitted to the facility, but has not been transferred to an inpatient unit (ACEP, 2011)." Patients are boarded inside the ED (e.g., an ED room or ED hallways), or outside the ED (e.g., observation units or hallways in inpatient units).

Prolonged boarding is a nationwide issue in the United States According to the Centers for Disease Control and Prevention (CDC), 78% of visits occurred in 2009 in EDs that reported boarding admitted patients in hallways and in other spaces until an inpatient bed becomes available (Hing & Bhuiya, 2012). In the United States, the median time boarded patients spent in the ED before being admitted to the hospital is approximately 79 minutes (Pitts, Vaughns, Gautreau, Cogdell, & Meisel, 2014), while the average boarding time was much higher for high-volume hospitals, at 253 minutes (Centers for Medicare & Medicaid Services, 2015).

Blocked patient transitions to inpatient units affects patient flow in the ED and throughout the hospital. Many studies have indicated that boarding is the primary factor contributing to ED overcrowding and frequent ambulance diversion (Olshaker & Rathlev, 2006; Schull, Lazier, Vermeulen, Mawhinney, & Morrison, 2003; Trzeciak & Rivers, 2003). Due to holding admitted patients in ED beds, patients who have urgent conditions do not have appropriate care in a timely manner. According to a survey conducted by the CDC, patients experienced longer waiting times to see a care provider in EDs with boarding (61.3 minutes), compared to those waiting times in EDs with no boarding (44.1 minutes) (Hing & Bhuiya, 2012). Also, boarding tends to increase the total length of stay in the hospital, which blocks access to inpatient care (Bullard et al., 2009).

Long boarding times have negative impacts on patient outcomes and satisfaction. Boarded patients are less likely to receive required medical attention and care in a timely manner. This delayed transition has significant potential to put patients in danger, especially critically ill patients (Trzeciak & Rivers, 2003). Patients who stay long in the ED may not receive care by continuity teams, which increases the possibility of undesirable events such as preventable adverse events (Liu, Thomas,

Gordon, Hamedani, & Weissman, 2009). Studies showed that long boarding times can increase hospital mortality rates (Forero et al., 2010; Singer, Thode, Viccellio, & Pines, 2011). Not surprisingly, long boarding times lower patient satisfaction (Pines et al., 2008), which can impact hospital reputation and incentives.

The primary cause of boarding is the lack of inpatient beds. The obvious solution to reducing boarding is to add capacity to inpatient units. One institution increased the number of ICU beds from 47 to 67. The capacity expansion led to a decrease in the length of stay of patients who were admitted to the ICU and a reduction in ambulance diversion hours (McConnell et al., 2005). An alternative solution is to open a new area remote from the main ED, in which admitted patients are held until they are transferred to a hospital floor. Studies showed that opening this kind of short-stay space (e.g., observation unit, holding unit, acute care unit) resulted in reducing the number of boarding patients in the ED and the length of stay of both admitted and regular patients in the ED (Bazarian, Schneider, Newman, & Chodosh, 1996; Gómez-Vaquero et al., 2009). Those areas also contributed to mitigating the overall ED overcrowding, which was measured by decreased ambulance diversions and LWOS rates (Kelen, Scheulen, & Hill, 2001).

OR Scheduling

Despite the system-wide operational and quality impact of the surgical schedule, in most cases, the OR alone determines the surgical schedule. Most of the OR schedules are "block" based, where individual service lines or surgeons have reserved block times for elective surgeries. In addition, a few trauma rooms with no blocks assigned are reserved to ensure patient access to care for emergent/urgent cases. The elective blocks are allocated to surgeons based on historical caseload and utilization rates. Surgical cases are typically added to the blocks by surgeons or service lines. Occasionally, emergent/urgent cases are performed within the elective blocks assigned to a surgeon to ensure access to care. Due to the stochastic nature of the surgical demand, there can be end-of-the-day unscheduled block time available for which surgeons may add on cases to achieve higher utilization during the day of the surgery or day prior. In general, block-scheduling practice creates considerable variability and low predictability of both number of cases performed and the hours of block time utilized. Such variability negatively impacts the expensive staffing and resource allocation in ORs.

In addition to reduced efficiency, the variability in surgical caseload may negatively affect the quality of care in the OR. On days with high numbers of emergent surgeries exceeding the capacity of trauma rooms, elective surgeries are delayed or put on hold to accommodate the emergent cases. Such delays force elective patients to wait long periods in perioperative units, causing patient and staff dissatisfaction. When demand exceeds capacity, staff may often become stressed and overworked, creating an opportunity for medical errors. As evident, OR design through reengineered scheduling policies could make significant improvements to patient flow in OR operations and overall system performance.

Effective scheduling is one key to smoothing flow (i.e., reducing variability) in the OR. In the recent past, some hospitals have adopted a scheduling policy based on separating elective cases from emergency cases and dedicating an appropriate number of ORs each day solely for emergent/urgent cases. While this policy claims to enhance efficient scheduling with high predictability and staff satisfaction, no research has been done on sufficient and appropriate conditions to adopt such a policy. Some of the ORs in the country may not be good candidates to adopt this policy based on various factors, including current OR utilization, room turnover times, inpatient bed occupancy, type of surgeries performed, and staffing policies. Blindly adopting such policies can impose extreme stress on the OR that could negatively impact the entire system performance.

Thus, the development of optimal OR scheduling policies to maximize total system performance based on a given set of system metrics can help improve patient flow. These optimal policies would guide the selection of a scheduling policy to reengineer the ORs based on current system metrics and parameters. Multiple criteria should be considered when seeking to optimize performance of this complex system, such as maximizing OR utilization and minimizing patient boarding time and ensuring timely access to care. Given the OR demand variability and limited healthcare resources impacting patient flow, health systems engineers must (1) determine the set of optimal scheduling policies that satisfy conflicting system objectives, and (2) quantify their respective impacts on the overall system performance.

Hospitals struggle with overcapacity because of the widespread practice of scheduling elective surgeries on specific days, rather than spreading them evenly throughout the week.

Variability

Normal variability is inevitable, but other sources of variability can be manageable. Compared to other healthcare settings, acute care hospitals are exposed to more variability for several reasons. Care providers in a hospital deal with a wide range of different diseases with different levels of severity. Since they have different capabilities to treat patients and manage cases, the speed and quality of work may vary. Many times, patient arrivals are unexpected and hard to control. This variability and uncertainties cause most of the flow problems. Variability in patient flow also has a significant effect on hospitals with respect to staffing levels, workload of physicians and nurses, and bed occupancy (Litvak, 2012).

The OR accounts for a significant proportion of the inpatient admissions to a hospital. The inherent variability of emergent/urgent surgical demand coupled with variability associated with scheduling elective surgical demand impacts the daily operations of the OR, including perioperative units and ancillary services and inpatient hospitals.

3.3 CARE TRANSITIONS

The term *care transitions* refers to a process in which a patient's care shifts between healthcare practitioners and across settings as the patient's health condition and care needs change (Center for Improving Value in Health Care, 2012). Poorly managed transitional care causes incomplete education on patient self-management, inappropriate follow-up care, and medical errors, which result in decreasing quality of care (Coleman, Parry, Chalmers, & Min, 2006). These patients in the vulnerable transition period may end up being readmitted to the hospital (Peikes, Chen, Schore, & Brown, 2009). Fragmented care transitions can also lead to financial burdens for individuals and their society. For example, duplicate tests in different care settings cause unnecessary costs and increased utilization of expensive resources (Lonowski, n.d.). It was estimated that \$25 to \$45 billion was spent in 2011 in the United States for avoidable complications and preventable hospital readmissions (Burton, 2012). In this respect, timely and high-quality care transitions are important for improving patient outcomes while containing healthcare costs.

In this section, we review various care transitions patients may experience in a healthcare system during the course of their illnesses. Transitions between different care units affect patient flow, staff, and operations.

Transfers within a Hospital

Patients frequently move through a hospital to receive different levels of care by different care providers, from the ED to an OR to an *intensive care unit (ICU)* to a step-down unit to a general medical-surgical unit (Naylor & Keating, 2008). Similar to what was discussed in the admission and discharge process, these transfers can be as critical. As not all care provisions are explicitly evidence based, there is interpretation of signs, symptoms, and tests that go into the diagnostic process. With the transfer inside the hospital, the transition from one hospitalist to the next may result in a care plan change based on interpretation. Similarly, the condition of the patient may require an upgrade in care (to ICU) or a downgrade (to floor), signaling a transition in care. Standard data transmission and clear communications during these transitions have suggested improved safety, but standards of process may not be in place.

Patient Transfer from the ED to Inpatient

One of the significant care transitions within a hospital is an admission of a patient from the ED to an inpatient unit. The majority of patients who visit the ED are sent home after receiving care. However, some patients who need further observation or intensive treatment are admitted to a hospital. During this care transition from the ED to inpatient units, patients frequently experience delays.

Increasing capacity may be effective but costly. Other potential solutions for rapid transfer of admitted patients include reducing variability in hospital admissions and improving patient flow in inpatient units. As explained in the previous section, managing and eliminating variability is critical to improving patient flow. Smoothing the flow of patients in and out of institutions can help to reduce wide fluctuations in occupancy rates and prevent surges in patient visits that lead to delays in care. Admitted patients through the ED compete with elective admissions for hospital beds. Appropriate scheduling for elective admissions (surgical and nonsurgical) during the course of a week can decrease the variability in demand for hospital beds (Olshaker & Rathlev, 2006). A study conducted by Levin et al. (2008) showed the potential impact of rescheduling an elective catheterization case on a reduction in boarding times. Vermeulen et al. (2009) showed that variability in the admission-discharge ratio of inpatient units affected ED length of stay. Another way hospital systems attempt to improve the balance between the inflow and the outflow of inpatients is by discharging patients early in the day when appropriate. Using simulation techniques, studies showed the potential impact of the different discharge patterns on reducing boarding times and improving patient flow (Powell et al., 2012; Shi, Chou, Dai, Ding, & Sim, 2015).

Other approaches to reducing boarding times include holding admitted patients in hallways (Garson et al., 2008; Richards et al., 2011), cancelling elective surgeries (Asplin et al., 2008), having a transition team who care for inpatients boarded in the ED (Schneider et al., 2001), and better communication between inpatient units and the ED (Richards, Linden, & Derlet, 2014).

Primary Care to Specialty Care

Patients who visit primary care are frequently referred to specialty care for various reasons. Each year, more than one-third of patients in the United States are referred to a specialist (Forrest, Majeed, Weiner, Carroll, & Bindman, 2002). The number of visits resulting in referrals to specialty care increased 159% nationally from 1999 to 2009 (Barnett, Song, & Landon, 2012). The patients who have specialty care are sent back to primary care for follow-up. In other words, patients continuously move back and forth between these two care settings.

To make these care transitions more smoothly, an effective referral system and information sharing is crucial. However, studies have indicated that a referral process poses some structural and operational challenges, and key patient information for care is commonly missed during this transition (Kim-Hwang et al., 2010; Kinchen, Cooper, Levine, Wang, & Powe, 2004; Mehrotra, Forrest, & Lin, 2011). These challenges often lead to delays in specialty care referrals, duplicated testing, and increased risk of adverse events (Lin, 2012). To overcome the challenges, it is required to promote adequate communication between referring clinicians and specialists, and a standardized referring process supported by information technology.

Inpatient Setting to Long-Term Care Facilities

Discharge from the hospital requires care transitions to other care settings, including home, primary care, specialty care, and long-term care facilities. Patients are vulnerable during this transition period, which increases the potential risk of medical errors (Kripalani, Jackson,

Schnipper, & Coleman, 2007). Considering the volume and medical impact of these care transitions, it is essential to make this patient flow effectively and smoothly.

However, there are several gaps in care transition from a hospital to other care settings. One of the barriers to care transitions to primary care is incomplete or lack of discharge orders that misses key information about patients' hospitalizations such as treatment, medications, and tests (Burton, 2012). Often, patients do not have timely and consistent follow-up care by their primary care physicians after they leave the hospital, which increases the possibility of readmission or ED visits ("Implementing health reform: Community based care transitions program," 2011).

Care transitions to other post–acute care facilities involve more complicated processes and multiple players. During a discharge-planning phase, the appropriate post-hospital discharge destination is determined based on patients' care needs and physicians' medical decision. Patients and their caregivers may choose one from a list of available facilities, with aid from a care manager or social worker. The decision should also comply with federal and state health and safety standards. Then a discharge coordinator team assesses whether the facility has appropriate equipment, care teams, and environment that meet patients' care needs. Also, it should be ensured that the patient transition meets the facilities' admission policies and requirements. If these conditions are satisfied and the facility agrees to admit the patient, required arrangements are made. This complicated process makes this transition vulnerable to medical errors and safety challenges (Sandvik, Bade, Dunham, & Hendrickson, 2013). For safe transitions to post–acute care facilities, it is key to improve communication between hospital and the receiving facilities and effectively convey information about the patients (King et al., 2013). Also, appropriate assessment of the patient is required prior to transfer (American Medical Directors Association, 2010).

Inpatient Setting to Home

Home is also an important care venue where patients are cared for after leaving hospital. Upon discharge, effective education of patients and their caregivers is key to preventing complications and adverse outcomes. The key education components include but are not limited to: instructions for medications (what, when, and how to take), symptoms to monitor, self-care instructions, follow-up plans, and how to contact if problems occur (Kripalani et al., 2007). Sufficient information and effective physician-patient communication help increase patients' adherence to recommend medications and care guidelines. Also, it is important to assess the needs of patients and their families and provide appropriate support to meet the needs for safe transition to home.

3.4 PROCESS MAPPING

Efficient patient flow is crucial for safe and quality care and effective use of healthcare resources. Patient waits, delays, bottlenecks, and backlogs are typically not the result of lack of effort or commitment on the part of the providers and staff, nor can these issues be solved by working harder. Variability in processes causes delays in patient flow. Root causes of poor care transitions across different care providers and settings include lack of communication, standardized procedures, and information system. For more effective and smooth patient flow, these challenges should be overcome at a system level, not an individual part of a system. The key for improving flow of patients stem in the redesign of the overall, system-wide work process that engender the patient flow problems.

Thus, *process mapping* is a cornerstone for improving patient flow, since optimal care can be delivered only when the right patient is with the right provider in the right place with the right information at the right time. By reducing waits and delays in hospital care, the flow of patients and information through the health system can be greatly improved; this improvement of flow also can lead to better patient/provider satisfaction, increased access to care, better patient outcomes, and lower costs.

Benefits of Process Mapping

A process is a series of steps or actions performed to achieve a specific purpose. Process mapping is a graphical depiction of a process showing inputs, outputs, and steps in a sequential manner. It can be used many times to understand what goes into a process, how a process interacts with other processes, and how to optimize different processes. In simple terms, it helps to identify the flow of events in a process.

A process map can be used as a business tool, a communication tool, or an organization tool. It can be used to capture workflow of an entire business unit to bring information technology and operations together. It is important to note that at a basic level, a process map is not more than a flow chart outlining a specific process. There are many benefits to using process mapping:

- Graphically identify the steps in a process.
- Visually show the complexity of a process and identify sources of non-value-added tasks (i.e., waste).
- Identify the key process input variables that go into a process step and the resultant key output variables.
- Builds on basic flowcharting by adding more process information.

In other words, process mapping can help visualize, analyze, and improve on processes by finding steps that you can combine, and discover processes in the map that can be eliminated to improve efficiency. Further, it is an effective way to identify bottlenecks and redundancies necessary to save time and simplify projects.

Note that a value-added task must be performed to meet customer needs, add form or feature to a service, enhance service quality, and be one in which the customer is willing to pay for. On the other hand, waste is non-value-added task, such as handling beyond what is minimally required to move work, rework to fix errors, duplicative efforts, wait and idle times, delays, unnecessary motion, and overprocessing (taking too many steps to complete the job).

Creating a Process Map

Creating a process map may involve the following steps:

1. Define the scope of the process by clearly defining where the process starts and stops (i.e., boundaries).
2. Document all of the steps in the process. Be sure to list and sequence all intermediate steps. This should be done by walking through the process to identify activities and decision points. Remember to map the process as it actually happens, not how it should happen.
3. List all inputs and outputs at each process step. Be sure to label each input as controllable (can be changed), standard operating procedure (standard method), or noise (cannot be controlled). Each step must be classified using the appropriate symbol.
4. Finalize the process flow and ask several questions. What are the redundancies that could be removed? Are there steps to appear

out of order? Are there steps that appear to be missing that could improve the process? Where is time lost in the process? What are the non-value-added steps?

When making a process map, there are several things to consider. First, ensure that there is a logical start and finish, as well as a general direction that a reader can follow. Second, be sure to use the correct symbol in representing the process step, and remember that each symbol should represent only one step.

An oval is used to show the materials, information, or action to start the process or to show the results at the end of the process. A box/rectangle is used to show a task or activity performed in the process. Although multiple arrows may come into each box, usually only one output or arrow leaves each activity box. A diamond shows those points in the process where a yes/no decision is required or question is being asked. Arrows show the direction or flow of the process. Additional process map symbols are depicted in Figure 3.2. Note that process mapping uses the same symbols and rules as flowcharting but is applied to a specific process.

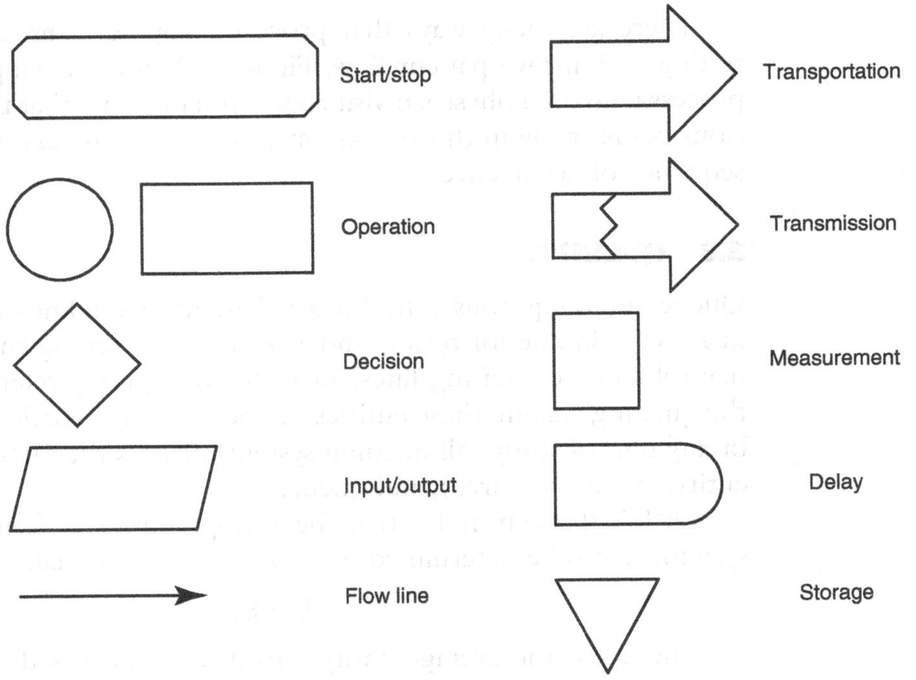

Figure 3.2 Process Map Symbols

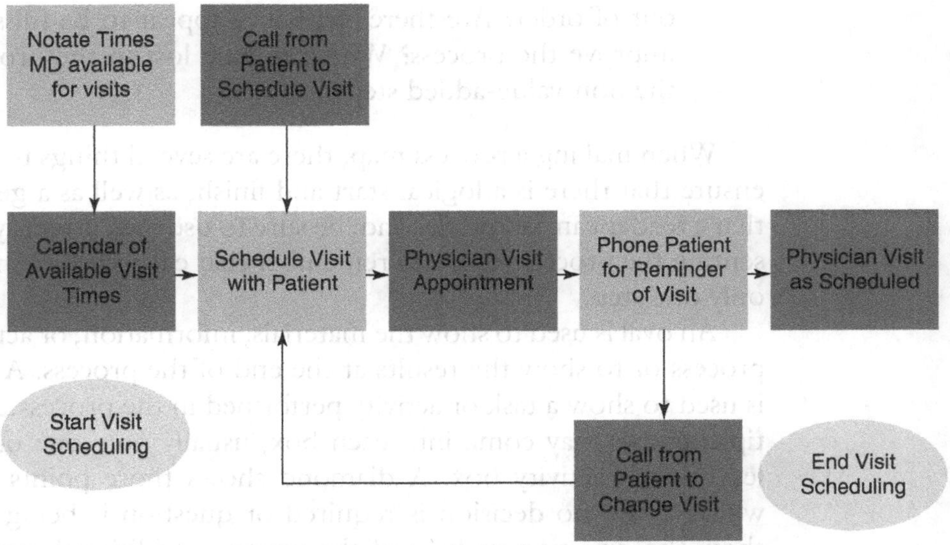

Figure 3.3 Process Map for a Physician Visit Appointment
Source: **Cousins M., Follow the Map2003; 2003.**
Available at: http://saferpak.com/process_mapping_art2.htm

There are many ways that process mapping is used in a healthcare setting to improve patient flow. Figure 3.3 shows a simple example of a process map for a physician visit appointment. Note that there are no decisions being made in this process map, but each process step is mapped in sequence of occurrence.

3.5 QUEUING

Queues are ubiquitous in healthcare. Patients get in line for care; radiology images are in line for review and assessment. Queuing theory is a mathematical study of waiting lines. Broadly, we typically refer to "entities" in the queuing system; these entities can be customers, patients, data packets, or any unit of study. All queuing systems possess the same basic elements: entity, server (resource), and queue.

Little's theorem holds that the average number of entities in a stable system, L, can be determined from the following equation:

$$L = \lambda T$$

where λ is the average entity arrival rate and T is the average service time for an entity. If we consider, for example, a walk-in clinic where

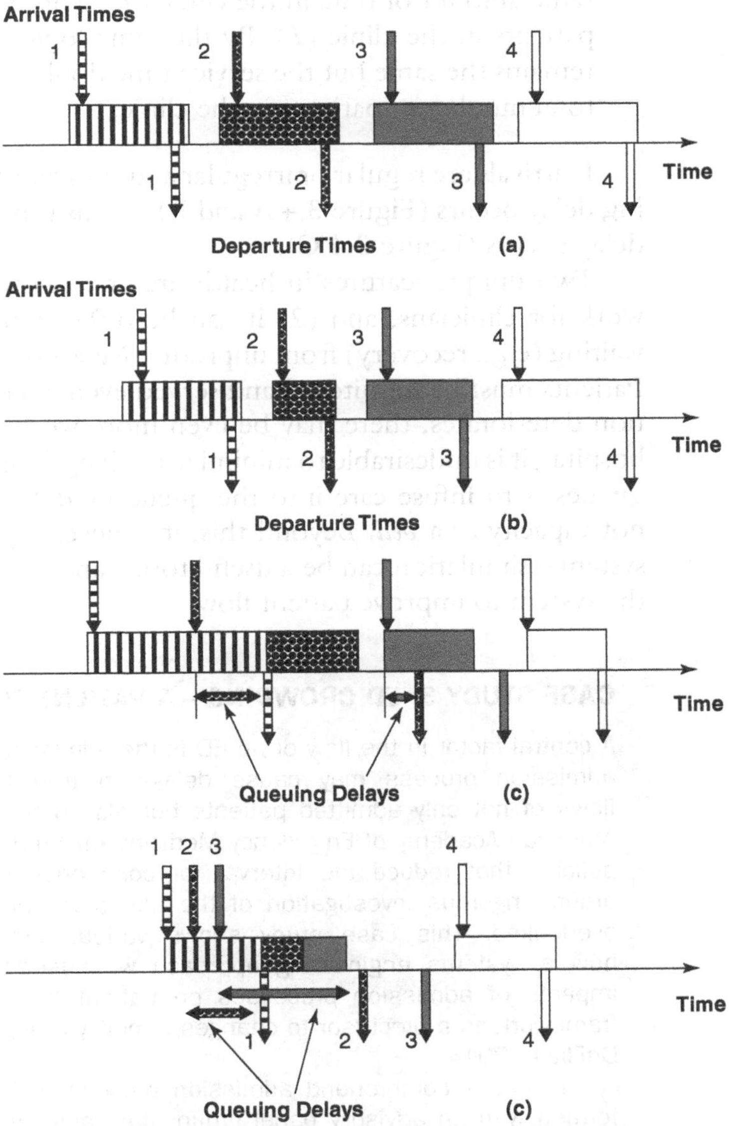

Figure 3.4 (a) Regular arrivals and departures spaced so there's no queue. (b) Irregular arrivals and departures spaced so there's no queue. (c) Regular arrivals with service length variation cause queuing delays. (d) Irregular arrivals queues with queuing delays

the customer arrival rate (λ) doubles but the patients still spend the same amount of time in the clinic (T). This will double the number of patients in the clinic (L). By the same logic, if the patient arrival rate remains the same but the service time doubles, this will also double the total number of patients in the clinic.

If arrivals are regular or irregular and sufficiently spaced apart, no queuing delay occurs (Figure 3.4 A and B). If this is not the case, then queuing delay occurs (Figure 3.4 C).

Two unique features in healthcare are: (1) waiting creates additional work for clinicians, and (2) it can be difficult to distinguish productive waiting (e.g., recovery) from unproductive waiting (e.g., waiting for tests). Patients must be monitored and served even while waiting. If their condition deteriorates, there may be even more work once they are seen. In a hospital, it is undesirable to minimize the length of stay. One way to address queues is to infuse care into the queue by defining the capacity to care, not capacity as a *bed*. Beyond this, it is necessary to analyze the queuing system—simulation can be a useful tool—and use these results to redesign the system to improve patient flow.

CASE STUDY 3: ED CROWDING – A PATIENT FLOW SOLUTION

A central factor in the flow of an ED is the admission process. An inefficient admission process may cause delays in transition and affect patient flows of not only admitted patients but also other ED patients. While the American Academy of Emergency Medicine encourages hospitals to develop policies that reduce the interval for completion of inpatient admissions orders, rigorous investigation of the admission process has been largely overlooked. This case study shows various admission processes and how a systems engineering approach was used to investigate and the impacts of admission processes on patient flow in the ED, providing a framework as a precursor to changes in policy (Kang, Nembhard, Rafferty, & DeFlitch, 2014).

To better comprehend admission process policies, a survey was performed with an advisory panel group that consisted of eight physicians in the ED and internal medicine. The survey asked the physicians to "describe the admission process carried out in a hospital in which you had previously worked (not Hershey Medical Center)." Based on literature and a survey, admission process policies (APPs) are classified into four types, as indicated in Table 3.2.

Table 3.2 Classification of Admission Process Policies (APPs)

		Provider Making Admission Decisions	
		Admitting Service	**Emergency Medicine**
Admission decision maker	Physician group decision making (Attending, resident, intern and/or physician extender)	Type 1: Decision by group of physicians on the admitting service	Type 3: Decision by group of physicians in the ED
	Attending physicians Exclusively	Type 2: Decision by attending physicians on the admitting service	Type 4: Decision by attending physicians in the ED

A case study was conducted in a tertiary care, suburban academic medical center in Pennsylvania with a level 1 trauma designation where the ED has 55 beds and 64,000 patients a year. In order to understand the current admission system of the hospital, a time-motion study and interview with physicians, nurses, and a bed management manager was conducted. Figure 3.5 shows patient flow from the ED to an inpatient unit in the study hospital.

Using discrete event simulation based on the hospital data, this study assessed the impact of six different admission processes on patient flow for discharged and admitted patients. The simulation results suggested several important implications. First, the flow of the internal medicine (IM) admitted patients was strongly dependent upon the APP. When compared to the currently employed policy (Type 1), their ED length of stay decreased by 14% to 21%. This means that patients may spend 1.4 to 2 hours less on average in the ED before being admitted to IM under a new APP. This improvement will likely have a positive impact on patient satisfaction, staff workflow, and performance measures for evaluations.

This study also showed that hospital APPs from the ED to an inpatient ward affect both admitted and discharged ED patients. For instance, a small change in procedures can lead to a substantial reduction in length of stay and patient flow. Through the classification of existing APPs and the analysis of simulation results, this study contributed to demonstrating the potential value of leveraging APPs and developing a framework for pursuing these policies.

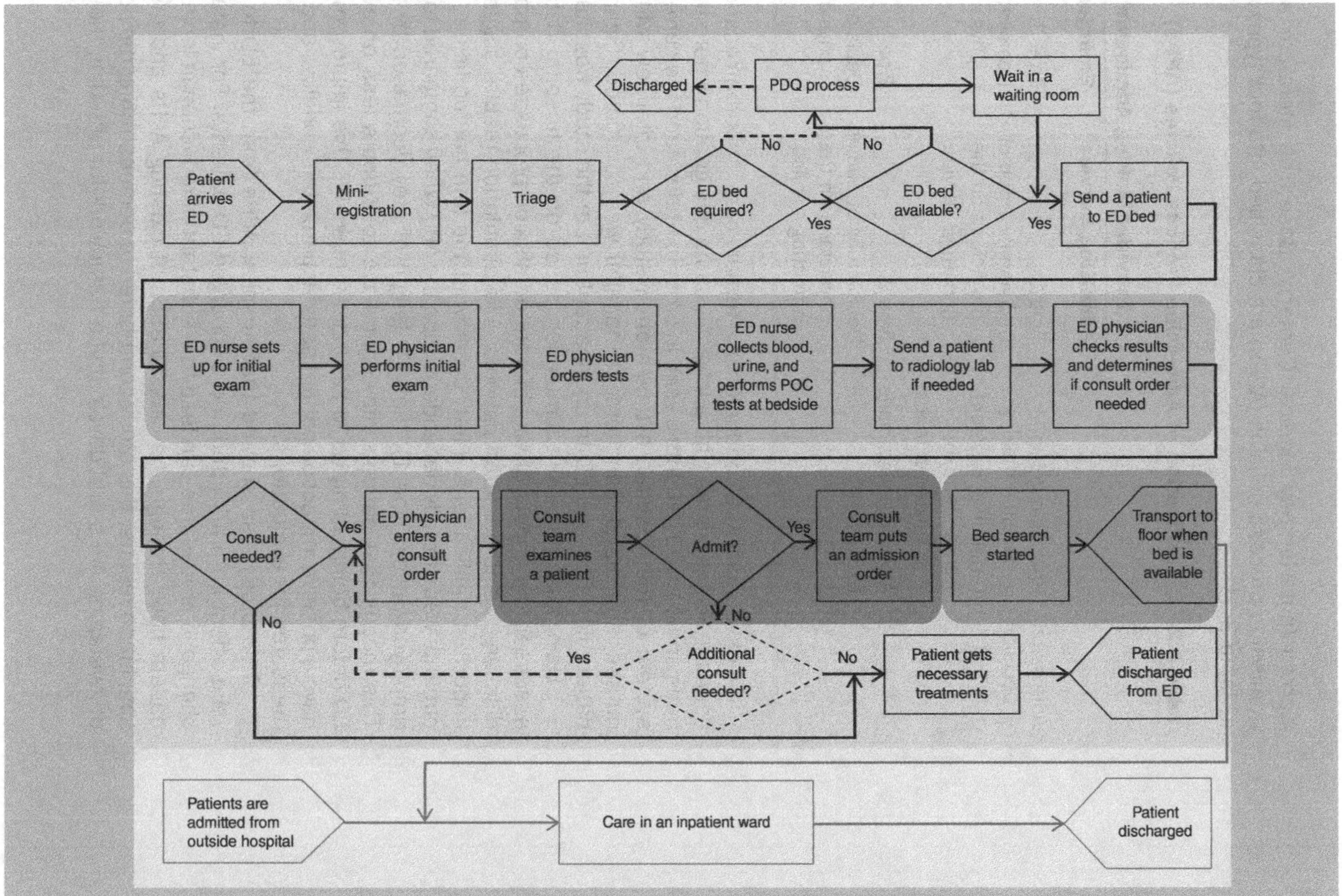

Figure 3.5 A Process Map of Patient Flow from the ED to Inpatient Ward

QUESTIONS AND LEARNING ACTIVITIES

1. What are the similarities and differences between the main healthcare settings?
2. What patient flow issues exist in each of the different healthcare settings?
3. Explain the problems involving ED crowding and discuss potential ways to reduce variability in an effort to improve patient flow.

REFERENCES

American College of Emergency Physicians (ACEP). (2011). Definition of boarded patient. Retrieved from http://www.acep.org/Content.aspx?id=75791

American Hospital Association. (2012). Prepared to care. Retrieved from http://www.aha.org/content/12/preparedtocare.pdf.

American Medical Directors Association. (2010). Improving care transitions between the nursing facility and the acute-care hospital settings. Retrieved from https://www.nhqualitycampaign.org/files/Transition_of_Care_Reference.pdf

Amini, M., Otondo, R. F., Janz, B. D., & Pitts, M. G. (2007). Simulation modeling and analysis: A collateral application and exposition of RFID technology. *Production and Operations Management*, *16*(5), 586–598. Retrieved from http://search.proquest.com/docview/228766772

American College of Emergency Physicians (2008). Emergency department crowding: High-impact solutions. ACEP Task Force Report on Boarding. Retrieved from http://www.acep.org/workarea/DownloadAsset.aspx?id=50026

Asplin, B., Magid, D. J., Rhodes, K. V, Solberg, L. I., Lurie, N., & Camargo, C. A. (2003). A conceptual model of emergency department crowding. *Annals of Emergency Medicine*, *42*(2), 173–180. Retrieved from http://doi.org/10.1067/mem.2003.302

Barnett, M. L., Song, Z., & Landon, B. E. (2012). Trends in physician referrals in the United States, 1999–2009. *Archives of Internal Medicine*, *172*(2), 163–170. Retrieved from http://doi.org/10.1001/archinternmed.2011.722

Bazarian, J. J., Schneider, S. M., Newman, V. J., & Chodosh, J. (1996). Do admitted patients held in the emergency department impact the throughput of treat-and-release patients? *Academic Emergency*

Medicine, *3*(12), 1113–1118. Retrieved from http://doi.org/10.1111/j.1553-2712.1996.tb03370.x

Billings, J., Parikh, N., & Mijanovic, T. (2000). *Emergency department use in New York City: A substitute for primary care*? Retrieved March 22, 2013, from http://www.commonwealthfund.org/Publications/Issue-Briefs/2000/Mar/Emergency-Department-Use-in-New-York-City--A-Substitute-for-Primary-Care.aspx

Bullard, M. J., Villa-Roel, C., Bond, K., Vester, M., Holroyd, B. R., & Rowe, B. H. (2009). Tracking emergency department overcrowding in a tertiary care academic institution. *Healthcare Quarterly*, *12*(3), 99–106.

Burton, R. (2012). Health policy brief: Care transitions. *Health Affairs*. Retrieved from http://www.healthaffairs.org/healthpolicybriefs/brief.php?brief_id=76

Center for Improving Value in Health Care (2012). Improving care transitions: A strategy for readmissions. Policy Issue Brief. Retrieved from http://www.civhc.org/getmedia/475510fc-ff67-4655-a915-c9539117eb2f/Care-Transitions-Policy-Brief-FINAL_9.2012.pdf.aspx

Centers for Medicare & Medicaid Services. (2015). Official hospital compare data. Retrieved May 1, 2015, from https://data.medicare.gov/data/hospital-compare

Coleman, C., Hammerschmith, M., & Duvall, D. (2013). The path to enterprise locating. *Health Management Technology*, *34*(6), 22–23. Retrieved from http://search.proquest.com/docview/1399962555/abstract?accountid=13158

Coleman, E. A., Parry, C., Chalmers, S., & Min, S.-J. (2006). The care transitions intervention: Results of a randomized controlled trial. *Archives of Internal Medicine*, *166*(17), 1822–1828. Retrieved from http://doi.org/10.1001/archinte.166.17.1822

Cooke, M. W., Wilson, S., & Pearson, S. (2002). The effect of a separate stream for minor injuries on accident and emergency department waiting times. *Emergency Medicine Journal*, *19*(1), 28–30. Retrieved from http://www.pubmedcentral.nih.gov/articlerender.fcgi?artid=1725754&tool=pmcentrez&rendertype=abstract

DeFlitch, C., G. Geeting, & H. L. Paz. (2015). Reinventing emergency department flow via healthcare delivery science. *Health Environments Research & Design Journal*, *8*(3), 105–115.

Derlet, R. W., & Richards, J. R. (2000). Overcrowding in the nation's emergency departments: Complex causes and disturbing effects. *Annals of Emergency Medicine*, *35*(1), 63–68. Retrieved from http://www.ncbi.nlm.nih.gov/pubmed/10613941

Derlet, R. W., Richards, J. R., & Kravitz, R. L. (2001). Frequent overcrowding in U.S. emergency departments. *Academic Emergency Medicine*, *8*(2), 151–155. Retrieved from http://www.ncbi.nlm.nih.gov/pubmed/11157291

Fernandes, C. M., Christenson, J. M., & Price, A. (1996). Continuous quality improvement reduces length of stay for fast-track patients in an emergency department. *Academic Emergency Medicine*, *3*(3), 258–263. Retrieved from http://www.ncbi.nlm.nih.gov/pubmed/8673783

Forero, R., Hillman, K. M., McCarthy, S., Fatovich, D. M., Joseph, A. P., & Richardson, D. B. (2010). Access block and ED overcrowding. *Emergency Medicine Australasia : EMA*, *22*(2), 119–135. Retrieved from http://doi.org/10.1111/j.1742-6723.2010.01270.x

Forrest, C. B., Majeed, A., Weiner, J. P., Carroll, K., & Bindman, A. B. (2002). Comparison of specialty referral rates in the United Kingdom and the United States: retrospective cohort analysis. *BMJ*, *325*(7360), 370–371. Retrieved from http://www.pubmedcentral.nih.gov/articlerender.fcgi?artid=117891&tool=pmcentrez&rendertype=abstract

Garson, C., Hollander, J. E., Rhodes, K. V, Shofer, F. S., Baxt, W. G., & Pines, J. M. (2008). Emergency department patient preferences for boarding locations when hospitals are at full capacity. *Annals of Emergency Medicine*, *51*(1), 9–12, 12.e1–3. http://doi.org/10.1016/j.annemergmed.2007.03.016

Georgiou, A., Prgomet, M., Paoloni, R., Creswick, N., Hordern, A., Walter, S., & Westbrook, J. (2013). The effect of computerized provider order entry systems on clinical care and work processes in emergency departments: A systematic review of the quantitative literature. *Annals of Emergency Medicine*, *61*(6), 644–653.e16. Retrieved from http://doi.org/10.1016/j.annemergmed.2013.01.028

Gómez-Vaquero, C., Soler, A. S., Pastor, A. J., Mas, J. R. P., Rodriguez, J. J., & Virós, X. C. (2009). Efficacy of a holding unit to reduce access block and attendance pressure in the emergency department. *Emergency Medicine Journal: EMJ*, *26*(8), 571–572. Retrieved from http://doi.org/10.1136/emj.2008.066076

Grumbach, K., Keane, D., & Bindman, A. (1993). Primary care and public emergency department overcrowding. *American Journal of Public Health*, *83*(3), 372–378. http://doi.org/10.2105/AJPH.83.3.372

Hing, E., & Bhuiya, F. (2012). Wait time for treatment in hospital emergency departments: 2009. Atlanta, GA. Retrieved from http://www.cdc.gov/nchs/data/databriefs/db102.htm

Hoot, N. R., & Aronsky, D. (2008). Systematic review of emergency department crowding: causes, effects, and solutions. *Annals of Emergency Medicine*, *52*(2), 126–136. Retrieved from http://doi.org/10.1016/j.annemergmed.2008.03.014

Implementing health reform: Community based care transitions program. (2011). *Family Practice News*, *41*(6), 67.

Institute for Healthcare Improvement. (2003). *Optimizing patient flow: Moving Patients Smoothly Through Acute Care Settings*. IHI Innovation Series white paper. Boston: Institute for Healthcare Improvement.

Kang, H., Nembhard, H. B., Rafferty, C., & DeFlitch, C. J. (2014). Patient flow in the emergency department: A classification and analysis of admission process policies. *Annals of Emergency Medicine*, *64*(4), 335–342.e8. Retrieved from http://doi.org/10.1016/j.annemergmed.2014.04.011

Kelen, G. D., Scheulen, J. J., & Hill, P. M. (2001). Effect of an emergency department (ED) managed acute care unit on ED overcrowding and emergency medical services diversion. *Academic Emergency Medicine: Official Journal of the Society for Academic Emergency Medicine*, *8*(11), 1095–1100. Retrieved from http://www.ncbi.nlm.nih.gov/pubmed/11691675

Kim-Hwang, J. E., Chen, A. H., Bell, D. S., Guzman, D., Yee, H. F., & Kushel, M. B. (2010). Evaluating electronic referrals for specialty care at a public hospital. *Journal of General Internal Medicine*, *25*(10), 1123–1128. Retrieved from http://doi.org/10.1007/s11606-010-1402-1

Kinchen, K. S., Cooper, L. A., Levine, D., Wang, N. Y., & Powe, N. R. (2004). Referral of patients to specialists: factors affecting choice of specialist by primary care physicians. *Annals of Family Medicine*, *2*(3), 245–252. Retrieved from http://www.pubmedcentral

.nih.gov/articlerender.fcgi?artid=1466676&tool=pmcentrez&rendertype=abstract

King, B. J., Gilmore-Bykovskyi, A. L., Roiland, R. A., Polnaszek, B. E., Bowers, B. J., & Kind, A. J. H. (2013). The consequences of poor communication during transitions from hospital to skilled nursing facility: a qualitative study. *Journal of the American Geriatrics Society*, *61*(7), 1095–1102. Retrieved from http://doi.org/10.1111/jgs.12328

Kripalani, S., Jackson, A. T., Schnipper, J. L., & Coleman, E. A. (2007). Promoting effective transitions of care at hospital discharge: A review of key issues for hospitalists. *Journal of Hospital Medicine*, *2*(3), 314–323.

Levin, S. R., Dittus, R., Aronsky, D., Weinger, M. B., Han, J., Boord, J., & France, D. (2008). Optimizing cardiology capacity to reduce emergency department boarding: a systems engineering approach. *American Heart Journal*, *156*(6), 1202–1209. Retrieved from http://doi.org/10.1016/j.ahj.2008.07.007

Lin, C. Y. (2012). Improving care coordination in the specialty referral process between primary and specialty care. *North Carolina Medical Journal*, *73*(1), 61–62.

Litvak, E. (2012). Bridging the gap: Operations management science, patient safety and healthcare cost. Retrieved from http://www.iom.edu/~/media/Files/Activity Files/Quality/VSRT/IC Meeting Docs/SAIHIC 12-14-12/Eugene Litvak.pdf

Liu, S. W., Thomas, S. H., Gordon, J. A., Hamedani, A. G., & Weissman, J. S. (2009). A pilot study examining undesirable events among emergency department-boarded patients awaiting inpatient beds. *Annals of Emergency Medicine*, *54*(3), 381–385. Retrieved from http://doi.org/10.1016/j.annemergmed.2009.02.001

Lonowski, S. (n.d.). Improving care transitions: A strategy for reducing readmissions. Retrieved from http://www.civhc.org/getmedia/475510fc-ff67-4655-a915-c9539117eb2f/Care-Transitions-Policy-Brief-FINAL_9.2012.pdf.aspx/

McConnell, K. J., Richards, C. F., Daya, M., Bernell, S. L., Weathers, C. C., & Lowe, R. A. (2005). Effect of increased ICU capacity on emergency department length of stay and ambulance diversion. *Annals of Emergency Medicine*, *45*(5), 471–478. Retrieved from http://doi.org/10.1016/j.annemergmed.2004.10.032

Mehrotra, A., Forrest, C. B., & Lin, C. Y. (2011). Dropping the baton: Specialty referrals in the United States. *The Milbank Quarterly*, *89*(1), 39–68. Retrieved from http://doi.org/10.1111/j.1468-0009.2011.00619.x

Muntlin Athlin, A., von Thiele Schwarz, U., & Farrohknia, N. (2013). Effects of multidisciplinary teamwork on lead times and patient flow in the emergency department: A longitudinal interventional cohort study. *Scandinavian Journal of Trauma, Resuscitation and Emergency Medicine*, *21*(1), 76. Retrieved from http://doi.org/10.1186/1757-7241-21-76

Naylor, M., & Keating, S. A. (2008). Transitional care: Moving patients from one care setting to another. *American Journal of Nursing*, *108*(9 Suppl), 58–63; quiz 63. Retrieved from http://doi.org/10.1097/01.NAJ.0000336420.34946.3a

Niska, R., Bhuiya, F., & Xu, J. (2010). National hospital ambulatory medical care survey: 2007 emergency department summary. *National Health Statistics Reports*, *26*, 1–31. Retrieved from http://www.ncbi.nlm.nih.gov/pubmed/20726217

Olshaker, J. S., & Rathlev, N. K. (2006). Emergency department overcrowding and ambulance diversion: The impact and potential solutions of extended boarding of admitted patients in the emergency department. *Journal of Emergency Medicine*, *30*(3), 351–356. Retrieved from http://doi.org/10.1016/j.jemermed.2005.05.023

Peikes, D., Chen, A., Schore, J., & Brown, R. (2009). Effects of care coordination on hospitalization, quality of care, and healthcare expenditures among medicare beneficiaries. *JAMA*, *301*(6), 603. Retrieved from http://doi.org/10.1001/jama.2009.126

Pines, J. M., Iyer, S., Disbot, M., Hollander, J. E., Shofer, F. S., & Datner, E. M. (2008). The effect of emergency department crowding on patient satisfaction for admitted patients. *Academic Emergency Medicine*, *15*(9), 825–831. Retrieved from http://doi.org/10.1111/j.1553-2712.2008.00200.x

Pitts, S. R., Vaughns, F. L., Gautreau, M. A., Cogdell, M. W., & Meisel, Z. (2014). A cross-sectional study of emergency department boarding practices in the United States. *Academic Emergency Medicine: Official Journal of the Society for Academic Emergency Medicine*, *21*(5), 497–503. Retrieved from http://doi.org/10.1111/acem.12375

Powell, E. S., Khare, R. K., Venkatesh, A. K., Van Roo, B. D., Adams, J. G., & Reinhardt, G. (2012). The relationship between inpatient discharge timing and emergency department boarding. *Journal of Emergency Medicine*, *42*(2), 186–196. Retrieved from http://doi.org/10.1016/j.jemermed.2010.06.028

Richards, J. R., Linden, C. van der, & Derlet, R. W. (2014). Providing care in emergency department hallways: demands, dangers, and deaths. *Advances in Emergency Medicine*, *2014*. doi:10.1155/2014/495219

Richards, J. R., Ozery, G., Notash, M., Sokolove, P. E., Derlet, R. W., & Panacek, E. A. (2011). Patients prefer boarding in inpatient hallways: Correlation with the national emergency department overcrowding score. *Emergency Medicine International*, *2011*, 840459. Retrieved from http://doi.org/10.1155/2011/840459

Rowe, B. H., Guo, X., Villa-Roel, C., Schull, M., Holroyd, B., Bullard, M., … Innes, G. (2011). The role of triage liaison physicians on mitigating overcrowding in emergency departments: a systematic review. *Academic Emergency Medicine*, *18*(2), 111–120. Retrieved from http://doi.org/10.1111/j.1553-2712.2010.00984.x

Sandvik, D., Bade, P., Dunham, A., & Hendrickson, S. (2013). A hospital-to-nursing home transfer process associated with low hospital readmission rates while targeting quality of care, patient safety, and convenience: A 20-year perspective. *Journal of the American Medical Directors Association*, *14*(5), 367–374. Retrieved from http://doi.org/10.1016/j.jamda.2012.12.007

Schneider, S., Zwemer, F., Doniger, A., Dick, R., Czapranski, T., & Davis, E. (2001). Rochester, New York: A decade of emergency department overcrowding. *Academic Emergency Medicine: Official Journal of the Society for Academic Emergency Medicine*, *8*(11), 1044–1050. Retrieved from http://www.ncbi.nlm.nih.gov/pubmed/11691666

Schull, M. J., Lazier, K., Vermeulen, M., Mawhinney, S., & Morrison, L. J. (2003). Emergency department contributors to ambulance diversion: a quantitative analysis. *Annals of Emergency Medicine*, *41*(4), 467–76. Retrieved from http://doi.org/10.1067/mem.2003.23

Shi, P., Chou, M. C., Dai, J. G., Ding, D., & Sim, J. (2015). Models and insights for hospital inpatient operations: Time-dependent ED boarding time. *Management Science*. http://dx.doi.org/10.1287/mnsc.2014.2112

Singer, A. J., Thode, H. C., Viccellio, P., & Pines, J. M. (2011). The association between length of emergency department boarding and mortality. *Academic Emergency Medicine: Official Journal of the Society for Academic Emergency Medicine*, *18*(12), 1324–1329. Retrieved from http://doi.org/10.1111/j.1553-2712.2011.01236.x

Stahl, J. E., Drew, M. A., Leone, D., & Crowley, R. S. (2011). Measuring process change in primary care using real-time location systems: Feasibility and the results of a natural experiment. *Technology and Health Care*, *19*(6), 415–421.

Svirsky, I., Stoneking, L. R., Grall, K., Berkman, M., Stolz, U., & Shirazi, F. (2013). Resident-initiated advanced triage effect on emergency department patient flow. *Journal of Emergency Medicine*, *45*(5), 746–751. Retrieved from http://www.sciencedirect.com/science/article/pii/S0736467913003351

Surgical Safety Checklist [Online image]. (2008). Retrieved June 9, 2015, from http://www.who.int/patientsafety/safesurgery/tools_resources/SSSL_Checklist_finalJun08.pdf

Trzeciak, S., & Rivers, E. P. (2003). Emergency department overcrowding in the United States: An emerging threat to patient safety and public health. *Emergency Medicine Journal: EMJ*, *20*(5), 402–405. Retrieved from http://www.pubmedcentral.nih.gov/articlerender.fcgi?artid=1726173&tool=pmcentrez&rendertype=abstract

Vermeulen, M. J., Ray, J. G., Bell, C., Cayen, B., Stukel, T. A., & Schull, M. J. (2009). Disequilibrium between admitted and discharged hospitalized patients affects emergency department length of stay. *Annals of Emergency Medicine*, *54*(6), 794–804. Retrieved from http://doi.org/10.1016/j.annemergmed.2009.04.017

Williams, S. J., & Torrens, P. R. (2002). *Introduction to health services* (6th ed.). Albany, NY: Thomson Learning.

Chapter 4

Healthcare Financing

"The art of medicine consists of amusing the patient while nature cures the disease."
—Voltaire

Overview

The financing of healthcare systems is a very complicated process, and varies significantly by country. Further, hospitals have the difficult task of handling multiple patient types that pay through different mechanisms, through often heterogeneous information technologies, in a changing regulatory environment. In this chapter, we present an introduction on how health services are financed and how healthcare providers can properly allocate costs and make investment decisions.

4.1 FINANCING MODELS FOR HEALTH SERVICES

One of the complicating features in healthcare delivery is the relationship between the major stakeholders. This relationship varies by country, but essentially a patient receives care from a provider who is paid by a third party. This third party, or payer, is typically one of four forms: (1) out-of-pocket payment from the patient, (2) private insurance, (3) government-provided insurance, and (4) universal care. The complication arises from the payment mechanism. Since in most cases patients don't directly pay for their care, their consumption of care may be in excess of their need for care. Further, the payer only has information about the patient's needs through the provider. The payer, therefore, needs to protect itself against unnecessary services given to the patient.

Insurance and Third-Party Payment

Health insurance is provided in a variety of forms, including employer-based, patient-purchased, and through a government exchange. The role of health insurance is to reduce risk to the patient. We can illustrate with a simple example. Suppose a patient has an annual income I. Further, the utility that they receive from that income is $U(I)$, and there is diminishing marginal utility to income as shown in Figure 4.1. Suppose the patient has a 50% chance of medical expenses of $0.2I$, then their income could drop to $0.8I$, and on average their income would be $0.9I$ with utility U_1. Suppose further that $U(I) = 100$, $U(.9I) = 95$, and $U(.8I) = 80$. This gives $U_2 = 0.5\,U(I) + 0.90\,U(.8I) = 90$. The average medical expenses would be $0.1I$. Note that if an insurance provider offered insurance that cost (i.e., a premium of) $0.1I$, the patient would purchase the insurance since it would increase their utility from U_2 to U_1. In fact, the patient would pay up to A since their utility would be at least U_2. An insurer charging between $0.1I$ and A would cover their average expenses, and would maximize their revenue at A.

Agency Issues

Two potential problems in the insurance market are *adverse selection* and *moral hazard*. Both problems occur because patients, providers, and payers have different information; that is, there is an information

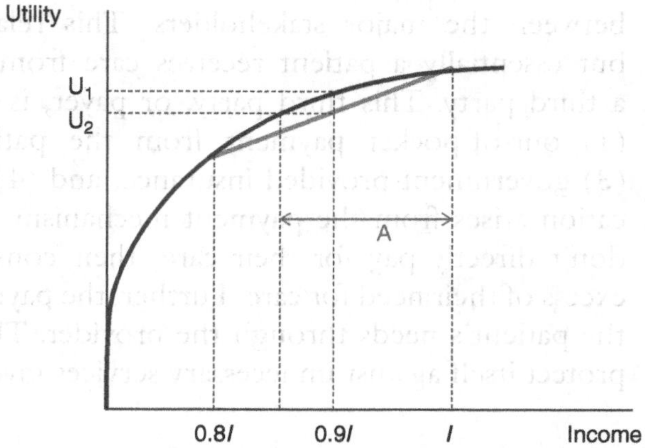

Figure 4.1 Utility versus Income for Health Insurance Example

asymmetry. When these lead to a potential conflict, they are known as agency problems.

Adverse selection is an insurance market failure due to the information asymmetry. Consider the following example: Suppose 25% of a population smokes and their average health costs are $1,500 per year greater than nonsmokers. Further, suppose nonsmokers have average health costs of $2,000 per year and are willing to pay an insurance premium of $2,300/year. Smokers are willing to pay a higher premium of $2,800 per year. If the insurer had full information about whether an individual were a smoker or nonsmoker, then they could offer a premium of $2,500 to a nonsmoker, and all nonsmokers in the population would buy the premium. If, however, the insurer did not know whether a person was a smoker or nonsmoker, their average costs would be $0.25(3500) + 0.75(2000) = \$2,735$. In order to make a profit, the insurer would need to charge a premium in excess of their costs. However, if they offered insurance at a premium at any value above this, only smokers would purchase it. The average cost of a smoker, however, is greater than this amount. In fact, at any premium up to $2,800, smokers would purchase it and the insurer would lose money. In order for the insurer to remain profitable, they would need to charge at least $3,500, but this is in excess of the willingness to purchase from smokers. In other words, the market would fail in this case due to lack of information from the insurer. In order to counter adverse selection, insurers try to remove the informational asymmetry through either detailed questionnaires or requiring physicals or some other form of testing. This process is known as ***underwriting***.

Moral hazard occurs when a party takes an action because they do not incur the cost of that action. As in adverse selection, it occurs due to the information asymmetry. In healthcare, a common form of moral hazard is from supplier-induced demand, where a healthcare provider gives more services than necessary to the patient since they will be paid for those services by the payer. This problem can be particularly prevalent when providers are given a "fee for service." In order to try to reduce moral hazard, insurers have developed various incentive-based contracts that allow the insurer and provider to share both the risks and the benefits. One example is a pay for performance scheme where a bonus is paid to the providers if they are able to meet certain quality targets. We will discuss this topic in more detail later in the chapter.

Government-Provided Insurance

The government plays a major stakeholder in healthcare, since in almost all countries it pays a significant component of the cost. In the United States, the government provides Medicaid for individuals below an income threshold and Medicare for individuals over 65 years of age, or roughly 31% of the population. In Japan, National Health Insurance is provided to persons not eligible for employer-based insurance. In Mexico, public care is provided for all citizens, though the amount of care that is subsidized is a function of employment status.

Often times, different levels of government contribute to care. For example, in the United States, Medicare, which was initiated in 1965, is completely provided at the federal level. It is largely funded by a payroll tax paid by employees as well as premiums paid by enrollees. Medicaid, however, is paid by both the state and federal governments, and management by the states. States set eligibility requirements, though after passage of the Affordable Care Act, a national eligibility standard was set at 133% of the federal poverty line.

Governments are also responsible for determining what acceptable healthcare outcomes are. This can also come in the form of penalties. An example is the "two-midnight" rule for Medicare, established by the Centers for Medicare & Medicaid Services (CMS). Medicare reimburses at a higher rate for inpatient services than for outpatient services. Under the two-midnight rule, a patient is only designated to inpatient status if they are admitted for a minimum of a two-night stay. Financial penalties are applied to the hospital by CMS if the rule is violated.

Universal Healthcare

Universal healthcare refers to a system where all citizens are provided a specified package of benefits by the government. Currently, over 50 countries supply some form of universal healthcare, including Argentina, Australia, France, Greece, Spain, and the United Kingdom. There are significant variations in these systems, however. These include the way they are funded, who is covered, what is covered, and the role of private insurance. Funding is primarily done through of mix of public funding (taxes) and private funding (employer/employee contributions). The private funding is compulsory in most cases.

A variation on universal healthcare is a compulsory insurance model. In this case, the government requires all citizens to purchase insurance. The insurance can either be a single source, such as done in Canada, or the individual may be able to select form a government provided or private source. The United States has moved toward this later model, though the purchase of insurance is not truly compulsory; a citizen can pay a penalty in lieu of purchasing insurance. One of the benefits of compulsory insurance is that it pools risk across the population, which helps to overcome adverse selection mentioned earlier.

4.2 COMPENSATION MODELS FOR PROVIDERS

In the previous section, we discussed general financing of the healthcare system. In this section we discuss how providers are compensated for their services. Although the list below is not complete, it does cover the major classes of reimbursements (Nowicki, 2011).

Fee for Service

In the *fee for service (FFS)* model, a provider is reimbursed for each service it provides. This may include an office visit, a blood test, or a tracheotomy. Payment is received after the service is performed.

One of the criticisms of this approach is that the provider is incentivized to provide more services than needed. This is called supplier-induced demand and is a type of moral hazard. Further, if patients are insured, they also have no incentive to consider the cost of treatment. This can particularly be an issue for end-of-life care, where very low probability treatments may be performed far beyond their expected usefulness.

Another key problem with FFS is that it can lead to fragmented care. A patient may need several provider types such as primary care and a set of specialists. In the FFS model, there is no coordination that occurs across the providers.

In order to help to address some of the problems of FFS, insurers specify which services they will reimburse and use a billing review process. Government programs also define not only which services they will cover, but what rate they will reimburse at. In many cases, the reimbursement rate from programs such as Medicaid and Medicare may be lower than the cost to provide the service. As a result, providers will use "cost-shifting" as a response. This topic will be discussed later in the next section.

Bundled Payments—Diagnostic Related Groups

In the *bundled payment* approach, a set of services is bundled into a *diagnostic-related group (DRG)* for an episode of care. DRGs are defined for both medical and surgical procedures and are also based on patient groupings (e.g., age, gender, the presence of a complication, the presence of comorbidities). An example is revision of a revision of hip or knee replacement with complication for an adult (DRG 467).

DRGs have become a popular form of reimbursement for acute inpatient care in many countries with high gross domestic product (GDP) per person. The potential benefit of DRGs is that they help to coordinate the services for a particular episode of care. In addition, it eliminates the incentive to add services needlessly, since the provider will receive the same reimbursement regardless of the actual services provided. Several studies have found that DRGs have a positive impact on hospital efficiency (primarily through reducing length of stay) and that case volumes tend to increase.

Pay-for-Performance Plans

Pay-for-performance (P4P) programs provide financial incentives to providers that achieve specified quality metrics with the goal of improving health outcomes at a lower cost. The outcomes measured include process outcomes such as the rate of central line–associated bloodstream infections (CLABSIs) per 1,000 central line days, or patient outcomes such as reduction of average A1c level for a diabetic patient. Since patient outcomes are more difficult to measure, the more common approach is to use process measures.

The basic idea of a P4P program is to incentivize the provider to improve health outcomes for a patient population. A set of goals is set by the payer (e.g., insurance company), and if the target goals are met, the provider is financially compensated. If, however, the targets are not met, then there is an associated penalty.

P4P programs have been implemented in several countries, including the United Kingdom and the United States. While much of the literature shows a positive effect on outcomes, it is unclear what the corresponding costs to achieve these outcomes were; that is, whether these programs are actually cost saving is at best unclear.

Shared Savings Plans

Shared savings plans (SSPs) are an approach to encourage providers to reduce the cost of providing care. There are two components to SSPs. First, a set of target health outcomes is established, which are primarily process based. Second, a baseline for the cost-of-care provision is established. Providers that are able to meet the target health outcomes at a cost lower than baseline are reimbursed a portion of the cost savings. In a country such as the United States that has historically relied on FFS payments, there has been little reason to consider the cost of care. With the use of SSPs, however, providers are now incentivized to focus on reducing costs in their system.

A key user of SSPs is Medicare in the United States. In this program, there is at present no financial risk for a provider to participate in an SSP. The caveat is that SSPs are offered only to accountable care organizations (ACOs). As a result, there has been very rapid growth over the past several years in the formation of ACOs and in hospital mergers. Further, SSPs can help to coordinate care in ACOs since cost saving often is a result of looking across the system of providers in an ACO.

Capitation

Capitation is a payment arrangement where providers (typically nurse and physician practices) are paid a fixed amount per person enrolled in their system per period of time (e.g., annually). Even if an individual received no service, the providers are still reimbursed. The potential benefit of capitation is that it helps eliminate supplier-induced demand, since physicians and nurses will want to limit their expenses. The drawback is that it also incentivizes providers to "cherry-pick" healthy patients. This can reduce access for at-risk patients and also increases the likelihood for adverse selection for those programs (e.g., government insurer) that do enroll the at-risk patients.

4.3 COST ALLOCATION AND CHARGES

One of the key functions in a hospital or physician's office is billing for services provided. This is complicated by the fact that their patients are typically covered in many ways (including uninsured). This has also provided hospitals with an opportunity to shift costs from one type of provider that reimburses at a lower rate (e.g., publicly insured) to another that reimburses at a higher rate (e.g., privately insured). In this section, we discuss the basics of cost allocation and charge setting as well as the billing process.

Costs

In healthcare organizations such as hospitals, there are revenue-generating functions such as radiology and cost (or non-revenue)-generating functions such as laundry services. In billing a patient, the charge must be allocated over both revenue-generating and non-revenue-generating functions. There are several approaches that can be used for cost allocation, and we present a few simple examples for illustration.

The first step in cost allocation is apportioning direct and indirect costs. One approach to this is called the ***reciprocal allocation method.*** In this approach, costs from non-revenue-generating functions and revenue-generating functions are allocated simultaneously. This is best illustrated by an example.

Example 4-1

Consider a very simple hospital system that consists of four functions: A = Facilities, B = Administration, C = Obstetrics, and D = Diagnostics. A and B are non-revenue-generating and C and D are revenue-generating. Suppose that over a month, the services provided by the non-revenue-generating functions to all departments are given in the following table. Note that the preallocated costs are also provided.

Source	Services Allocated			
	A	**B**	**C**	**D**
A	0%	20%	40%	40%
B	10%	0%	60%	30%
Preallocation costs	$75,000	$50,000	$250,000	$140,000

The costs can be allocated by solving a simple set of simultaneous equations. For Obstetrics, for example, its cost equals \$250,000 plus 40% of the final allocated costs from Facilities and 60% of the final allocated costs from Administration. If this is set up for all four departments, we get the following set of simultaneous equations:

$$A = 75000 + 0.1B \tag{4-1}$$

$$B = 50000 + 0.2A \tag{4-2}$$

$$C = 250000 + 0.4A + 0.6B \tag{4-3}$$

$$D = 140000 + 0.4A + 0.3B \tag{4-4}$$

Solving the system yields: A = \$81,632.65, B = \$66,326.53, C = \$322,449, and D = \$192,551. Note that the sum of C and D after applying the method equals the sum of all preallocation costs. For this example, for Obstetrics, the direct cost is \$250,000 and indirect cost is \$322,449 – \$250,000 = \$72,449.

Two other methods that are commonly used in practice are the step-down method and the double apportionment method. However, both methods do not as accurately allocate costs as the reciprocal method.

A second important costing function is to determine the product cost based on patient volumes and the allocated costs. Although alternative approaches exist, we will assume here that direct and indirect costs are assembled directly for each of the revenue-generating departments. This is called *full costing*. A popular approach for determining product cost is *activity-based costing* (ABC). ABC uses a driver to assign indirect costs for resources. For example, consider a diagnostic lab that uses radiology equipment, and has three types of costs: labor, materials/supplies, and equipment. In this case, labor and materials/supplies are direct costs since they are required for each procedure. However, the equipment is a shared resource across procedures. The key question is how to allocate the cost of the shared resource. A common driver is time required on the shared resource. The driver can then define the relative value that the resource has for each procedure. We illustrate the ABC approach through a simple example.

Example 4-2

Three procedures are performed in a diagnostic laboratory (P1, P2, and P3). Based on the reciprocal allocation method, direct costs are projected to be $400,000 and indirect costs are projected to be $200,000. The minutes of use on the diagnostic equipment will be the ABC driver. Data for the three procedures are:

Procedure	Patient Volume	Direct Costs ($/procedure) (labor + materials/supplies)	Equipment Time (min)
P1	400	100	20
P2	200	80	25
P3	100	90	30

We can convert the direct cost into direct relative value units (RVU), by dividing by the greatest common denominator (GCD). The GCD for direct costs is 10, which gives RVUs for P1 to P3 equal to 10, 8, and 9, respectively. Similarly for indirect costs, the GCD for equipment time is 5, which gives indirect RVUs for P1 to P3, respectively, of 4, 5, and 6, respectively. For each case, RVUs can be converted to total RVUs by multiplying by patient volume. This gives 10(400) = 4,000, 8(200) = 1,600, and 9(100) = 900 for direct and 4(400) = 1,600, 5(200) = 1,000, and 6(100) = 600 for indirect. The sum of the total direct RVUs is 6,500. Since the given projected direct costs are $400,000, then the direct cost per direct RVU is $400,000 / 6,500 = $61.54. Similarly, the indirect cost per indirect RVU is $200,000 / 3,200 = $62.50. The direct cost per procedure is calculated by the product of direct cost/RVU and the direct RVUs and the indirect cost per procedure is calculated by the product of the indirect cost/RVU and the indirect RVUs. The total cost per procedure is the sum of direct and indirect cost per procedure. This is summarized in the following table.

Procedure	Direct Cost/ Procedure	Indirect Cost/ Procedure	Total Cost/ Procedure
P1	$615.40	$250.00	$865.40
P2	$492.32	$312.50	$804.82
P3	$553.86	$375.00	$928.86

The important point in this example is that the minutes of equipment served as the cost driver by defining how the indirect costs were spread per procedure. If revenue generated for P1 and P2 were the same, for example, even though P1 has higher direct costs than P3, it would be more profitable to perform P1 over P3 due to the high indirect costs (i.e., equipment requirements) of P3.

Setting Charges

There are several ways that hospitals can set charges for services. It is worth noting, however, that since government insurance determines the rate they will pay, the charge set will likely not be met in billing. We will discuss a practice that some hospitals use to deal with this issue in the next section.

One approach for setting charges builds off of the relative value units used in ABC. Charges can simple be set as the total cost/procedure determined by the analysis. If the procedures are bundled into a DRG, then the sum of the total costs/procedure can be used. A basic markup defined by management (e.g., 10%) can also be directly applied to the ABC costs.

There are several opportunities to use revenue management strategies to improve profitability as well. For example, with the increase in use of bundled payments through DRGs, hospitals are incentivized to be selective in their admissions, as opposed to FFS, which encourages hospitals to keep occupancy rates high. In bundled payments, each admission generates revenue, though patients with different conditions have different levels of profitability. Hospitals therefore target the most profitable admissions and avoid the least profitable, which they can influence for their physician mix, services offered, and marketing. The opportunities are even greater for patients with multiple chronic conditions. A good survey of revenue management practices may be found in Talluri and van Ryzin (2014).

Cost Shifting

As mentioned previously, hospitals are "price-takers" for government programs such as Medicaid and Medicare. These programs pay based on a published charge schedule. The charge schedule does not necessarily match the actual costs. In fact, for both programs, the reimbursement can be significantly below cost. Further, uninsured patients are often not able to pay their bill, which is called ***uncompensated care***. On average, private insurance reimburses at a rate that is roughly 30% higher than average (Coughlin, 2014). This practice of charging higher rates to insurers to make up for losses from uninsured and government populations is called ***cost shifting***. We illustrate the practice by an example.

Example 4-3

A hospital uses a fee-for-service reimbursement scheme. The cost of performing the procedure is $1,000. There are three patient populations: government insured, privately insured, and uninsured. The government reimburses at $800 per procedure, and it is estimated that 80% of the uninsured population will be uncompensated care. The estimated demand for the service is 100 government-insured patients, 200 privately insured patients, and 80 uninsured patients. We will assume that we can shift costs to the privately insured patients and the uninsured patients that will pay.

The loss from government insured patients is $100(\$1,000 - \$800) = \$20,000$. The loss from uncompensated care is $0.2(80)(1000) = \$16,000$. Therefore, for the hospital to break even, it must shift $36,000 of costs to the private and paying uninsured. Since the total population of these groups is $200 + 0.8(80) = 264$, then the charge per procedure will need to be at least $\$1,000 + \$36,000 / 264 = \$1,136.36$.

Figure 4.2 shows the inverse relationship between hospital payment-to-cost ratios for private payers compared to Medicare and Medicaid from 1993 to 2013, which implies the practice of cost shifting. Although the somewhat controversial practice is widely acknowledged in practice, some recent literature has argued that the amount done may not be so great (Frakt, 2011; White, 2013). If hospitals could easily cost shift, then the

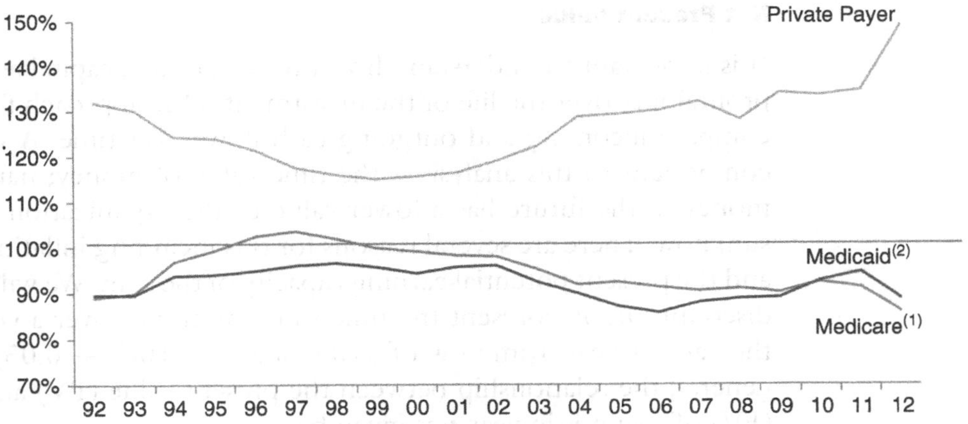

Source: Avalere Health analysis of American Hospital Association Annual Survey data, 2012, for community hospitals.
(1) Includes Medicare Disproportionate Share payments.
(2) Includes Medicaid Disproportionate Share payments.

Figure 4.2 Aggregate U.S. Hospital Payment-to-Cost Ratios for Private Payers, Medicare, and Medicaid, 1993–2013 ***Source:*** **American Hospital Association, Trendwatch Chartbook 2015**

private reimbursement rates should increase as the number of uninsured patients increases or as government reimbursement rates decrease. Some studies have shown, properly controlling for external factors, that this is not the case. The results imply that in most cases, hospitals simply do not have the market power to arbitrarily raise rates on private insurance. However, it is the case that smaller insurance companies with less market power do typically pay higher rates than larger companies with more market power.

It is worth noting that charging different parties different rates for the same service is called *price discrimination*. Price discrimination is a result of differences in market power, and there is no causal relationship between the different prices charged for the same service. In cost shifting, there is a direct relationship. Cost shifting implies there is price discrimination, but price discrimination does not necessarily imply cost shifting.

4.4 CAPITAL BUDGETING

Capital budgeting is the process of determining which investments in capital assets (land, buildings, and equipment) are appropriate to acquire. We discuss three important topics in this area, namely, cash flow analysis through the use of net present value, lease versus buy decisions, and project selection.

Net Present Value

It is important to understand how investment in a capital asset will impact profitability over the life of the investment. One approach for doing this is compare incoming and outgoing cash flows over time. A very important component to this analysis is the time value of money; namely, a sum of money in the future has a lower value to the organization than the same sum now. There are several reasons for this including inflation, default risk, and the present potential earning capacity of the sum. We will let the annual discount rate r represent the time value of money over a year. Therefore, the value 1 year from now of \$10 would be \$10(1 + 0.05) = \$10.50. In general, the relationship between the present value (PV) and future value (FV) of a sum is in year n is given by:

$$FV = PV(1 + r)^n \tag{4-5}$$

The *net present value (NPV)* approach sums future discounted cash flows. If we let cf_i equal the incoming minus outgoing cash flow for year i, then the NPV for a project life of N years and a discount rate of r is defined as:

$$NPV(r, N) = \sum_{i=0}^{N} \frac{cf_i}{(1 + r)^i} \tag{4-6}$$

If the NPV is positive, then the capital asset investment is profitable. We illustrate with an example.

Example 4-4

A hospital can take a loan of \$150,000 to purchase a piece of imaging that will last for 6 years. The loan is paid off over 10 years with an interest rate of 4%, and payments start in year 1. In this case, the yearly payments (A) is given by the annuity formula:

$$A = P\left(\frac{i(1 + i)^n}{(1 + i)^n - 1}\right) \tag{4-7}$$

where P is the loan amount, i is the interest rate, and n is loan life. Using our data gives A = \$18,493.64.

The equipment has a salvage value of $20,000, which would be taken at the end of year 6. The annual discount rate for the hospital is 6%. In this example, we will not consider replacement of the equipment. The hospital estimates net revenue (revenue minus direct costs of labor and supplies) for the next 6 years are $15,000 for years 1 to 3 and $20,000 for years 4 to 6. The cash flows are shown in the following table:

Year	Loan Payments	Net Revenue	Salvage	Cash Flow	Discounted Cash Flow
1	18,493.64	20,000	0	1,506.36	1,421.09
2	18,493.64	20,000	0	1,506.36	1,340.66
3	18,493.64	20,000	0	1,506.36	1264.77
4	18,493.64	30,000	0	11,506.36	9,114.12
5	18,493.64	30,000	0	11,506.36	8,598.22
6	18,493.64	30,000	20,000	31,506.36	22,210.74
7	18,493.64	0	0	−18,493.64	−12,299.30
8	18,493.64	0	0	−18,493.64	−11,603.10
9	18,493.64	0	0	−18,493.64	−10,946.40
10	18,493.64	0	0	−18,493.64	−13,326.80

Summing the discounted cash flows gives the NPV $(6\%, 10) = -\$1,225.98$. Since the investment yields a negative NPV, it should not be made based on the financials.

In general, investments that have a positive NPV achieve the organizations discount rate and hence should be made. This assumes that the potential investments are independent and that the firm has or can acquire the capital needed to make the investments. Often, there is a budget that cannot be exceeded, and the best projects from the feasible set should be chosen. We discuss this in the next sections. Further, the fact that an investment yields a negative NPV does not necessarily imply the investment should not be made. It is possible, for example, that the investment is a requirement. The imaging device in Example 4-4, for instance, may be needed for the hospital to perform other services.

There are several limitations to NPV analysis. First, the choice of a discount rate is typically a subjective one. Further, organizations are typically not able to project future inlays from demand with much certainty. It is typically wise, therefore, to perform sensitivity analysis on the results.

Project Selection

In the previous section, we discussed an approach to evaluate a single capital investment project. There are many cases where a set of projects need to be evaluated. Consider the case where a set of *n* investment projects are considered simultaneously and the NPV of each have been determined. The goal is to choose a set of projects that maximizes the total NPV. There are often constraints that must be satisfied. Examples include:

- *Budget:* The total number of selected projects (or total investment) must be a certain value.
- *Mutually exclusive:* If project A is chosen, B cannot be.
- *Dependency:* If project A is chosen, then B must be.

Integer programming can be used to determine the best set. We illustrate with an example.

Example 4-5

There are five potential projects to choose from (P_1 to P_5). The NPV of each is \$500, \$900, \$600, \$600, and \$400, respectively. If project P_2 is selected, then project P_5 must be selected. Finally, no more than three projects can be selected. This can be formulated as a binary program, where $x_i = 1$ if project P_1 is selected and 0 otherwise:

$$\max 500x_1 + 900x_2 + 600x_3 + 600x_4 + 400x_5$$

subject to:

$$x_2 \leq x_5$$

$$x_1 + x_2 + x_3 + x_4 + x_5 \leq 3$$

$$x_i \in \{0, 1\} \ \forall i$$

Solution of the above gives $x_2 = x_3 = x_5 = 1$.

Recent changes in healthcare financing and reimbursements in many countries have led to a significant interest from hospitals in understanding and reducing their costs. There are significant opportunities for operations managers and industrial engineers to impact this area.

QUESTIONS AND LEARNING ACTIVITIES

1. In what ways is health insurance different than dental insurance? What are the ramifications?
2. Suppose the utility curve in Figure 4.1 were a straight line. What would be the implications? What does this imply about the individual's view of "risk"?
3. Suppose the utility function for an individual is given below (the units are scaled). Further, the individual's income is \$7. There is a 50% chance that over the year an illness will occur that would cost the individual \$5. What is the most that an insurance firm could charge this individual where they would still purchase the insurance?

$$U(I) = I^{0.3}$$

4. In the adverse selection example, suppose 40% of the population smokes instead of 25%. How would the results change, if at all?
5. In what ways is cost allocation similar and different than cost allocation for a manufacturing plant? What do you believe is the most challenging part?
6. List examples of types of moral hazard that can occur in healthcare. Discuss what could be done to prevent it.
7. What do you see as the key drawback to the use of a shared saving's plan?
8. Discuss why there are such differences in financing mechanisms across countries.
9. In example 4-1, suppose the services allocated from A are 10% to A, 20% to B, 40% to C and 30% to D. How would the results change?
10. The step down method mentioned after Example 4-1 works as follows: the non-revenue department with the highest percentage of costs is allocated first to other non-revenue and revenue department. The second highest non-revenue department is then allocated in a similar way.

This process is repeated until all non-revenue costs are allocated. Apply this method to Example 4-1. How are the results different? Compare and contrast the step-down and reciprocal allocation methods.

11. Consider a CT-scanner that is used for three procedures. Use ABC analysis to determine the cost/procedure assuming that minutes of use on the equipment are the driver.

Procedure	Patient Volume	Direct Costs ($/procedure) (labor + materials/supplies)	Equipment Time (min)
P1	100	10,000	80
P2	200	8,000	55
P3	100	9,000	130

12. What might be other drivers for ABC analysis other than equipment time? What is the implication of choosing the driver?
13. Discuss the ethics of the practice of cost shifting by hospitals.
14. Discuss the factors that go into how a provider should choose a discount rate for capital budgeting.
15. Discuss how taxes can be taken into account in the NPV method.
16. Repeat Example 4-4 using discount rates of 5%.
17. A hospital is considering expanding their ICU by adding 5 beds. The cost of the expansion is $3M, which they will finance through a loan with a 5% interest rate. The lifetime of the expansion is 20 years, which is the same length of the loan. The discount rate that the hospital uses is 10%. If the same revenue is obtained each year from the expansion, what is the minimum it can be that would yield a positive NPV?
18. Discuss of a firm can estimate their revenues. What are the challenges? What are the implications?
19. How does risk aversion on the part of the provider impact capital budgeting? Think of some ways that could account for this.

REFERENCES

Coughlin, T. A., Holahan, J., Caswell, K., & McGrath, M. (2014). *Uncompensated care for the uninsured in 2013: A detailed examination.* Washington, DC: Urban Institute. https://kaiserfamilyfoundation

.files.wordpress.com/2014/05/8596-uncompensated-care-for-the-uninsured-in-2013.pdf

Frakt, A. B. (2011). How much do hospitals cost shift? A review of the evidence. *Milbank Quarterly*, *89*(1), 90–130.

Nowicki, M. (2011). *Introduction to the financial management of healthcare organizations* (5th ed.). Chicago, IL: Health Administration Press.

Talluri, K. T., & Van Ryzin, G. J. (2014). *The theory and practice of revenue management*. New York, NY: Springer Science+Business Media, Inc.

White, C. (2013). Contrary to cost-shift theory, lower Medicare hospital payment rates for inpatient care lead to lower private payment rates. *Health Affairs*, *32*(5), 935–943.

Chapter 5

Health Data and Informatics

"V97.33XD: Sucked into jet engine, subsequent encounter.
W61.62XD: Struck by duck, subsequent encounter.
Z63.1: Problems in relationship with in-laws."

—Actual ICD10 Codes

Overview

In 1984, Marsden S. Blois offered the following observation: "During the past few decades the volume of medical knowledge has increased so rapidly that we are witnessing an unprecedented growth in the number of medical specialties and subspecialties. Bringing this new knowledge to the aid of our patients in an economical and equitable fashion has stressed our system of medical care to the point where it is now declared to be in a crisis. All these difficulties arise from the present, nearly unmanageable volume of medical knowledge and the limitations under which humans can process information" (Blois, 1984).

Today, we recognize the evolution and importance of **health informatics**, which is the science of combining healthcare data into information, to derive knowledge and wisdom. Figure 5.1 provides a schematic of the discipline of health informatics.

In particular, health informatics is a "scientific field that deals with resources, devices and formalized methods for optimizing the storage, retrieval and management of biomedical information for problem solving and decision making... understanding, skills and tools that enable the sharing and use of information to deliver healthcare and promote health" (Hoyt, 2014).

Health informatics is a combination of **bioinformatics** and **medical informatics.** Bioinformatics is the collection, classification, storage, and

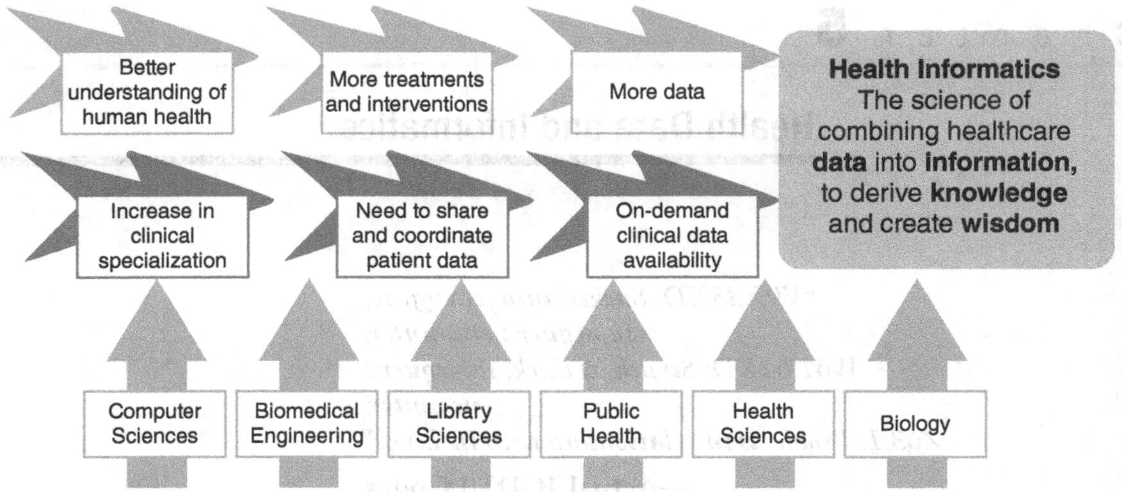

Figure 5.1 Schematic of the Discipline of Health Informatics

analysis of biological and biochemical information using computers. Frequently associated with the laboratory and genomics world, bioinformatics is the basis for which the electronic exchange of information for care of the patient was born. We may think of bioinformatics as the microscopic birth of medical data that, at its core, is inherent to our being. Many of the bioinformatics systems are within laboratories (laboratory information systems [LISs]) or in research facilities and were the first information technology (IT) systems to be introduced to facilitate medical care. In contrast, medical informatics is a more recent term. The Healthcare Information and Management Systems Society (HIMSS) defines medical informatics as the interdisciplinary study of the design, development, adoption, and application of IT for the planning, delivery, and management of healthcare. In contrast to bioinformatics, medical informatics is the use of information on the macro-systems of healthcare. Electronic medical records (EMRs) or electronic health records (EHRs) are the base systems of medical informatics. In both bio and medical informatics, these records generate healthcare data, which are converted into knowledge and wisdom and are the basis of health informatics.

Health informatics provides information that enables providers to make decisions, and better information leads to better decisions. At the end of the day, healthcare is the function of people taking care of people. This function includes a number of opportunities to make a big difference in a patient's life with the right information at the right time. For example, the most meaningful interaction for a patient in the doctor's office might be being greeted by name upon registration. This is especially important in our fast-paced healthcare system since healthcare is no longer delivered by one primary care physician in one office, with the patient seeing that same provider every time. Similarly, it is rare for the primary physician to also be the physician responsible for care in the hospital. The reality of healthcare is that patients have a "medical home," and that medical home is staffed with an interdisciplinary group of providers (physicians, nurse practitioners, physician assistants) as well as the support staff that are involved with the patient. As a result, the only true constant is the health informatics system that binds all of the data.

Healthcare delivery, and health informatics, is a multidisciplinary approach that involves how people transform technology and how technology transforms people. Health informatics lies at the intersection of people, processes, and technology as information systems, and focuses on the expanding relationship between information systems and the daily lives of real people. Further, it deals with the resources, devices, and methods required to optimize acquisition, storage, retrieval, and use of information in health.

Health informatics helps develop new/better uses for IT in order to design solutions that reflect the way people create, as well as how they use and find information. It also takes into account the social, cultural, political, and organizational settings in which those solutions will be used. Informatics includes the use of computers for change management, human-computer interactions, communications, risk management, organizational behavior, workflow redesign, productivity improvement, and organizational culture, safety, and quality. Tools include not only computers but also clinical guidelines, formal medical terminologies, and information and communication systems.

Although they may have similarities, it is important to note that health informatics is not health IT. Informatics is the collection, manipulation, and use of information, whereas IT is the hardware and software that informatics uses. IT infrastructure must "support evidence-based practice, including the provision of more organized and reliable information sources on the Internet for both consumers and clinicians and the development and application of decision support tools" (Kohn, Corrigan, & Donaldson, 2001).

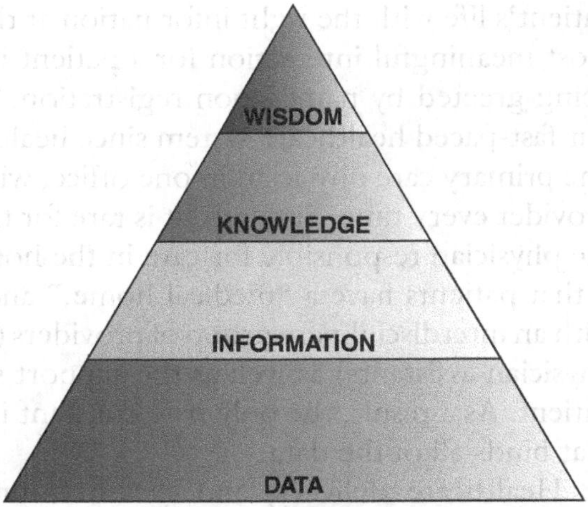

Figure 5.2 Information Hierarchy Used in Health Informatics ***Source:*** **Hoyt (2014)**

Decision making within the health informatics framework consists of data that are synthesized into information, which are analyzed and become knowledge and then abstracted to become wisdom. Figure 5.2 depicts this hierarchy.

Let's define these elements depicted in Figure 5.2. **Data** are unorganized and unprocessed facts—a set of discrete facts about events. No meaning is directly attached to data and, as a result, it may have multiple meanings. **Information** is the aggregation of data that makes decision making easier. Meaning is attached and contextualized, which is the *fundamental problem and challenge in informatics.* **Knowledge** includes facts about real-world entities and the relationship between them; it is an understanding gained through experience. **Wisdom** embodies principles, insight, and morals by integrating knowledge.

The key elements of health informatics include acquisition, storage, communication, manipulation, and display. We will now define these five key elements. **Acquisition** is the capture of data, taking care to strive for quality (accurate, timely, reliable, and complete). **Storage** is saving the data so that they can be retrieved (the key term here is *retrieval,* as the trash bin will store data but they are difficult to retrieve). **Communication** is such that data need to be moved from point of collection to storage for analysis, and finally to point of use (this may be a very short distance and time where data are used near the collection point, or these may be widely separated points). **Manipulation** is such that data usually need to be manipulated in some way, combined with other data, aggregated, or compared. Finally, **display** is how the data can be best displayed so that they can be easily understood and acted upon.

The display of data cannot be overemphasized. Data do not become information without context, and information cannot become knowledge without interpretation. The ultimate wisdom that is derived from knowledge is imparted to healthcare providers and consumers based on their own context. To make this practical, a cardiologist is mostly focused on the cardiovascular system (heart, blood vessels, blood pressure, etc.). Orthopedists are primarily concerned with musculoskeletal system (bones, muscles, joints, etc.) Data presented (displayed) to a provider in the clinical context is essential. Simply, an x-ray of a bone would not help the cardiologist care for a patient with a heart attack, while the orthopedist would not benefit from the knowledge of a cholesterol level in a patient with a dislocated shoulder.

Both health informatics and IT help connect information (from external/internal data), technology (networks, databases), and healthcare functions (patient care, medical records, ancillary services, administrative, research). The traditional perspective is that health informatics and IT consists of:

1. **Architectures** for electronic medical records and other health information systems used for billing, scheduling, and research;
2. **Standards** to facilitate the exchange of information among healthcare information systems—these specifically define the means to exchange data, not the content;
3. Controlled **vocabularies** used to allow a standard, accurate exchange of data content among systems and providers; and
4. **Software** for specialist services and devices.

The driving forces behind health informatics are to increase efficiency of healthcare (decrease medical costs, improve physician productivity), improve quality (patient outcomes) of healthcare, improve patient safety, provide information brokerage (the sharing of a variety of information back and forth between people and healthcare entities), and the EHRs (to be used for evidence-based decisions).

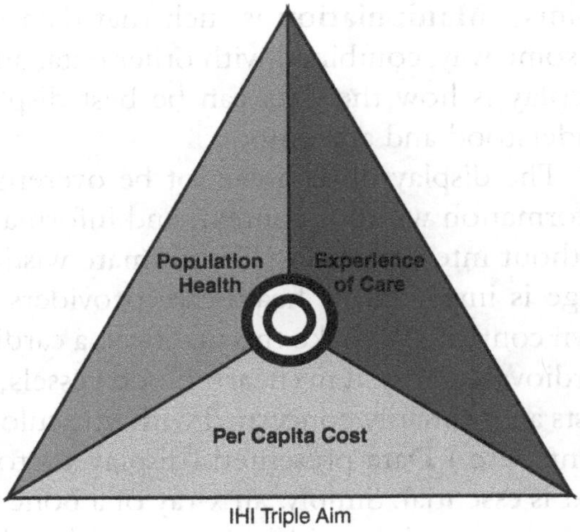

Figure 5.3 IHI Triple Aim

The Institute for Healthcare Improvement (IHI) has promoted the "triple aim" as a framework to approach simultaneous healthcare systems optimization to:

1. Improve the patient experience of care (quality and satisfaction),
2. Improve the health of a population, and
3. Reduce the costs of healthcare.

This is represented graphically in Figure 5.3.

This framework provides the goal of the healthcare delivery system. At the center of the triple aim is transaction-level healthcare data. The relationship between the data and decisions in the context of improving costs, patient experience, and overall health is the basis of health informatics.

The diagram in Figure 5.4 shows an example of the relationship between decisions and data.

5.1 HEALTHCARE DATA

As already seen, **healthcare data** are the central component of health informatics. Recall that data are observations, whereas information is meaningful data to draw conclusions, and knowledge is information justifiably believed to be true. Figure 5.5 depicts various levels of data, from nonelectronic to unstructured to structured to computable.

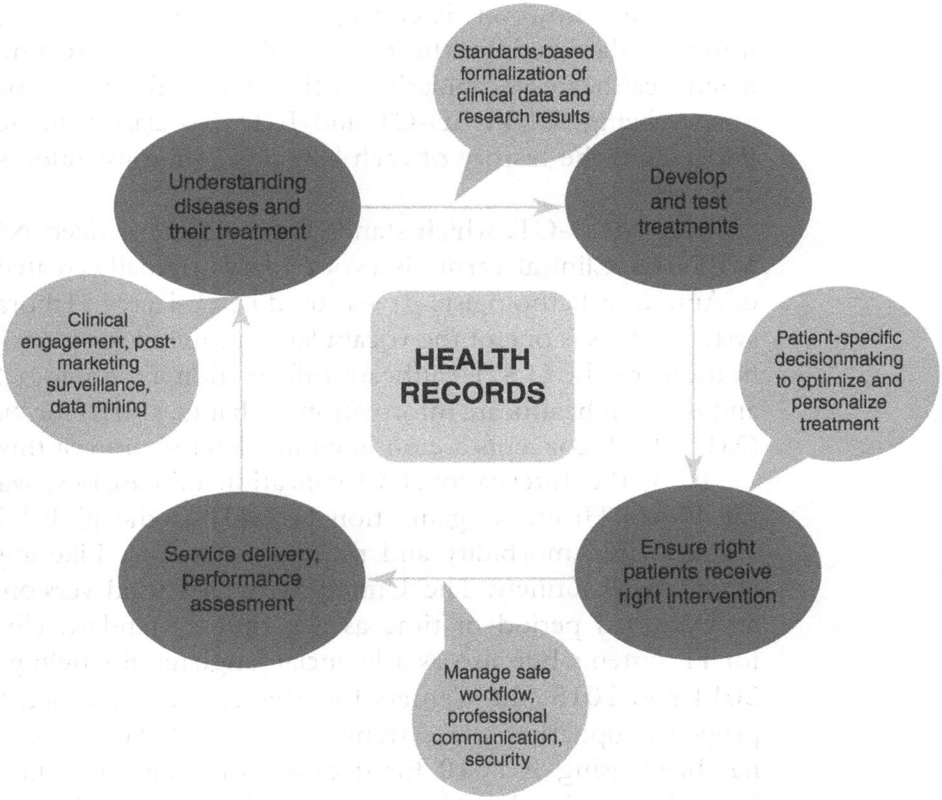

Figure 5.4 Relationships between Decisions and Data

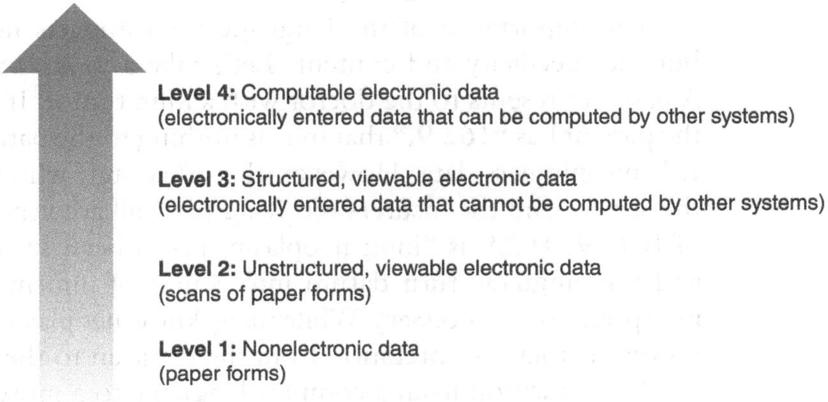

Figure 5.5 Increasingly Sophisticated and Standardized Data

The key, however, is converting data to information, which is done using standardized healthcare vocabularies. There are a number of recognized healthcare vocabularies, with some of the two most common languages being SNOMED-CT and ICD. To clarify the acronyms and to understand the history of each language, we must understand the origin and use.

SNOMED-CT, which stands for the Systematized NOmenclature of MEDicine Clinical Terms, is a vocabulary originally created by the College of American Pathologists. It was used extensively in laboratory-pathology systems. This is one of the vocabularies required for interoperability specifications by the U.S. Healthcare Information Technology Standards panel and is use in healthcare information exchange (HIE) for "meaningful use" (MU). Both concepts are outlined in a later section of this chapter.

ICD, the International Classification of Diseases, was generated by the World Health Organization (WHO) as the global health standard for reporting morbidity and mortality statistics. Like any vocabulary, it is updated routinely. The United States has used version 9 (ICD-9) for an extended period of time as not only a standard clinical vocabulary for IT systems but also as a financial language for billing and coding. In 2014 and 2015, the Centers for Medicare & Medicaid Services (CMS) proposed updating the systems to ICD-10. While most of the world has been using ICD-10 for decades, this was for clinical reasons. The United States is relatively unique in using the ICD language for financial structure. The release of subsequent versions of ICD are consistent and expected (ICD-11 in 2015).

The importance of the language used impacts not only the content but the specificity and content. Let's take a look at an example in ICD. A patient presents to the doctor with a lung tumor. If the EMR notes that the patient has "162.9," that means nothing to the patient or the physician. It is meaningless data. However, the office staff who generate the bill for the service and the insurers receiving that bill are very aware that in terms of ICD-9, 162.9 is "lung neoplasm, not otherwise specified." The term and the language turn datum into a unit of information. Thus, human interpretation is necessary. While many know neoplasm equals cancer, there is even further interpretation from the clinician to the patient.

The transition from a complex language to a more specific vocabulary is important when outlining details of healthcare data. Table 5.1 outlines the comparable number of codes for procedures and diagnosis between versions 9 and 10.

Table 5.1 ICD-9 to 10 Comparison of Complexity and Specificity

Differences between ICD-9-CM and ICD-10 Code sets

Procedure	ICD-9-CM	ICD-10 code sets
Diagnosis	3,824 codes	71,924 codes
	14,025 codes	69,823 codes

ICD-10 Code Structure Changes (selected details)

Diagnosis Structure	**Old** ICD-9-CM	**New** ICD-10-CM
	• 3–5 characters • First character is numeric or alpha • Characters 2–5 are numeric	• 3–7 characters • Character 1 is alpha • Character 2 is numeric • Character 3–7 can be alpha or numeric
Procedure Structure	ICD-9-CM	ICD-10-PCS
	• 3–4 characters • All characters are numeric • All codes have at least 3 characters	• ICD-10-PCS has 7 characters • Each can be either alpha or numeric • Numbers 0–9; letters A–H, J–N, P–Z

Moving forward with advanced ICD language, version 10s and 11, the simple "162.9 = lung neoplasm" will be expanded in detail and specificity. An ICD-10 version might be c34.90, "Malignant neoplasm of unspecified part of unspecified bronchus or lung." The generic "lung cancer" may be a devastating but actionable term to the patient, but in the context of the healthcare provider, based on that healthcare data, it says nothing of location, type, extent, or other details. However, if both the clinician and the coder are working toward the triple aim, or patient experience, cost reduction, and population health, the specificity would allow for more accurate diagnosis, clearer treatment plans, and avoiding unnecessary detail in care, while gathering the data in enough detail to support the knowledge gained with a population of similarly diagnosed patients.

Previously alluded to but not explicitly called out is the importance of the language to be able to be transmitted and used by other systems. **Interoperability** refers to the standards that make it possible for diverse electronic health record systems to work compatibly in a true information network. Thus, a standardized healthcare vocabulary is essential for converting data into information. Further, data cannot be converted directly to knowledge, since information produces knowledge. In the clinical world, evidence exists that knowledge is true rather than proven fact.

As seen in Table 5.1, healthcare data are collected via electronic health records, which are composed of both structured data (billing codes, lab results, medication lists, etc.) and unstructured data (clinical notes, natural language, etc.). Structured data are in simple, repeatable formats that are easy to automate analytical processing. This is frequently entered as a "point-and-click" function for healthcare providers or office staff. Unstructured data are freeform text, making it difficult to automate analytical processing. Historically, clinicians always "told the story" of a patient while documenting the care, either in handwritten notes or letters, but more recently in electronic "word processing" components of the EMR. The "storytelling" about the care of the patient is important to the care of the patient and to the care team providing that care. Healthcare is a human-to-human interaction, and the story in human terms is the preferred language of patient and physicians. However, many clinicians believe that because it is electronic, it is easily used and accessible. Unless and until the promise of converting unstructured to structured data comes to fruition (natural language processing and/or optical character recognition), the unstructured formatting does lead to interoperability and secondary use of data challenges for any data warehouse or registry.

The ***clinical data warehouse (CDW)*** is a shared database that collects, integrates, and stores healthcare data from a variety of sources, including electronic health records and other health information systems (e.g., laboratory). These systems will frequently receive data from registration systems, financial systems, and clinical systems as well as logistics/operational systems. The CDW does more than archive data, as it must "make sense" of clinical data to turn data into information and information into knowledge. The CDW enables healthcare organizations to monitor quality and trends in a given patient population, conduct practice-based research, track pathogens within hospitals, and conduct epidemiological surveillance. The use and output of these data warehouses will be directly related to the type of data included in the warehouse and the integrity of the data. Again, the fact that it is electronic does not mean it is usable. Healthcare has recently adopted the general approach that other industries have employed with ***Business intelligence (BI)*** tools. The BI tools are used to analyze and report data from the CDW in the context of that healthcare business query. Again, data are converted to knowledge only in context.

There are many challenges posed by healthcare data. Some are inherent to the patient or disease itself (e.g., the first instance of disease in the health record may not be when it was first manifested). More frequently, healthcare data generated can be inaccurate or incomplete based on who or how the data were entered originally. For decades there were no standards in the industry, leading to variable and nonstandard approaches to data entry. As standards developed, the widespread adoption of the standard was slow to the industry. Some were related to practical concerns, such as a change in what the clinician or nurse does on a daily basis, while others were more related to the vendor or EMR developer being able to maintain those standards.

The lack of early standards and variable input consistency impacts the ability to research and discover with the data, especially with the identified population. "Dirty data" only allow for observational rather than experimental studies (raising the issue of cause and effect of findings discovered). In the research and discovery with the use of these "secondary data," there are ethical concerns over how the healthcare data for individuals are used (who owns the data and who has privileges to use it). Given these challenges, the role of health informatics professionals is to serve as human experts to build systems, capture healthcare data, put them into usable form, and apply the results of analysis. It is not just about healthcare data science, but it requires an understanding of the healthcare data, types, and how to manipulate and leverage them.

5.2 ELECTRONIC HEALTH RECORDS

Electronic health records (EHR), also known as electronic medical records (EMRs), continue to be one of the most controversial yet important topics in health informatics. As an essential component of the national health information technology (HIT) strategy and healthcare reform, widespread adoption of EHRs will likely help reform the delivery and quality of medical care. However, EHRs also present many new challenges. There are mixed reports as to whether the adoption of EHRs will consistently produce improved medical quality and patient safety or reduce healthcare costs. Some question whether improvement in medical care due to EHR use will take many years to occur and whether only large, integrated delivery network–type healthcare organizations will experience these gains.

There is no universally accepted definition of the electronic health record. Informally, the EHR is a digital collection of patient health information compiled at one or more meetings in any care delivery setting. A patient's record typically includes patient demographics, progress notes, problems, medications, vital signs, past medical history, immunizations, laboratory data, and radiology reports. The term *EHR* is often used to refer to the software platform that manages patient records maintained by a hospital or medical practice. Formally and historically, the EMR is the EHR in one location or office and is owned and operated by the healthcare provider or system. *Personal medical records* (PMRs) is a term used that represents electronic medical records that are owned and maintained by the individual patients themselves. Other terms such as *computerized medical record* (CMR), *electronic clinical information system* (ECIS), and *computerized patient record* (CPR) are rarely used.

In 2008, the National Alliance for Health Information Technology released more concrete definitions of EHR and EMR to better standardize terms used in HIT. The EHR is "an electronic record of health-related information on an individual that conforms to nationally recognized interoperability standards and that can be created, managed and consulted by authorized clinicians and staff across more than one healthcare organization," while the EMR is "an electronic record of health-related information on an individual that can be created, gathered, managed and consulted by authorized clinicians and staff within one healthcare organization" (National Alliance for Health Information Technology, 2008). For further discussion, and colloquially, EMR and EHR can be used interchangeably.

EHRs are, at their simplest, digital (computerized) versions of patients' paper charts. But EHRs, when fully up and running, are so much more than that. EHRs are real-time, patient-centered records. They make information available instantly, and most importantly are universally and securely accessible to authorized persons. These systems bring together in one place everything about a patient's health.

EHRs can (1) contain information about a patient's medical history, diagnoses, medications, immunization dates, allergies, radiology images, and lab and test results; (2) offer access to evidence-based tools that providers can use in making decisions about a patient's care; (3) automate and streamline providers' workflow; (4) increase organization and accuracy of patient information; and (5) support key market changes in payer requirements and consumer expectations.

One of the key features of an EHR is that it can be created, managed, and consulted by authorized providers and staff across more than one healthcare organization. A single EHR can bring together information from current and past doctors, emergency facilities, school and workplace clinics, pharmacies, laboratories, and medical imaging facilities. Increasingly, EHRs have an associated or linked patient portal. This patient portal is the patient's direct access to view information released by that health system and a mechanism to communicate with that healthcare provider or office. Controversy exists in "what" and "when" to release information to the patient portal, but increasingly systems are becoming more transparent. Advanced systems are completely transparent with patients, engaging them in the care by releasing all information to the patient portal at the same time it is posted within the EHR.

The history of the EHR dates back to 1976, when the Medical Center Hospital of Vermont developed the Problem Oriented Medical Information System (POMIS). Soon after, other EHR systems started appearing throughout the United States; for example, the Wishard Memorial Hospital in Indianapolis, Indiana, developed the Regenstreif Medical Record System (RMRS). In 1991, the Institute of Medicine (IOM) recommended EHRs as a solution for many of the problems facing modern medicine. The American Recovery and Reimbursement Act (ARRA) of 2009 provided both and incentive and threat of future revenue loss for systems, which in turn functionally "requires" the use of a "certified" EHR. Figure 5.6 provides a visual display of the early history of the EHR.

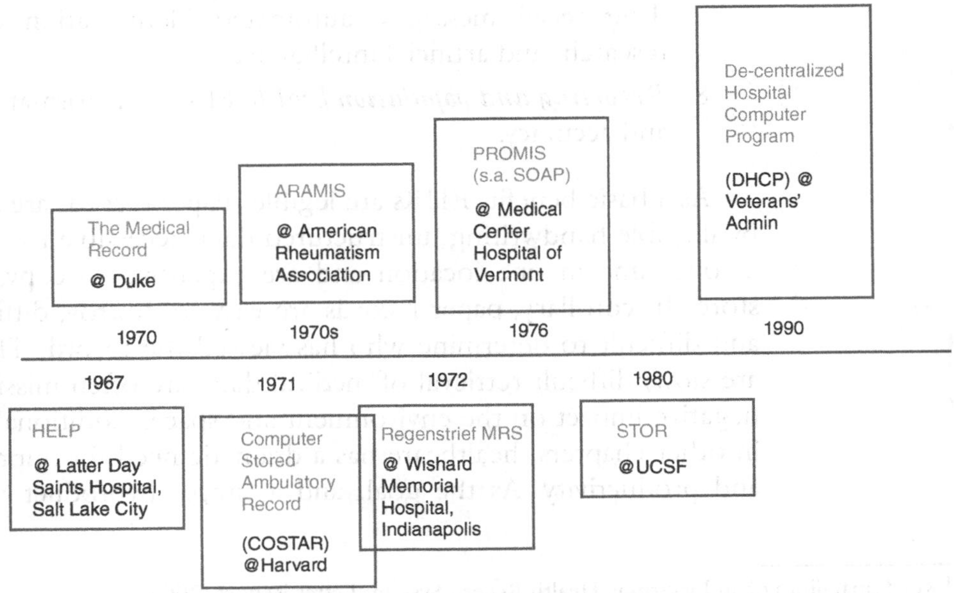

Figure 5.6 History of the Electronic Health Record

The need for EHRs is multifold. According to the IOM's vision for EHRs[1], there are eight core functions:

1. *Accuracy*: Accurate health information and data.
2. *Result management*: Quick access to lab, x-ray and consult results saves time and money and prevents redundancy and improves care coordination.
3. *Order management*: Computerized physician order entry (CPOE) should reduce order errors from illegibility for medications, lab tests, and ancillary services and standardize care.
4. *Decision support*: Should improve medical care quality by providing alerts and reminders.
5. *Electronic communication and connectivity*: Communication among disparate partners is essential and should include all tools such as secure messaging, text messaging, web portals, health information exchange, and so on.
6. *Patient support*: Recognizes the growing role of the Internet for patient education as well as home telemonitoring.
7. *Administrative processes and reporting*: Electronic scheduling, electronic claims submission, eligibility verification, automated drug recall messages, automated identification of patients for research, and artificial intelligence.
8. *Reporting and population health*: Electronic format improves speed and accuracy.

As a basic benefit, EHRs are legible. Paper records are severely limited by illegible handwriting, unstructured data, accessible to only one person at one time in one location and are expensive to copy, transport and store. In corollary, paper records are easy to destroy, difficult to analyze and difficult to determine who has viewed the record. The paper charts are slow/difficult retrieval of medical data, are often missing, and have a negative impact on the environment and space. Additionally, as discussed in other chapters, healthcare has a dramatic need for improved efficiency and productivity. As the goal, and a simple byproduct of making data

[1] Key Capabilities of an Electronic Health Record System: Letter Report (2003).

electronic, it allows patient information to be available to anyone who needs it, when they need it, and where they need it. EHRs can reduce duplication or redundant paperwork and have the ability to facilitate electronic review and transmission of the business office function of healthcare (billing, claims submission, etc.). Based on the understanding of processes in healthcare and the implementation of a given EHR, there is an improvement to overall office productivity, but it increases work of clinicians, particularly with regard to data entry.

As a key component to the triple aim, EHRs can help improve the quality of care and patient safety through a number of mechanisms:

1. Improves legibility of clinical notes;
2. Improves access anytime and anywhere;
3. Reduces duplication secondary to access to historical results;
4. Supports preventative care, though clinical decision support (CDS) such as ensure reminders of tests or preventive services are due;
5. Provides structure for standard dosing of medications; and
6. Displays the electronic problem summary at a glance.

While not comprehensive, the list provides a tangible example of quality of care impact.

Public expectations and perceptions of EHRs are that they decrease medical errors, reduce healthcare costs, facilitate better customer satisfaction (through fewer lost charts, faster refills, and improved delivery of patient educational material), and positively influence decisions about selecting a personal physician. While many of these perceptions are true, the medical community can point to additional risks identified with "clicking on the wrong chart" or ordering a study because it was "part of the order set." While EHRs are here to stay, the explicit and detailed benefits continue to be described. However, it is very clear that there are governmental and insurance company expectations that EHRs are imbedded in the health system as they are integral to healthcare reform (ARRA). The goal of a healthier population requires interoperable data collected at the bedside, resulting, to a degree, in an interoperable healthcare delivery system.

EHRs have the potential to engender major financial savings (savings from eliminated chart rooms and record clerks, fewer callbacks and pharmacists, reduced labor for copying, faxing and mail expenses, and perhaps even decrease in malpractice) but also incur a cost of vendor selection/purchase, change in process, decreased provider productivity, and maintenance. The consensus is that the benefit to the population outweighs the cost of implementation and adoption. Technological advances (Internet, computer speed, memory and bandwidth, and mobile technologies) make EHRs' usability and use the way of the future. While EHRs facilitate the point of care collection of data, there is need for aggregated data. In order to make evidence-based decisions, clinicians need high-quality data that should derive from multiple sources: inpatient and outpatient care, acute and chronic care settings, and urban and rural care and populations at risk. This can be accomplished within some EHRs and discrete structured data.

The need for data integration in context is clear. EHRs can foster integration with health information organizations; analytical software for data mining to examine optimal treatments genomic data, and the like; with local, state, and federal governments for quality reporting and public health issues; and with algorithms and artificial intelligence. The EHR can be used as a transformation tool for improvement in standardization of care, care coordination, and population health. Further, having more than one physician mandates good communication among the primary care physician, the specialist, and the patient.

As noted, the needs for EHRs are multifactorial and the components of an EHR are varied. Most include physician documentation, nursing and/or allied health (ancillary) documentation, pharmacy documentation, CPOE, educational/reference material, and decision support. Frequently, there are additional components such as tracking systems, niche modules for specific care areas, and patient/provider access portals. Supporting systems such as radiology, laboratory, cardiovascular, operating rooms, emergency department, supply management, registrations, or billing systems can be components of the core EHR or supporting systems with some interaction/interface with the EHR.

We next discuss the two key EHR components, where systems can be leveraged for cost and quality outcomes: CPOE and *Clinical decision support (CDS)*.

Computerized Physician Order Entry

CPOE is a feature of EHRs that processes orders for medications, lab tests, imaging, consults, and other diagnostic tests. CPOE has the potential to significantly reduce medical errors through a variety of mechanisms, as well as reduce variation of care, length of stay, and overall costs (in addition to decreased medication costs) (Hoyt, 2014).

Given that the CPOE process is electronic, users can embed clinical decision support "rules" to set up alerts, such as checking for allergies. There are several advantages of CPOE compared to paper-based systems for patient safety (Koppel et al., 2005):

- Overcomes the issue of illegibility
- Fewer errors associated with ordering drugs with similar names
- More easily integrated with decision support systems than paper
- Easily linked to drug-drug interaction warning
- More likely to identify the prescribing physician
- Able to link to adverse drug event reporting systems
- Able to avoid medication errors like trailing zeroes
- Creates data that are available for analysis
- Can point out treatment and drugs of choice
- Can reduce under- and overprescribing
- Prescriptions reach the pharmacy more quickly

Although CPOE may appear to be an "easy" thing to do, the CPOE represents a significant change in clinical workflow and process. Thus, the adoption of CPOE requires training, planning, and leadership on the part of the clinical team, and continuous improvement efforts should be implemented alongside to ensure that CPOE is enhancing the workflow (rather than hindering it). CPOE can be enhanced by prepopulating details of an order, for ease of use, but still relies on the clinician to choose the appropriate care and diagnostics for an individual patient. Order sets (groups of individual orders generated at the same time) are also an efficiency and evidence-based approach to care provision.

Clinical Decision Support Systems

Clinical decision support systems (CDSSs) are systems designed to influence clinicians by providing health knowledge and patient situational context to positively influence the choices to improve care. The CDSS may be inherent to the EHRs or may be an additional system(s) or set of rules.

Mature systems will leverage both the internal CDS within the EHR and supplement with additional CDSSs. CPOE is one of the natural points of interaction with the clinician EMR workflow for CDS. A few classic examples would include drug-allergy checking while prescribing a medication, or reminders to perform a needed test based on previous results or the lack or previous testing. More recently, CDSSs have become more robust in their capabilities. They are used for automatic knowledge support, calculators, flow sheets, patient lists and registries, graphs of captured metrics, medication ordering support (detect known allergies, drug-drug interactions, excessive dosages, formulary checking, etc.), computerized reminders and alerts (e.g., radiology, laboratory, public health), order sets (groups of preestablished inpatient orders that are related to a symptom or diagnosis), and differential diagnoses (diagnostic possibilities based on patient symptoms) (Hoyt, 2014).

Meaningful Use

"Meaningful use" (MU) is a phrase from the ARRA, which references the meaningful use of electronic records. Colloquially, in the United States, it has become a program that is defined by using certified EHR technology to (1) improve quality, safety, and efficiency and reduce health disparities; (2) engage patients and family; (3) improve care coordination and population and public health; and (4) maintain privacy and security of patient health information. The objective is that meaningful use compliance will result in better clinical outcomes, improved population health outcomes, increased transparency and efficiency, empowered individuals, and more robust research data on health systems (HealthIT.gov, 2015). In other words, meaningful use means that providers must show that they are using certified EHR technology in ways that can be measured significantly in quality and quantity. MU had jump-started the acquisition and use of the EHRs for some and validated the previous work for others. One component that MU has allowed is the measure of success, with new standards for components of the EHRs (independent of vendor). MU also has components for three key constituencies: inpatient care (under the eligible hospital [EH]), ambulatory care (via eligible providers [EPs]), and critical access facilities.

According to the ARRA of 2009, the three main aspects of meaningful use are (1) the use of a certified EHR in a meaningful way, (2) the electronic exchange of health information to improve quality of healthcare, and (3) the use of certified EHR technology to submit clinical quality and other measures. In 2010, the Centers of Medicare and Medicaid Services (CMS) published a final rule announcing the establishment of three phases of an EHR adoption incentive program. These three stages are designed to support eligible professionals and hospitals with implementing and using EHRs in a meaningful way to improve the quality and safety of the nation's health system (Health Resources and Services Administration [HRSA], 2015). Table 5.2 (HealthIT.gov, 2015) displays these three stages of meaningful use.

As of 2015, Medicare requires that all Medicare eligible professionals and hospitals meet meaningful use or they may be subject to a financial penalty. On the other hand, Medicaid eligible professionals participating in the EHR incentive program can elect to adopt, implement, or upgrade

Table 5.2 Stages of Meaningful Use for EHR Implementation

Stage 1	Stage 2	Stage 3
2011–2012	**2014**	**2016**
Data capture and sharing	**Advance clinical processes**	**Improved outcomes**
Stage 1: Meaningful use criteria focus on:	**Stage 2: Meaningful use criteria focus on:**	**Stage 3: Meaningful use criteria focus on:**
Electronically capturing health information in a standardized format	More rigorous health information exchange (HIE)	Improving quality, safety, and efficiency, leading to improved health outcomes
Using that information to track key clinical conditions	Increase requirements for e-prescribing and incorporating lab results	Decision support for national high-priority conditions
Communicating that information for care coordination processes	Electronic transmission of patient care summaries across multiple settings	Patient access to self-management tools
Initiating the reporting of clinical quality measures and public health information	More patient-controlled data	Access to comprehensive patient data through patient-centered HIE
Using information to engage patients and their families in their care		Improving population health

Table 5.3 Medicare and Medicaid EHR Incentive Programs

Medicare EHR Incentive Program	Medicaid EHR Incentive Program
Run by CMS	Run by your state Medicaid agency
Maximum incentive amount is $44,000	Maximum incentive amount is $63,750
Payments over 5 consecutive years	Payments over 6 years, does not have to be consecutive
Payment adjustments will begin in 2015 for providers who are eligible but decide not to participate	No payment adjustments for providers who are only eligible for the Medicaid program
Providers must demonstrate meaningful use every year to receive incentive payments	In the first year providers can receive an incentive payments for adopting, implementing, or upgrading EHR technology. Providers must demonstrate meaningful use in the remaining years to receive incentive payments.

to an EHR system in the first year and still receive the incentive payments. Table 5.3 (CMS, 2015) details the Medicare and Medicaid EHR incentive programs.

5.3 HEALTH INFORMATION EXCHANGE

The **health information exchange (HIE)** is a critical element of meaningful use, and it is integral to the future success of healthcare reform in the United States. The term *HIE* can be considered a "noun" or a "verb." HIE as a noun is an intermediary repository when healthcare data can be ingested and released in a secure fashion to appropriate individuals or systems. HIE as a verb is the process of exchanging health information. In either definition, the efficient exchange of health-related data is important to all healthcare organizations and insurance companies to improve the disability process, continuity of medical care issues, bio-surveillance, research, and natural disaster responses (Hoyt, 2014). Further, the U.S. federal government has been a major promoter of the HIE and the development of data standards to achieve interoperability. As a result, the HIE is a major component of a comprehensive game plan to share health information among disparate partners. Table 5.4 displays the common types of health-related data exchanged.

Table 5.4 Common Types of Health-Related Data Exchanged

Clinical results	Lab, pathology, medication, allergies, immunizations, and microbiology data
Images	Images radiology reports; scanned images of paper documentation
Clinical summaries	Documents office notes, discharge notes, emergency room notes
Documents	Continuity of Care Documents (CCDs), personal health record extracts
Financial informations	Claims data, eligibility checks
Medication data	Electronic prescription, formulary status, history
Performance data	Quality measures like blood pressure, cholesterol levels
Case management	Management of the underserved/emergency room utilization
Public health data	Infectious diseases outbreak data, immunization records
Case management	Management of the underserved/emergency room utilization
Referral management	Management of referrals to specialists

More formally, the HIE is the electronic movement of health-related information among organizations according to nationally recognized standards. On the other hand, the **health information organization (HIO)** is an organization that overseas and governs the exchange of health-related information among organizations according to nationally recognized standards. Finally, the **health information service provider (HISP)** is an organization that provides services and support for the electronic exchange of health information. In the 1990s, there were community health information networks, but these all failed due to lack of perceived value and sustainable business plans and immature technology. However, in April 2004, U.S. President George W. Bush created the Office of National Coordinator for Health Information Technology, while calling for interoperable EHRs within the next decade; this led to the establishment of the Nationwide Health Information Network (NHIN), which is a decentralized architecture built using the Internet. Figure 5.7 (Hoyt, 2014) depicts the NHIN.

The prototype architecture for the NHIN exchange (NwHIN) was built in 2005 and pilot demonstrated in 2007, where participating organizations utilized technology known as NHIN-Connect Gateway. The core capabilities tested include: 1) look-up, retrieval and secure exchange of health information, 2) application of patient preferences and permissions for sharing of data, and 3) use of NwHIN for other business purposes as

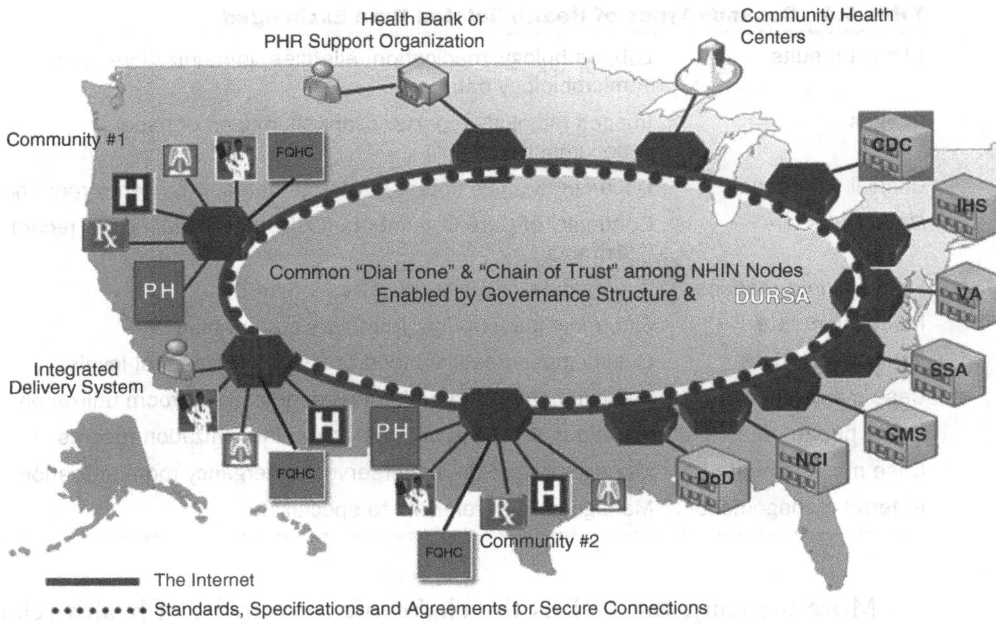

Figure 5.7 Overview of the National Health Information Network

authorized by consumers (Hoyt, 2014). The NwHIN was formally established in 2009, and in 2012 it was renamed eHealth Exchange.

In March 2010, the NHIN Direct project was launched to supplement traditional fax and mail methods of exchanging health information between known and trusted recipients with a faster, more secure, Internet-based method. The Direct project helps provider A transmit to provider B patient summaries, reconciliation of medications, and lab and x-ray results. This system is based on secure messaging that is managed by an HISP (which can be a healthcare organization, HIO, or IT entity). The role of the HISP is to provide user authentication, message encryption, and maintenance of system security for sending and receiving organizations or clinicians (see Figure 5.8) (Hoyt, 2014).

5.4 PUBLICLY REPORTED HEALTHCARE DATA

Transparency in healthcare data to develop public knowledge of clinical outcomes and "quality" of care is more prevalent and can lead to consumer

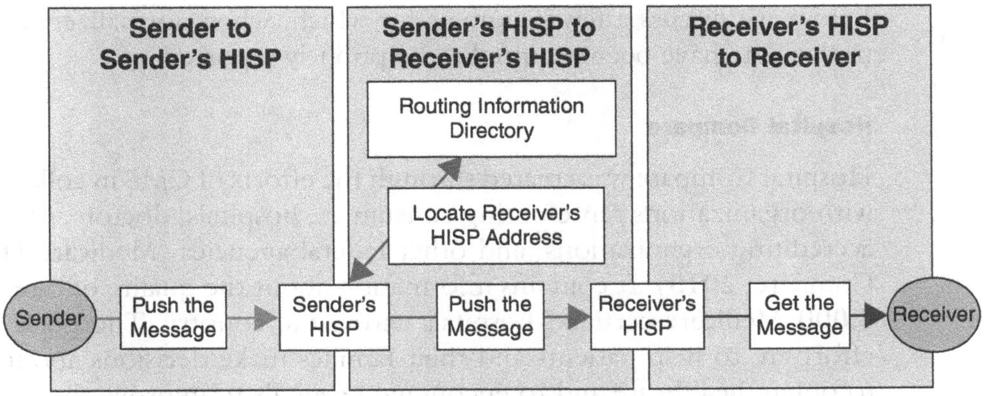

Figure 5.8 Relation between NHIN Direct and HISP

driven decisions based on data (knowledge). Unfortunately, the healthcare data that are used to report clinical quality are based in financial systems. The challenge of healthcare data and publicly reporting remains the source of the data, collection mechanism/intent, and the interpretation of the data. For example, the standard collection of clinical measures is not consistent within EHRs and therefore will be reported differently. Circa 2015, the data that are generally being reported are financial-driven data for insurers such as CMS or surveillance reporting to the state.

In a back-office process, clinical care is "coded" or defined by a specific vocabulary (frequently ICD). Additionally, if patients are "covered" for a service—an electrocardiogram (EKG) or blood test, for example—then the data that the patient had a payment/claim for an EKG or blood test are available. It says nothing about the actual test itself or its results (good or bad, normal or abnormal), let alone the quantification of a test. These claims for service, codes, or tests are then correlated and used as "clinical quality measures" for which, via financial systems data, are compared to other groups (hospitals, providers) who submit the same claims, and determine the "clinical quality." Ultimately, those financially generated data points, even if using clinical vocabularies, are adjusted for optimal financial outcomes, rather than clinical outcomes. Additionally, they may not have the specificity or results to determine "quality of care" in some cases. However, these are data that can be somewhat normalized, and through analysis, may suggest comparison opportunities, if not actual care quality.

Below, we discuss a few programs for which, when normalized, administrative data have been required to be publicly reported.

Hospital Compare

Hospital Compare was created through the efforts of CMS in collaboration with organizations representing consumers, hospitals, doctors, employers, accrediting organizations, and other federal agencies (Medicare Hospital Compare, 2015). It contains information about the quality of care at over 4,000 Medicare-certified hospitals across the country. The goals of this effort are to help patients and their families make decisions about where to obtain healthcare and to encourage hospitals to improve the quality of care they provide.

The data on Hospital Compare are updated annually, semiannually, or quarterly, depending on the measure, and include the following categories:

- Structural measures (e.g., cardiac surgery and stroke care registries)
- Hospital Consumer Assessment of Healthcare Providers and Systems Survey (HCAHPS) scores
- Timely and effective care measures on acute myocardial infarction (AMI), heart failure (HF), and pneumonia (PN)
- Surgical care improvement project (SCIP)
- Emergency department (ED) throughput
- Preventive care
- Children's asthma care
- Stroke care
- Blood clot prevention and treatment
- Pregnancy and delivery care
- Death and readmission rates
- Surgical complications
- Healthcare-associated infections (HAI) (including central line–associated bloodstream infection [CLABSI], catheter-associated urinary tract infection [CAUTI], surgical site infection [SSI], methicillin-resistant *Staphylococcus aureus* [MRSA])
- Medical imaging usage

- Medicare payments
- Number of Medicare patients

The data released by CMS through Hospital Compare comes from a number of sources. These include the CMS Certification and Survey Provider Enhanced Reporting (CASPER) system (i.e., self-reported by hospitals), the Joint Commission, the Centers for Disease Control and Prevention (CDC) via the National Healthcare Safety Network (NHSN), and Medicare enrollment and claims data.

Health, United States

Health, United States is an annual report on trends in health statistics. It includes topics such as birth and death rates, infant mortality, life expectancy, morbidity and health status, risk factors, use of ambulatory and inpatient care, health personnel and facilities, financing of healthcare, health insurance and managed care, and other health topics. The report contains detailed tables and charts on health status and its determinants, healthcare resources, healthcare utilization, and health insurance and expenditures. *Health, United States* focuses primarily on trends over time in health statistics. Comparable data must be available for at least two points in time and available at the national level. In addition to national level data, data are commonly shown by age group, sex, race and Hispanic origin, geography, poverty status, and education level whenever possible. It is compiled by the CDC and is available on its website (www.cdc.gov).

Medicare Provider Charge Data

Hospitals determine what they will charge for items and services provided to patients. These charges are then billed to Medicare, private insurers, or patients. Medicare does not actually pay the amount a hospital charges but instead uses the Medicare Inpatient Prospective Payment System (IPPS) to reimburse hospitals for treating specific conditions. The core vocabulary used in this model is the diagnosis-related groups (DRGs). These codes (vocabulary) group a set of services that are provided for a patient for any given inpatient hospitalization and categorize them based on the primary discharging diagnosis and associated co-morbidities and/or complications associated with that clinical encounter. These codes are assigned by billing professionals based on the documentation and care recorded in the EMR. The Medicare provider charge data for inpatients gives hospital-specific charges for the more than 3,000 U.S. hospitals that receive Medicare IPPS payments for the top 100 most frequently billed discharges, paid under Medicare based on a rate per discharge using the Medicare Severity Diagnosis-Related Group (MS-DRG) for fiscal year (FY) 2011). These DRGs represent almost 7 million discharges or 60% of total Medicare IPPS discharges.

Similarly, the Medicare charge data for outpatients include estimated hospital-specific charges for 30 Ambulatory Payment Classification (APC) groups paid under the Medicare Outpatient Prospective Payment System (OPPS) for calendar year (CY) 2011. These data can be downloaded from the CMS website (https://www.cms.gov/).

State Departments of Health

In the Commonwealth of Pennsylvania, the *Pennsylvania Health Care Cost Containment Council (PHC4)* is an independent state agency charged with collecting, analyzing, and reporting information that can be used to improve the quality and restrain the cost of healthcare in the state. PHC4 collects over 4.5 million inpatient hospital discharge and ambulatory/outpatient procedure records each year from hospitals and freestanding ambulatory surgery centers in Pennsylvania. These data, which include hospital charge and treatment information as well as other financial data, are collected on a quarterly basis and then verified by PHC4 staff. The data collected are then shared through free public reports. PHC4 also produces customized reports and data sets through its Special Requests division, for users such as hospitals, policymakers, researchers, physicians, insurers, and other group purchasers.

CASE STUDY 5: HEALTH INFORMATICS AT A HOSPITAL—A 10-YEAR JOURNEY

An academic health center must make determinations as to what systems and structures they need for the practice of healthcare. The current demographics include the following:

# Beds:	550
Total Admissions:	30,000
Total Outpatient Visits:	1,000,000
ED Visits:	70,000
Births:	2,000
Attending Physicians:	750
Residents:	550
APCs (NP and PAs)	200
Medical Students:	550
Nursing:	2,000

Goals of the initiative were clear, with a standard strategic vision to have one EMR for all providers across all venues. The health system understood that not only would it be a challenge to complete the conversion from a hybrid paper and electronic format to one standard EMR, but it would require strategy and leadership. The team developed four guiding principles: (1) improve patient safety, (2) efficiently deliver high-quality care, (3) improve our ability to recruit and retain talented and dedicated personnel, and (4) increase patient satisfaction. Note that there was no mention of computers or electronic records anywhere in the guiding principles. Upon completion, there would be one record.

This journey initiated with a vendor search and selection in 2002–2003. The system had decided that inpatient care automations (including the emergency department) were the highest priority with a natural extension to ambulatory care. An extensive search that included academics, administrators, and frontline clinicians was undertaken. There were multiple demonstrations on-site, as well as numerous site visits across the United States. The team narrowed it down to two vendors, and negotiations were completed with both. After final selection, the vendor was selected and engaged. Now the work really began.

The healthcare team and vendor committed and partnered to a timeline and road map to automate the entire inpatient facility in two years. Significant time was spent diagramming and understanding the current state of process and practice. The team paid special attention to the high-clinical-risk and high-revenue-producing areas of the facility. Actively practicing clinicians, called SMEs (subject matter experts) were paid to engage in the build and design of the system.

With patient safety as the priority, the health system prioritized medication safety, automating the medication process and nursing intake to support medication safety first. Second, they implemented CPOE and all nursing and ancillary documentation. Initially planned as a one-month rollout, the system was so successful in the pilot units after three days that they expanded the use to *all* children in the hospital. After managing the complexities of those high-risk areas, the system required all care to be provided and ordered via the new EMR on day 14. The knowledge gained during the pilot was valuable, but a true "big bang," one-day implementation would have been preferable to all users of the system. Additionally, key leaders in both administration and clinical areas were very prominent and supportive of the conversion.

Over the next few years, additional modules, upgrades to the software, and safety enhancements allowed the automation of the operating rooms, intensive care units, anesthesia, medication reconciliation, physician documentation, and e-prescribing. The intent of the EMR team was to build an optimization team, to ensure that all users were able to take advantage of the ever-expanding capabilities of the EMR. However, due to tight financial conditions and the institution's large appetite to expand EMR and related capabilities (new modules), the resources were reallocated to new projects rather than optimization.

The next step was to automate the nearly 1 million outpatient visits over 60 different locations. With all due diligence, the health system engaged in a process to determine if the current vendor who partnered with the inpatient and ED could be used in the outpatient setting. Neutral third-party consultants were engaged to assist in the evaluation of the two major EMR vendors in the market. After an extended evaluation, and to ensure compliance/timing with meaningful use requirements, and a return on investment (ROI) determination, the health system determined that extending the current EMR to the ambulatory space was the most prudent approach.

Similar to the inpatient deployment, there were expanding complexities to the processes of care. Some had procedures, while others used the academic office secretary as the clinical secretary. We found that specialists practice differently than primary care providers and each physician (specialist or primary) practices slightly different. This variability in process was a new challenge for a system that seeks standardization. The EMR team developed new teams and SMEs for outpatient care, primary care, specialist, surgeons, proceduralists, and pediatricians. Once the process was broken down, there were a number of core processes that were automated across all clinic sites. Colloquially called DMOD (diagnosis, medications, orders, depart), each one of the four steps *must* be completed by every provider for every patient for every visit. Every patient would have a diagnosis made and documented in the EMR. Every patient would have the medication history collected and appropriately adjudicated by the provider (medication reconciliation). Every patient would have orders placed, even if the *only* order was for follow-up care. Finally, a departing document (instructions and education included) would be provided to the patient. Sticking to the four guiding principles and the four-step clinic process, the system was able to automate all clinical venues on the same platform.

There was a challenge after 67% of the clinics were completed. The entire EMR underwent an upgrade, and included in that upgrade was a new graphical user interface (GUI) that was more pleasant and user friendly, and brought data into context on one screen. Users thought this was the best thing since sliced bread. However, despite aggressive testing, when the new user interface was expanded to many thousand users, the system slowed down. Specifically, the length of each of transaction effectively tripled. The target time of every transaction is 0.4 seconds or below. The median transaction time expanded to 1.1 seconds. Any gains in process and standard efficiency were lost in the speed of the system. This is an example of one of the many individual challenges trumping the gains of the greatest majority of the EMR.

After stabilizing and improving the transaction speed to 0.35 seconds, the system was able to complete the EMR rollout. In parallel, the inpatient EMR expanded capabilities for quality and safety using data contextualization tools with the new GUI and CDS (rules and alerts) to support best practices.

Over recent years, the system has turned the prioritization to patient engagement through patient portals, exchange of information for clinical care, reporting to regional and national bodies for transparency, and, most importantly, aggregating and analyzing the data for population health. To measure success, the system has attained MU stage 2 incentives, both inpatient and outpatient.

This case study and timeline is typical of a health system that has gone "in to out"—that is, starting in the acute care hospital and expanding outward to the ambulatory setting.

QUESTIONS AND LEARNING ACTIVITIES

1. Compare and contrast the reasons for implementing an EHR.
2. Discuss and define the key terms that are essential for the success of the EMR and the subsequent secondary use of that data.
3. Discuss the need for EHRs to be interoperable with other EHRs. Enumerate why current EHRs are not interoperable and what can be done about it.

4. Discuss the implications of information sharing on efficiency, patient safety, care coordination, cost reduction, and improved quality.
5. Discuss why HIE is so important in medicine and why it must occur at local, state, and national levels. Provide examples of medical information that should be shared at these three levels.

REFERENCES

Blois, M. S. (1984). *Information and Medicine: The Nature of Medical Descriptions*. Berkeley, CA: University of California Press.

Centers for Medicare & Medicaid Services (CMS). (2015). EHR incentive programs. Retrieved April 12, 2015, from https://www.cms.gov/Regulations-and-Guidance/Legislation/EHRIncentivePrograms/index.html?redirect=/ehrincentiveprograms/

HealthIT.gov. (2015). Meaningful use definition and objectives. Retrieved April 12, 2015, from http://www.healthit.gov/providers-professionals/meaningful-use-definition-objectives

Hoyt, R. E. (2014). *Health informatics: Practical guide for healthcare and information technology professionals* (6th ed.). Pensacola, FL: Informatics Education.

Health Resources and Services Administration (HRSA). (2015). *What is "meaningful use"?* U.S. Department of Health and Human Services. Retrieved April 12, 2015, http://www.hrsa.gov/healthit/meaningfuluse/MU%20Stage1%20CQM/whatis.html

Kohn, L. T., Corrigan, J. M., & Donaldson, M. S. (2001). *Crossing the quality chasm: A new health system for the 21st century*. Washington, DC: Committee on Quality of Health Care in America, Institute of Medicine.

Koppel, R., Metlay, J. P., Cohen, A., Abaluck, B., Localio, A. R., Kimmel, S. E., & Strom, B. L. (2005). Role of computerized physician order entry systems in facilitating medication errors. *Journal of the American Medical Association, 293*(10), 1197–1203.

Medicare Hospital Compare (2015). Retrieved from https://www.medicare.gov/hospitalcompare/search.html.

National Alliance for Health Information Technology. (2008). Defining key health information technology terms. Retrieved from www.nahit.org (organization no longer active).

Chapter 6

Lean

"The most dangerous kind of waste is the waste we do not recognize."

—Shigeo Shingo

Overview

Lean is a philosophy, methodology, and toolset that can help improve quality and patient safety by reducing waste and inefficiencies. The lean approach can also help to improve employee satisfaction because it can lead to eliminating the roadblocks that interfere with a focus on providing care.

This chapter begins with some origin on the use of lean in manufacturing and in the health domain. We then discuss four drivers for lean in healthcare and four tools for eliminating wastes along with examples of their application. Lean has now become widely implemented, but it takes strong commitment from leadership to advance an organization toward lean thinking.

6.1 LEAN PHILOSOPHY AND METHODS

The practice of **lean** considers the use of resources for any purpose other than the creation of **value** for the end customer to be wasteful. Lean principles are derived from the manufacturing industry and the **Toyota Production System (TPS)**. However, lean is not an instant remedy to healthcare problems. It took Toyota and other manufacturers three decades of continued effort to achieve world-class success with lean methods.

Taichii Ohno, an industrial engineer and businessman considered to be the father of the TPS, stated that: "The most important objective of the Toyota system has been to increase production efficiency by consistently and thoroughly eliminating waste. The concept and equally important respect for humanity that has passed down from the venerable Toyoda Sakichi (1867–1930) … are the foundations of the Toyota production system" (Ohno, 1988).

The deployment of lean led to a revolution across the automotive and manufacturing industries to improve quality and lower costs. Holweg (2007) gives a historical account of how the term *lean* was coined, and many authors have addressed details and cases on lean in manufacturing, notably Womack and Jones (2003) and Womack, Jones, and Roos (1990).

However, a patient is not a car, and for a time the implied suggestion that they could be viewed similarly from the standpoint of process improvement turned people off from lean initiatives. But ways of thinking about waste reduction methods as useful across boundaries has a long history—from Benjamin Franklin's documented thoughts on not wasting time in *Poor Richard's Almanac* to Frank and Lillian Gilbreth's approaches to motion efficiency. The Gilbreths, sometimes known by the original version of the film *Cheaper by the Dozen*, published studies on factory work as well as on medicine. In 1914, Frank Gilbreth's article entitled "Scientific Management in the Hospital" declared that:

> … having spent years of study in the theory and practice of management of the industries, and having also spent years in observing hospital practice and in studying hospital problems, we are able to tell you with authority not only that the same laws which govern efficient shop practice also govern efficient practice in the hospital, but also that many of the problems involved are not only similar, but identical, and that many of the solutions which we found to these problems in the shop can be carried over bodily into the field of hospital management. The problems of transportation; problems of assembling; problems of enforcing and maintaining system, orders and discipline; problems of motion study, time study, and standardization; problems of teaching problems of synthesizing these elements into methods of least waste; all may make two apparently dissimilar and unrelated lines of activity so closely akin that a successful solution in one case can be applied with little or no change with equal success to the other. You can recognize at a glance that all of these problems stated are to be found in hospital management, and that it may be possible for you to save much time and effort by becoming acquainted with efficient practice in the industrial world.

Gilbreth's article entitled "Motion Study in Surgery" (1916) went on to identify the worker; the equipment, surroundings, and tools; and the worker's motion as the three needed divisions of study. He determined that "the surgeon furnishes in every respect the ideal example to be motion studied." The article was delivered before the Hospital Section of the American Medical Association in 1915.

6.2 DRIVERS FOR LEAN HEALTHCARE SYSTEMS

Ohno's statement on the objectives of the lean system suggests that there are two core elements that define lean: respecting people and eliminating waste. As we translate these core elements into healthcare we have to consider what are the important drivers to focus on in that environment. Although there will be many concepts and tools to deploy in order to achieve solutions, we can identify four drivers that need to be considered.

Driver 1: Define Value Based on the Voice of the Patient (Customer)

Value is a key concept in lean thinking. Value is achieved by delivering exactly the (customized) product or service a customer wants with minimal time between the moment the customer asks for that product or service and the actual delivery at a suitable price (Womack & Jones, 2003). The "voice of the customer" means to define what customers want. The "voice of the customer" is an essential connection to the lean element of respecting people. It focuses on the belief that the customer's perspective is what should drive all activities in the system. Given this information, the process involved in delivering it can be divided into value-adding steps and non-value-adding steps. Value-adding steps directly contribute to creating a product or service a customer wants. Non-value-adding steps do not contribute directly, no matter how necessary they may be.

In healthcare, value is achieved for patients in many ways. The closest alignment with delivering what a patient wants with minimal delay suggests that first and foremost, patients will not wait for healthcare services. For example, when they arrive at the emergency department or clinic, value is imparted by treating their injuries or illnesses expeditiously—not having them sit in a waiting room or by telling their condition several times to multiple providers. Similarly, lab results will be communicated directly and quickly—not several days later after returning a message left by another clinician.

Driver 2: Create an Environment That Supports the Staff

The healthcare environment has many challenges as a place of work. The fact that here is a wide range of biological, physical, chemical, and biomechanical/ergonomic hazards has long been understood (e.g., Moore & Kaczmarek, 1991; Sepkowitz, 1996). Another category of exposures includes nonphysical or psychosocial hazards such as the job stress that results from factors such as rotating shift work and heavy workload (Johnson et al., 1995; Revicki & May, 1989); these factors may contribute to burnout and depression. In addition, organizational culture and climate, which is created by contextual factors such as leadership style and institutional goals, may also influence hospital safety and workplace exposure incidents for caregivers (Gerson et al., 2000).

We recognize that patients are often vulnerable as illness or injury upends their very lives and/or the lives of their loved ones. And yet the people who provide the care are also vulnerable: a recent career survey showed that 69% of healthcare respondents reported feeling stressed in their current jobs, and another 17% reported feeling highly stressed (Sullivan, 2014). It is understood that stress sustained over a long period of time can be detrimental to employee health and ultimately stand in their way of providing quality care to patients. Furthermore, in industrial settings, employee perceptions regarding their organization's commitment to safety has been shown to correlate to both the adoption and maintenance of safe work practices and to workplace injury rates.

In other words, if we want an environment that is safer for patients, we must make that same environment safer for employees. It is crucial that both clinical and nonclinical staff that supports patients must be fulfilled in their work environment. In a sense, this is also a component of respecting people—in this case, the people who provide the care. Providing meaningful work, a reasonable workload, a safe work environment, and a culture of respect are important for having empathetic care providers. Healthcare leaders must provide staff with the resources and support they need to perform at their best.

Driver 3: Eliminate All Forms of Waste

Simply put, all steps in a process that add no value are considered waste. At first pass, it may seem easy to just say "eliminate waste." However, in healthcare, waste may come in many difficult forms. For example, the overtreatment of patients results in a variety of medications, tests, clinical, and surgical procedures that drive up cost unnecessarily while exposing patients to risks of complication. The underuse of evidence-based medicine or evidence-based practices means that providers routinely fail to administer a variety of tests and treatments known to be effective. Medical errors such as drug misuse, wrong site surgery, hospital-associated infections, and failures to coordinate care are, at the core, all waste.

Don Berwick, MD, former administrator of the Centers for Medicare and Medicaid Services (CMS) and former president and CEO of the Institute for Healthcare Improvement, said that 20% to 30% of health spending is "waste" with no benefit to patients, because of the overtreatment of patients, the failure to coordinate care, the administrative complexity of the healthcare system, and burdensome rules and fraud (Pear, 2011). Berwick has also been a leading expert on using evidence-based medicine and comparative effectiveness research (CER) and has suggested that the rational approach to reducing waste is to look at what interventions are effective and what they cost ("Rethinking Comparative Effectiveness Research," 2009).

Driver 4: Deliver Excellent Clinical Quality

In the United States, healthcare reform is having an impact on making healthcare available and affordable. Delivering a high quality of care in a safe environment is, in a sense, a combination of drivers 1, 2, and 3. In *Crossing the Quality Chasm*, the Institute of Medicine (IOM) outlined six specific aims that a healthcare system must fulfill to deliver quality care:

1. *Safe*: Care should be as safe for patients in healthcare facilities as in their homes.
2. *Effective*: The science and evidence behind healthcare should be applied and serve as the standard in the delivery of care.
3. *Efficient*: Care and service should be cost effective, and waste should be removed from the system.
4. *Timely*: Patients should experience no waits or delays in receiving care and service.
5. *Patient centered*: The system of care should revolve around the patient, respect patient preferences, and put the patient in control.
6. *Equitable*: Unequal treatment should be a fact of the past; disparities in care should be eradicated.

6.3 A TOOLSET FOR ELIMINATING WASTES

The 3Ms of waste in lean are **muda, mura,** and **muri.** Eliminating **muda** deals with eliminating processes or activities that add cost but do not add value. Eliminating **mura** deals with eliminating unevenness so as to achieve production or system leveling. Eliminating **muri** deals with eliminating overburdens or unreasonableness so as to avoid difficulties and achieve standardized work and a logical workflow with repeatable processing steps, in a reasonable time/takt time.

Muda wastes were categorized by Taichii Ohno into the seven forms of waste: transportation, inventory, unnecessary motion, waiting, overprocessing, overproduction, and defects. A few other categories of waste—underutilized intellect, resources, byproducts—have been added by some practitioners. Table 6.1 shows examples of these wastes that are relevant to healthcare.

Mura wastes of unevenness and inconsistency in healthcare mean that there is a variability and unpredictability to patient arrivals or other demands (such as lab tests). This often creates a system of high demands with periods of "just get through" followed by low demands with periods of "not a lot to do." This see-saw may adversely impact the staff, particularly with high stress levels that accompany the high-demand periods, which may in turn increase costs and perhaps contribute to a less safe

Table 6.1 Wastes in Healthcare

Type of Waste	Examples in Healthcare
Transportation	Movement of patients or equipment
Inventory	Excess supplies in patient exam rooms and nursing stations
Motion	Movement of staff and information
Waiting	Delays in diagnosis or treatment
Overprocessing	Duplication of patient information on multiple forms or fields Asking patients their demographics several times
Overproduction	Requesting or preparing unnecessary medications Preparing too many procedure kits
Defects	Medical errors
Underutilized intellect	Nurses mopping floors Pharmacists checking refrigerator temperatures
Resources	Failing to turn off lights or unused equipment
By-products	Not making use of byproducts of processes such as the steam water from sterilization

environment for patients. In other words, failing to smooth the process may actually create many of the muda wastes.

Muri waste is to cause undue burden on the employees and processes. Some of the causes or nature of muri include working on processes without the requisite training, lack of proper maintenance or unreliable equipment or processes, and unclear instructions or poor communication channels. In healthcare, standard work is often employed to address these types of wastes.

In many lean implementations, muda is used as an easy starting point for quality improvement efforts. However, in most truly successful lean implementations—for example, those going back to Japanese companies or manufacturing organizations that deployed TPS—the full evolution of dealing with waste includes a joint and sophisticated consideration of muda, mura, and muri. In time, it is often realized that dealing with mura and muri are more difficult but far more important to tackle.

Table 6.2 shows several tools and methods that have been used to implement lean. The first four tools form a complementary set that can be quickly learned and deployed in healthcare settings. In the sections of this chapter that follow, these four tools are described with respect to their foundations and with a view toward healthcare applications. Some of the other tools are briefly mentioned in relationship to these four tools and to lean thinking in healthcare. Several authors, including Womack and Jones (2003), have addressed these and the other tools in greater detail and/or other contexts.

Table 6.2 Lean Tools and Methods

Value stream mapping—a method of diagramming and describing the flow of products, people, information, materials and with identification of value and non-value adding steps	**5 Whys, root cause analysis**—an approach to focusing on finding and resolving an underlying problem by asking why five times—each time moving closer to the root cause
A3—a standardized worksheet and a structured problem solving method that starts with A3 (11" × 17") size paper	**Poka-Yoke (error proofing)**—a method for designing error detection and prevention into a process
5S—a method for organizing and standardizing workspaces	**Andon**—a visual feedback system that indicates production status and alerts when assistance or attention is needed
Kanban—a pull method of inventory control using signals throughout a decentralized system	**Kaizen**—short-cycle continuous improvement sessions used by employees to work together to proactively achieve incremental improvements
	Plan, Do, Check, Act (PDCA)—an iterative methodology for implementing improvements

6.4 VALUE STREAM MAPPING

Organizations are designed around specialized functions or departments. Problems and waste are often found in the interactions or handoffs between functions or departments. The chief reason for this is a lack of focus on the critical pathways, or what are called **value streams** in quality improvement.

Womack and Jones define a **value stream** as the set of all the specific actions required to bring a specific product (whether a good, service, or increasingly, a combination of the two) through the three critical management tasks of any business: the problem-solving task; the information management task; and the physical transformation task. A **value stream map (VSM)** is a structured diagram that originated with Toyota in the 1980s. Both process maps and VSMs document the steps in a process and the activities that take place, but VSMs take this further by capturing time elements.

Value stream maps, flowcharts, swim lane diagrams, logic models—these techniques can uncover and explain how activities happen in the system. There are five key VSM principles as follows:

1. Specify value from the standpoint of the end customer.
2. Identify the value stream for each product family.
3. Make the product flow.
4. So the customer can pull.
5. As you manage toward perfection.

Value stream maps are particularly helpful because they capture both the activity and the time it takes for completion of the activity thereby enabling a fast visual assessment on where delays occur in the process. VSMs identify the flows that lead to each step and thus expose constraints and conditions that may contribute to waste. VSM is best done by a team that involves representatives from all of the areas within the process being mapped.

Figure 6.1 shows some of the symbols commonly used in VSMs. The process symbol can be any step, operation, machine, or department through which a person, thing, or concept flows. The customer/supplier symbol typically represents the supplier when placed in the upper left and the customer when in the upper right. The remaining symbols are then pulled into the map as necessary to represent the system flow. It is likely

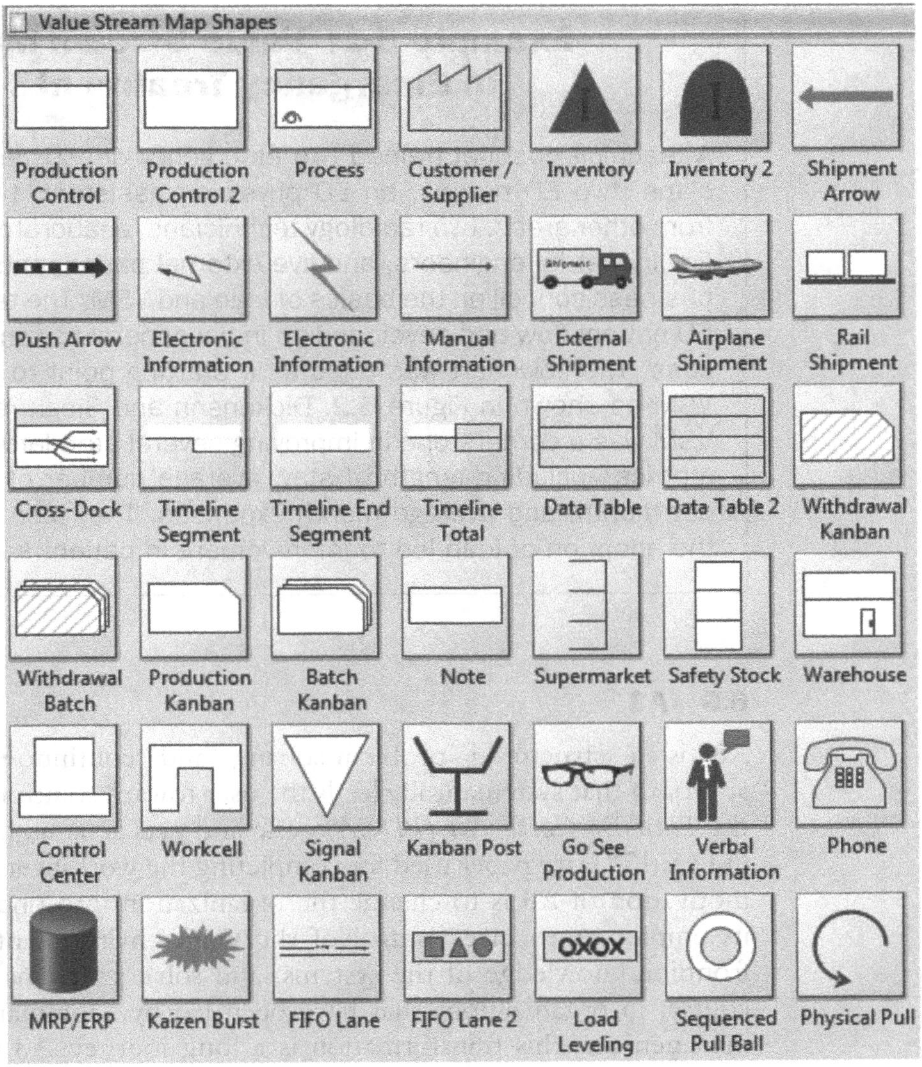

Figure 6.1 Value Stream Map Symbols

that data will have to be collected on elements like inventory, cycle time, change over time, and so forth.

The process of creating the VSM for the current state typically reveals several problems. The team should use this insight to create an ideal or future state VSM that represents a shared vision of where the process needs to be.

Example 6-1: Value Stream Map in Emergency Treatment

A teaching hospital trained two emergency department (ED) physicians, two ED nurses, an ED physician assistant, two physicians from other areas, two radiology technicians, a laboratory technician, five industrial engineers, and five external participants from a local business council on the basics of lean and VSM. The team observed ED patient flow and developed an initial general flowchart of the process. The flowchart was used as a starting point to construct the VSM as shown in Figure 6.2. Dickenson and Singh report that the VSM was a cornerstone in improving several standard ED operation metrics, including length of stay, average number of patient visits per month, and average month expenses. They also reported that the adoption of lean led to improvement in patient satisfaction.

6.5 A3

A3 is a structured problem-solving and continuous improvement approach that systematically leads the user toward solutions. It is based on the *Plan-Do-Check-Act (PDCA)* cycle and gets is name from the ISO A3 (11″ × 17″) size paper used for completing the worksheet. The underlying motivation of A3 is to change the organization into one that empowers its employees to take control of the work environment, leverage their frontline knowledge of the systems, and solve problems at the core. Of course, in organizations that have been led by "command-and-control" management, this transformation is a long journey. A3 provides a template, as well as a necessary buttress, to help in this transformation. Jimmerson (2007) reports on a speech given by Art Niimi, the CEO of Toyota America, which was accompanied by a single slide that simply read: THINK DEEPLY. Through a relentless effort to think deeply and move toward an ideal state, Toyota was able to manifest its success.

In "Decoding the DNA of the Toyota Production System," Spear and Bowen (1999) described the four rules that underlie the TPS and "guide the design, operation, and improvement of every activity,

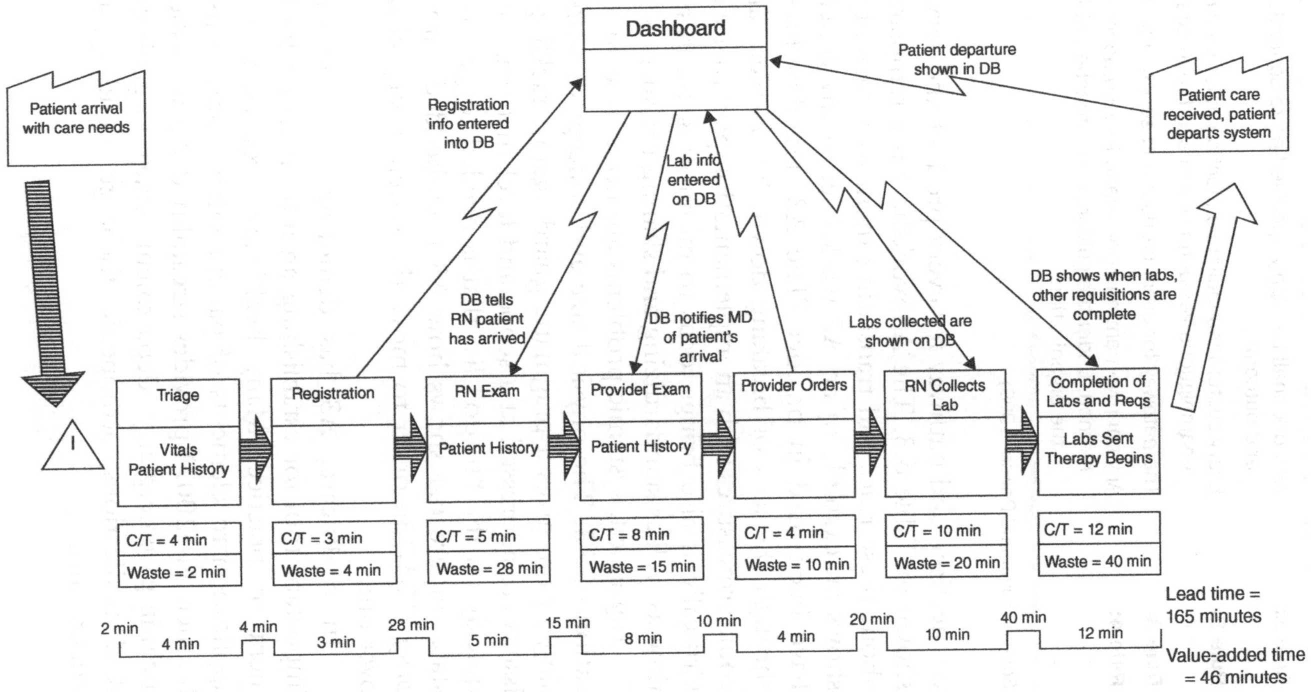

Figure 6.2 A Value Stream Map (VSM) for an Emergency Treatment Encounter
Source: **Dickenson and Singh (2009)**

Table 6.3 The Four Rules of the TPS

Rule 1:	All work shall be highly specified as to content, sequence, timing, and outcome.
Rule 2:	Every customer-supplier connection must be direct, and there must be an unambiguous yes-or-no way to send requests and receive responses.
Rule 3:	The pathway for every product and service must be simple and direct.
Rule 4:	Any improvement must be made in accordance with the scientific method, under the guidance of a teacher, at the lowest possible level in the organization.

Source: Spear and Bowen (1999)

connection, and pathway for every product and service." These rules are shown in Table 6.3. The A3 worksheet is a template for thinking deeply about these rules and translating them to a current problem. Figure 6.3 shows an example of an A3 worksheet; many other forms or templates have been used in practice. The A3 describes how the process works through the lens of breaking down the problem and provides a set of countermeasures with an implementation plan that will solve the problem. Critically, it also brings focus to monitoring the results and process over time as well as standardizing and sharing the success.

Ingrained, systemic problems are not solved in a single step or application of A3. The physical size and concept of working a problem on a single page gives a pragmatic guard against tackling too much. Instead, issues are addressed at a scope and level that make it possible to hone in quickly on the "to do" list that is needed to get to an improved future state. Figure 6.4 shows how A3 is embedded with *Value Stream Mapping (VSM)* in order to move the system a step closer to the ideal state over time.

In healthcare, A3 has demonstrated potential for becoming an important tool for establishing an infrastructure for healthcare improvement in practice. Many healthcare organizations have embraced its application to elements of patient safety, clinic visits, and staff support. Jimmerson (2007) provides streamlined A3 worksheets for several cases including emergency department chart organization, shock trauma ICU medications, orthopedic discharge rounding, and medical surgical stockrooms.

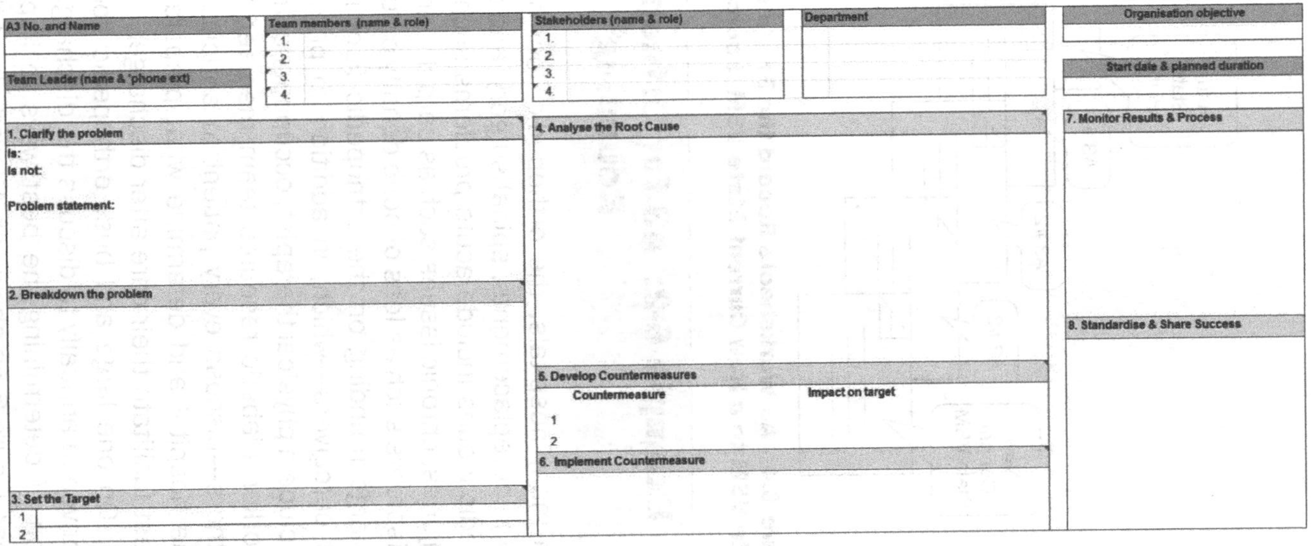
A3 No. and Name

Team Leader (name & 'phone ext)

Team members (name & role)
1.
2.
3.
4.

Stakeholders (name & role)
1.
2.
3.
4.

Department

Organisation objective

Start date & planned duration

1. Clarify the problem
Is:
Is not:
Problem statement:

2. Breakdown the problem

3. Set the Target
1
2

4. Analyse the Root Cause

5. Develop Countermeasures

Countermeasure	Impact on target
1	
2	

6. Implement Countermeasure

7. Monitor Results & Process

8. Standardise & Share Success

Figure 6.3 A3 Worksheet

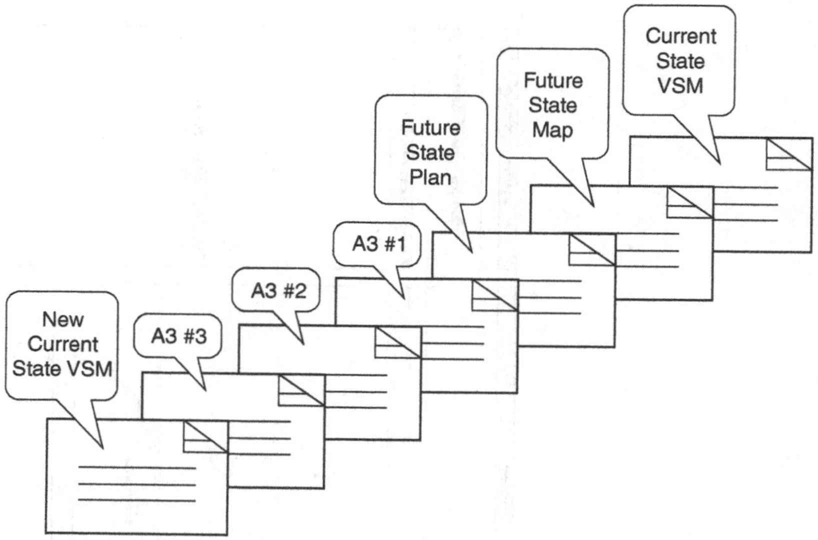

Figure 6.4 A3 Worksheets Record the Progress in Moving from a Current State VSM to a New Current State VSM ***Source:*** **Jimmerson (2007)**

Example 6-2: A3 for Orthopedic Discharge Rounding[1]

In many hospitals, the orthopedic floor handles patients in need of joint replacements, spinal surgery, and arthroscopy. Other orthopedic issues include acute problems such as fractures and sports injuries, chronic issues such as carpal tunnel, and chronic systemic disorders such as loss of bone density or lupus erythematosus. Discharge rounding on the orthopedic floor refers to having a range of caregivers—which, in addition to physicians and nurses, may include a physical therapist, occupational therapist, chaplain, social worker, diabetic resource team member, trauma coordinator, and others—discuss every patient to anticipate when they may leave the hospital and determine what preparations could be made to best facilitate their care after discharge.

On one large and busy orthopedic floor, the entire team was convened annually to discuss the discharge process with the intention of determining the best ways to improve it over the coming year. Figure 6.5a shows an A3 worksheet that arose from one such

[1]Based on Jimmerson (2007)

ISSUE *Not all patients receive complete discharge planning.*

BACKGROUND

Discharge planning is done in meetings with staff from all necessary departments on Mondays and Thursdays. 14 people attend the meetings. with an average wage of $55/hour. Meetings are an average of 1.5 hours long. 14 x $55 x 2 x 52 weeks = $80,080/year spent on discharge planning meetings.

CURRENT CONDITION

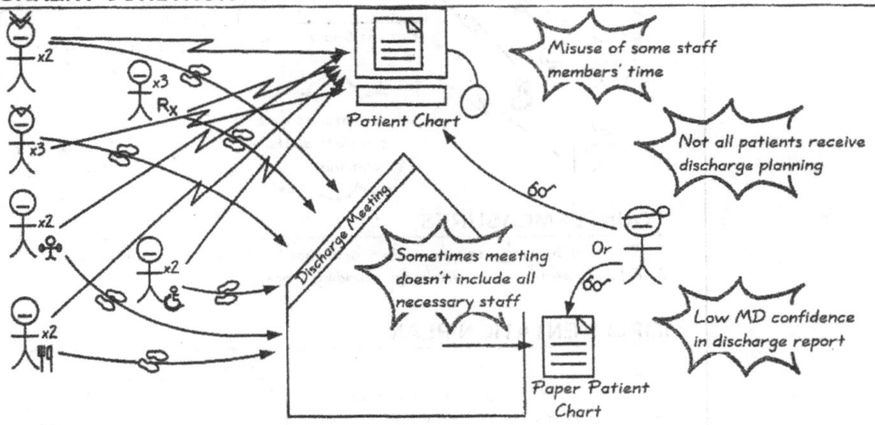

PROBLEM ANALYSIS

1. *Sometimes discharge meeting doesn't include all necessary staff.*
   - *Why? Attendees of meetings are not always the ones who have been treating the patient.*
     - *Why? Representatives are sent from each department.*
       - *Why? Staff from the department are at work and not available at the same time.*
2. *Not all patients receive discharge planning.*
   - *Why? Some patients check in and check out before they receive it.*
     - *Why? Discharge rounding is only done on Mondays and Thursdays.*
       - *Why? Staff can't be taken away from work every day.*
3. *Meeting is a misuse of some staff members' time.*
   - *Why? Staff members have to sit through the review of patients that they have not treated.*
     - *Why? Not all patients receive care from all departments present at the meeting.*
       - *Why? It is not always necessary.*
4. *There is low MD confidence in the discharge report.*
   - *Why? Discharge report is sometimes incomplete.*
     - *Why? The discharge report was not created by all the people who cared for the patient.*
       - *Why? Not all the staff was present at the discharge meeting.*
         - *Why? All the staff is not available to attend meetings.*
           - *Why? It would take too many people away from direct patient care.*

Figure 6.5a A3 for Orthopedic Discharge Rounding
***Source:* Jimmerson (2007)**

TARGET CONDITION

Title: Orthopedic Discharge Rounding

TO	Bob Johnson
BY	Elizabeth Smith
DATE	June 12, 2007

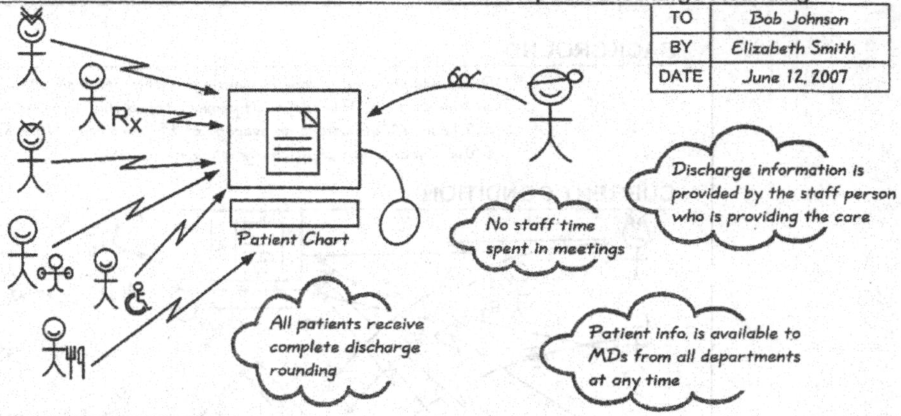
Rx
Patient Chart
Discharge information is provided by the staff person who is providing the care
No staff time spent in meetings
All patients receive complete discharge rounding
Patient info. is available to MDs from all departments at any time

COUNTERMEASURES

1. Create a discharge screen in electronic chart notes for all departments and MDs to access.
2. Educate staff on new discharge rounding process.

IMPLEMENTATION PLAN

What	Who	When	Outcome
Meet with staff to determine what info is necessary on discharge screen	Elizabeth	June 15	Discharge rounding screen criteria created
Meet with IT staff to create new discharge screen	Elizabeth	June 20	Comprehensive screen completed and online
Orient Staff	Elizabeth /IT	June 30	All staff understands new process and can use the new screen.

COST/BENEFIT

Cost	$$$
In-house development of discharge rounding screens	12 hours x $50 = $600
Benefit	$$$
Staff time saved from meeting elimination	$80,080
All patients receive accurate and timely discharge planning	Improved patient care

TEST

Each department uses the new screen in test mode for one week. Staff reviews discharge page and evaluates it for accuracy and quality of patient's discharge plan.

FOLLOW UP

New discharge planning screen has been used successfully for 6 months with the orthopedic staff. Three changes incorporated and tested. The same discharge rounding system will be implemented hospital-wide over the next 2 months.

Figure 6.5b A3 for Orthopedic Discharge Rounding
***Source:* Jimmerson (2007)**

team meeting. It showed that discharge planning was regularly conducted by a team of 14 people during 1.5-hour meetings. Even so, some necessary staff were not present, while some staff did not need to be present. Moreover, not all patients received discharge planning, and even when they did, physicians did not have great confidence in the discharge report. The A3 worksheet also showed how to move to an improved state through countermeasures that included better use of the electronic medical record and staff education. It also laid out an implementation plan and cost/benefit analysis. Follow-up showed that the new approaches were helpful six months later and that the methods would be exported to other hospital floors.

6.6 5S

5S (Five S) is the name of a workplace organization method that provides a basic fundamental, systematic approach for productivity, quality, and safety improvement. It uses five Japanese words translated into five English words that all start with the letter "S."

Seiri :: sort means to remove unnecessary or unwanted material from the workplace and dispose of it properly. **Seiton :: systematic arrangement :: set in order** means to arrange all items so they can be easily found and selected for use. **Seiso :: shine** means to clean the workplace so that it is safe and easy to work in. **Seiketsu:: standardize** means to have the best practices in each work area. **Shitsuke :: sustain** means to keep the workplace in good working order.

The 5S program aims to eliminate the waste that stems from a poorly organized work area. The organized space is typically more visually appealing and thus sets the stage establishing a **visual factory,** which uses visual indicators and displays and controls to improve the communication of information. In healthcare, even a partial deployment of visual factory concepts can be beneficial.

Example 6-3: Red Tag Event for Medical Supply and Equipment Rooms[2]

In healthcare, one of the typical applications is to use a rapid improvement team to apply 5S to areas that hold inventory such as supply rooms, equipment rooms, and so forth. A general approach is to form the team with a representative from each department or unit that uses the room, define a red-tag holding area to temporarily hold the excess material identified during the sort phase, and complete the set in order, shine, and standardize phases within one to two days. After two weeks (or similar limited time period), the team will decide what to do with the tagged items. If it is unneeded in the current department, the options are to recycle or dispose of it, move it to another department if needed there, return to vendor, or possibly sell it. A log book could be kept to show the final disposition of all items and serve to document how building more space was unneeded or how time and money were saved by not looking through unneeded items to find desired items.

Once the room is organized, it becomes a natural next step to continue the organization into the usage points it serves. For example, Figure 6.6 shows an anesthesia board with several items

Figure 6.6 An Anesthesia Board *before* the Use of Visual Factory Concepts From Pelletier (2015)

[2]Based on Chalice (2007)

from a supply room. Figure 6.7 shows the application of visual factory concepts—it becomes easier to visually establish which tray has all the parts.

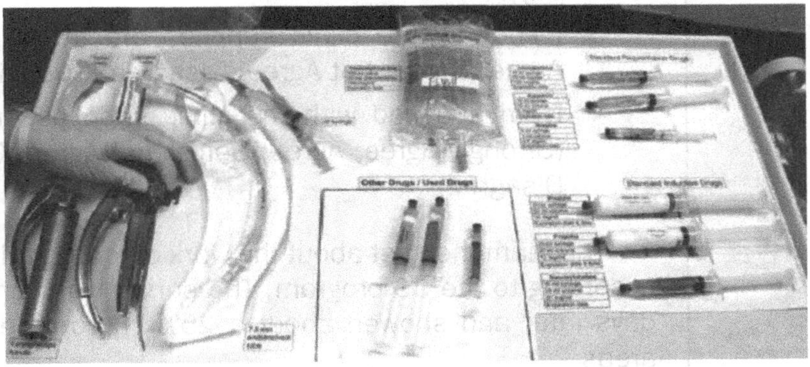

Figure 6.7 An Anesthesia Board *after* the Use of Visual Factory Concepts From Pelletier (2015)

Example 6-4: Improving the Patient Experience with Standardized Exam Rooms[3]

At Seattle Children's Hospital, a rapid improvement team consisting of nurses, physicians, and supporting healthcare providers used 5S to standardize exam rooms and office workspace. At the beginning of a three-day rapid improvement workshop, all of the steps of a surgical clinic visit were diagrammed in order to create a process map that identified the steps as value-added or non-value-added from the patient perspective. Through this review, searching for supplies appeared to be a ripe opportunity for improvement. A survey was then conducted that contained four questions/items:

1. How often do you leave the exam room to get supplies? (Never; once in a while; once a clinic/day; two to five times per day; more than five times per day.)

[3] Based on Waldausen et al. (2010)

2. How much time does it take you in an average day to search for supplies? (Zero—I have what I need; 1 to 10 minutes; 11 to 15 minutes; 16 to 30 minutes; more than 30 minutes per day).
3. I know where clinic supplies are stored. (Strongly Agree; Agree; Somewhat Agree; Disagree; Strongly Disagree).
4. I am satisfied with the clinic supplies in the exam room. (Strongly Agree; Agree; Somewhat Agree; Disagree; Strongly Disagree).

The team then set about the work of reorganizing the exam room according to the 5S program. The survey was conducted again two days later and showed about a 25% improvement in each of the areas.

6.7 KANBAN

Kanban means signboard or billboard in Japanese; more precisely, "kan" means "color" and "ban" means "card" or "ticket." A kanban provides a visual signal that an item requires replenishing or replacement. Beyond the board or card, kanban is a system to control inventory that was developed by Ohno in the 1950s. The aim of the kanban system is to provide a decentralized approach for regulating the flow of materials. As the implementation of the system highlights inventory problem areas, it also serves to become an effective catalyst in promoting process improvement. Kanban is almost a necessity in healthcare because of the abundance in storage locations such as central supply, department supply rooms, nursing stations, patient rooms, and exam and procedure rooms (Poole, Hinton, & Kraebber, 2010).

Traditionally, kanban cards contain information such as the item name and part number, picture or description of the item, process name where the kanban is to be used, number of units in the standard container or batch, and often a contact name and phone number to use in case of questions or problems. Figure 6.8 shows an example of a kanban card. Each container has a kanban card that always circulates with the actual material flow. Through the kanban system, employees come to understand

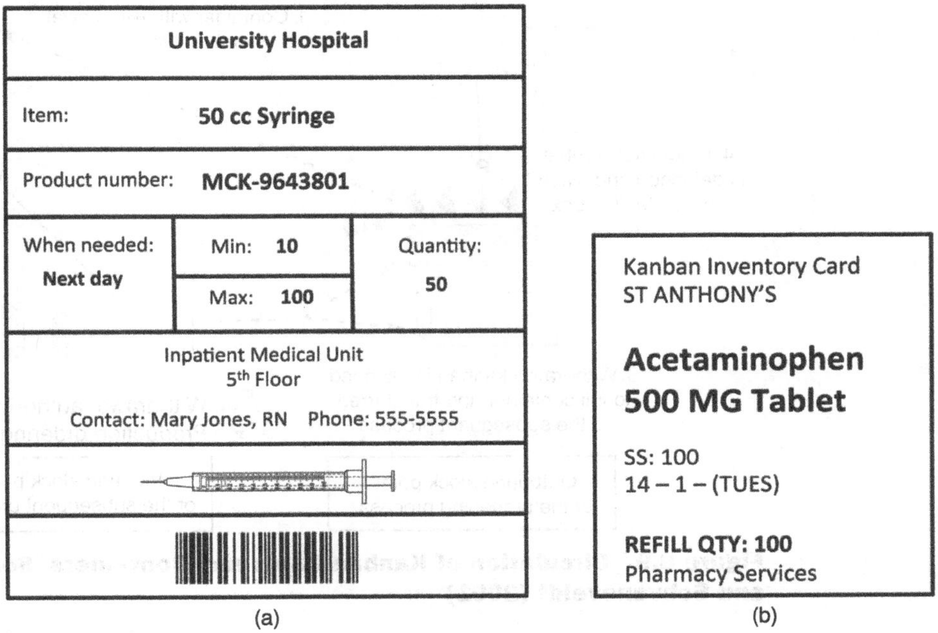

Figure 6.8 A Typical Kanban Card *Source:* Zidel (2006)

the operational procedures and standards and learn and share information needed for process control.

Figure 6.9 shows the circulation of kanban cards and containers. **Withdrawal kanbans** authorize the transfer of a standard container of an item from the preceding process where it was produced to the subsequent process where it is to be used. **Production kanbans** authorize the production/batching of one standard container of a specific item from the proceeding process. When an item from a container is used, the withdrawal kanban is taken to the outbound stock point of the preceding process with an available empty container. At that stock point, a full container of the needed items has its production kanban removed and replaced by the withdrawal kanban. The removed production kanban is placed in a collection box at the preceding process. When a new container is produced, the production kanban is attached to it and the container is placed in the outbound stock point to complete the cycle (Enkawa and Schvaneveldt (2001)).

The kanban delivery cycle is expressed as *a-b-c,* where *a* indicates the number of days until *b* deliveries, and *c* indicates the delivery delay factor (i.e., replenishment lead time). For example, 7-2-1 means that every

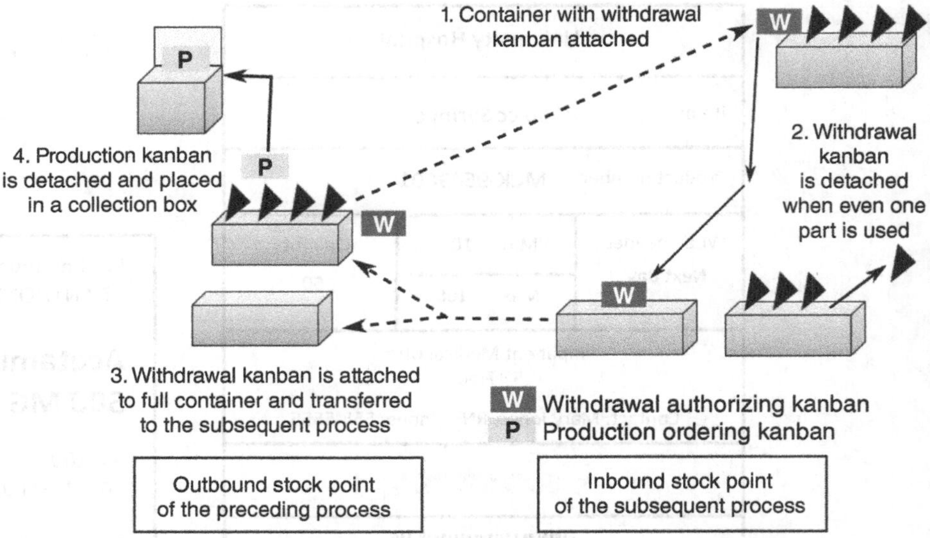
1. Container with withdrawal kanban attached
2. Withdrawal kanban is detached when even one part is used
3. Withdrawal kanban is attached to full container and transferred to the subsequent process
4. Production kanban is detached and placed in a collection box
W Withdrawal authorizing kanban
P Production ordering kanban
Outbound stock point of the preceding process
Inbound stock point of the subsequent process

Figure 6.9 Circulation of Kanban Cards and Containers ***Source:*** **Enkawa and Schvaneveldt (2001)**

7 days, 2 containers are delivered and that a new production order would be delivered by the next 1 subsequent delivery. Minimizing work-in-process inventory suggests that the number of units per standard container should be kept as small as possible, that the number of deliveries per day (b/a) should be as frequent as possible (so as to synchronize takt time of the subsequent process), and that the delivery delay factor c should be kept as short as possible. A safety stock factor s may be desired. The number of kanbans, k, may then be calculated as:

$$k = \frac{\textit{number of units required per day}}{\textit{number of units in standard container}} \times \frac{a}{b} \times (1 + c + s)$$

Circulation of the kanbans is triggered only by actual usage of items and only what is needed is withdrawn from the preceding process, and then only what is needed for replacement is produced. The kanban effectively establishes an upper limit to the work-in-progress inventory, avoids overloading the system, and reduces waste. The kanban demand signal immediately travels through the supply chain and ensures that intermediate stock held in the supply chain is better managed. Ultimately, kanban is a "pull" approach, where the resupply or production is determined according to the actual demand of the customer.

Example 6-5: A Lean Supply Room using a Kanban System[4]

In order to improve efficiency in a healthcare organization, changeover reduction is key. By implementing lean principles the critical process of changing over from patient to patient, test to test, and operating room to operating room can be improved, resulting in increased revenue, reduction of cost, and an increase in patient satisfaction. One way to achieve changeover reduction is through a lean supply room that uses a kanban system for inventory management.

When a stocked item such as syringes are depleted to a predetermined level, the kanban (we will assume the card shown in Figure 6.8 for this example) is sent to the supplier, indicating that the item must be replenished. When the supplier, the storeroom in this case, receives the withdrawal kanban card, a box of twenty-five 50cc syringes is scheduled for delivery to the unit no later than the following day. The storeroom returns the kanban card with the item when it is delivered. The card is kept with the item until it needs to be replenished again. The cycle is repeated as necessary.

The storeroom should have its withdrawal kanban to indicate that a box was pulled from stock and sent to the unit. This kanban signals the depletion of storeroom stock and the need for replenishment by the vendor.

Employees of Sisters of St. Francis Health Services Inc. (SSFHS) implemented a kanban system to improve changeover times and more effectively manage inventory. At SSFHS, a two-bin kanban system was implemented in the department supply room at many hospitals, with the empty bins acting as the kanban cards. Figure 6.10 shows the reorganized supply room. This technique aided in the elimination of daily inventory counts as well as the prevention of stock-outs. The system also used color coding to identify a specific type of supplies, thereby improving the staff's ability to find supplies more efficiently. The color coding also helped to keep processes

[4]Case based on Zidel (2006, p. 80)

efficient even for nurses who didn't generally work in the unit (such as nurses from outside the agency or student nurses).

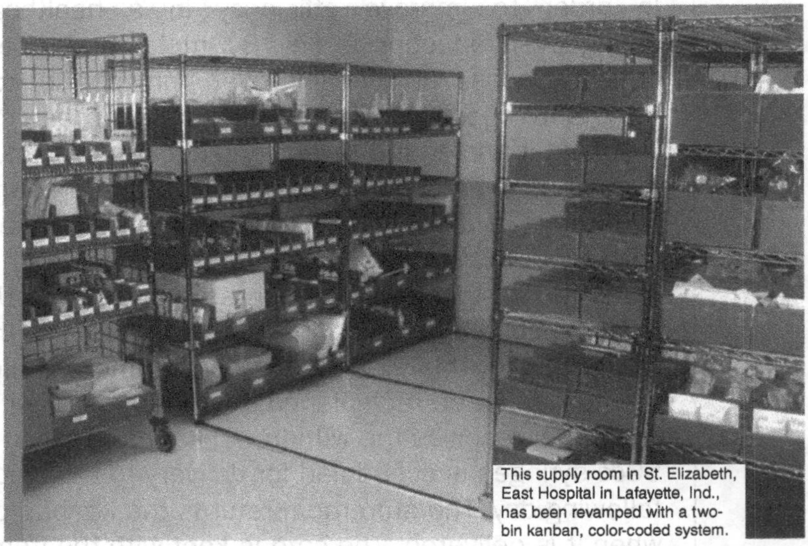
This supply room in St. Elizabeth, East Hospital in Lafayette, Ind., has been revamped with a two-bin kanban, color-coded system.

Figure 6.10 A reorganized hospital supply room after the application of kanbans. ***Source:*** **Poole et al. (2010)**

6.8 LEAN IMPLEMENTATIONS

The success of lean in manufacturing made it a natural approach to turn to as healthcare organizations worldwide started to answer the call to improved healthcare delivery and safety. Healthcare organizations are composed of a series of processes or sets of actions intended to create value for the customers/patients who use or depend on them. When applied rigorously, lean principles can have a positive impact on productivity, cost, quality, and timely delivery of services.

Although it would be difficult to validly assess how many organizations deploy lean there is plenty of evidence that lean can and does work in healthcare. Organizations such as the Institute for Healthcare Improvement (IHI) have embraced the application of lean to the delivery of healthcare. The Internet is replete with white papers and presentations about lean implementations. Grunden (2008) and Graban (2009) provide numerous case studies on lean deployments in hospitals.

A growing body of more formally disseminated studies examines the implementation of lean in various domains of healthcare. A literature review by Mazzocato (2010) provides an assessment of studies at health-care sites (e.g., family medicine clinics, rural community health providers, and outpatient clinics), clinical specialites (e.g., anesthesiology, emergency medicine, intensive care, surgery), diagnostic units (e.g., cytology, pathology, radiology), and departments (e.g., central sterile processing, supply room). In addition, Holden (2011) provides a critical review of lean in emergency departments. Collar et al. (2012) conducted a prospective quasi-experimental study wherein a multidisciplinary task force undertook a systematic implementation of lean in an operating room. Beck and Gosik (2015) look at using lean to lead improvements in a pediatric service. Poole et al. (2010) discuss how lean is gradually diffusing across health systems. The formal literature in lean implementations will continue to grow.

6.9 LEAN THINKING

Leadership and management skills are a critical component of lean. Leaders need to help employees understand why improvement is necessary and how lean methods will be used to initiate and sustain improvement. Healthcare providers need to embrace a mindset of developing a sense of trust, fair treatment, fact-based decision making, and long-term thinking.

In order for lean principles to take root, leaders must create an organizational culture that is receptive to lean thinking. Ultimately, lean thinking is a management strategy and as such cannot be done half-heartedly or by piecemeal. Inducing an organizational shift to lean thinking is a long journey and is certainly not for the faint of heart. There is no single "silver bullet" solution—there is no new computer system or short course or consultant visit that will accomplish lean thinking.

An IHI report, "Going Lean in Health Care" (IHI, 2005), summarized it this way:

> Implementing lean thinking requires major change management throughout an entire organization, which can be traumatic and difficult. Strong commitment and inspiring leadership from senior leaders is essential to the success of an effort this challenging. The CEO must be a vocal, visible champion

of lean management, create an environment where it is permissible to fail, set stretch goals, and encourage "leaps of faith." A senior management team that is aligned in its vision and understanding of lean is a critical foundation for "going lean."

QUESTIONS AND LEARNING ACTIVITIES

1. Reflecting on the definition of value from the lean perspective, how is value created in healthcare?
2. Critically read and analyze the article by Beck and Gosik (2015). What lean tools are used? How could the study be translated to a different or larger audience?
3. Recall your last trip to a hospital or clinic. Did you notice any wastes or inefficiencies? How could lean tools and methods be applied?
4. A sterile processing unit, also called central sterile services department (CSSD), central supply department (CSD), or central supply, is a location in hospitals and healthcare facilities that performs sterilization for medical equipment and devices. The sterilization processes usually consist of cleaning previously used items with a sterilizing liquid, drying them on a stand, and wrapping them in an aseptor bag with an expiry date. Kimsey (2010) reports on how A3 thinking was used to improve the CSSD. Use the descriptions provided therein to construct an A3 worksheet.

REFERENCES

Beck, M.J., and Gosik, K. (2015). Redesigning an Inpatient Pediatric Service: Using Lean to Improve Throughput Efficiency. *Journal of Hospital Medicine*, *10*(4), 220–227.)

Chalice, Robert. (2007). ***Improving healthcare using Toyota lean production methods*** (2nd ed.) Milwaukee, WI: ASQ Quality Press.

Collar, R. M., Shuman, A. G., Feiner, S., McGonegal, A.K. Heidel, N., Duck, M., ... Bradford, C. R. (2012). Lean management in academic surgery. *Journal of the American College of Surgeons*, *214*(6), 928–936.

Dickenson, E. W., Singh, S., Cheung, D. S., Wyatt, C. C., Nugent, A. S. (2009). Application of lean manufacturing techniques in the emergency department, *The Journal of Emergency Medicine*, *37*(2), 177–182.

Enkawa, T. and Schvaneveldt, S.J. (2001). "Chapter 17. Just-in-Time, Lean Production, and Complementary Paradigms" (pages 544–561) in Handbook of Industrial Engineering: Technology and Operations

Management, Third Edition, G. Salvendy, editor, John Wiley & Sons, Inc.

Gerson, R. R. M., Karkashian, C.D., Grosch, J. W., Murphy, L. R., Escamilla-Cejudo, A., Flanagan, P. A., ... Martin, L. (2000). Hospital safety climate and its relationship with safe work practices and workplace exposure incidents. *American Journal of Infection Control*, *28*(3), 211–221.

Gilbreth F. B. (1916). Motion study in surgery. *Canadian Journal of Medicine and Surgery*, *40*, 22–31.

Graban, M. (2009). *Lean hospitals*. New York, NY: Productivity Press.

Grunden, N. (2008). The Pittsburg way to efficient healthcare: Improving patient care using Toyota-based methods. New York, NY: Healthcare Performance Press.

Holweg, M. (2007). The genealogy of lean production. *Journal of Operations Management*, *25*(2), 420–437. doi:10.1016/j.jom.2006.04.001 DOI:10.1016%2Fj.jom.2006.04.001

Institute for Healthcare Improvement. (2005). Going lean in healthcare. Retrieved from https://www.entnet.org/sites/default/files/GoingLeaninHealthCareWhitePaper-3.pdf

Jimmerson, C. (2007). *A3 problem solving for healthcare*. Boca Raton, FL: CRC Press, Taylor and Francis Group.

Johnson, J. V., Hall, E. M., Ford, D. E., Mead, L. A., Levine, D. M., Wang, N.-Y., & Klag, M. J. (1995). The psychosocial work environment of physicians: The impact of demands and resources on job dissatisfaction and psychiatric distress in a longitudinal study of Johns Hopkins Medical School graduates. *Journal of Occupational and Environmental Medicine*, *37*(9), 1151–1159.

Kimsey, D. B. (2010). Lean methodology in healthcare. *AORN Journal*, *92*(1), 53–60.

LaGanga, L. R. (2010). Lean service operations: Reflections and new directions for capacity expansion in outpatient clinics. *Journal of Operations Management*, *29*, 422–433.

Mazzocato, P., Savage, C., Brommels, M., Aronsson, H., and Thor, J. (2010). Lean thinking in healthcare: a realist review of the literature. *Quality and Safety in Health Care 19* (no number), 376–382.

Moore, R. M., & Kaczmarek, R. G. (1991). Occupational hazards to health care workers: Diverse, ill-defined, and not fully appreciated. *American Journal of Infection Control*, *18*, 316–327.

Ohno, T. (1988). *Toyota production system: Beyond large-scale production*. New York, NY: Productivity Press.

Pear, R. (2011, December 3). Health official takes parting shot at "waste." *New York Times*, http://www.nytimes.com/2011/12/04/health/policy/parting-shot-at-waste-by-key-obama-health-official.html

Pelletier, L. (2015). "AD – The basic problem solving tool." Unpublished lecture.

Poole, K., Hinton, J., & Kraebber, K. (2010). The gradual leaning of health systems. *Industrial Engineer*, *42*(4), 50–55.

PRA website (n.d.). *Poor Richard's almanac.* https://en.wikipedia.org/wiki/Poor_Richard%27s_Almanack

Rethinking comparative effectiveness research. (2009, June). Biotechnology Healthcare. Retrieved from http://www.ncbi.nlm.nih.gov/pmc/articles/PMC2799075/pdf/bth06_2p035.pdf

Revicki, D. A., & May, H. J. (1989). Organizational characteristics, occupational stress, and mental health in nurses. *Behavioral Medicine*, *15*(1), 30–36.

Sepkowitz, K. A. (1996). Occupationally acquired infections in health care workers: Part I. *Annals of Internal Medicine*, *125*(10), 826–834.

Spear, S., & Bowen, H. K. (1999, September–October). Decoding the DNA of the Toyota production system. Harvard Business Review, (September-October issue) 96–108.

Sullivan, K. (2014). Retrieved from http://www.fiercehealthcare.com/story/healthcare-workers-report-highest-stress-levels/2014-02-12

Value Stream Map Symbols and their Usage [Online image]. (n.d.). Retrieved May 19, 2015, from https://www.edrawsoft.com/valuestreammapsymbols.php

Waldhausen, J. H.T., Avansino, J. R., Libby, A., and Sawin, R. S. (2010). Application of lean methods improves surgical clinic experience, *Journal of Pediatric Surgery*, *45* (no number), 1420–1425.

Womack, J., & Jones, D. (2003). *Lean thinking*. New York, NY: Simon & Schuster.

Womack, J. P., Jones, D. T., & Roos, D. (1990). *The machine that changed the world*. New York, NY: Simon & Schuster.

Yousri, T. A., Khan, Z., Chakrabarti, D., Fernandes, R., & Wahab, K. (2011). Lean thinking: Can it improve the outcome of fracture neck of femur patients in a district general hospital. *Injury: International Journal of the Care of the Injured*, *42*, 1234–1237.

Zidel, T. (2006). A lean guide to transforming healthcare: How to implement lean principles in hospitals, medical offices, clinics, and other healthcare organizations. Milwaukee, WI: ASQ Quality Press.

Chapter 7

Six Sigma

"Only the truly educated can be moved to tears by statistics."

—G. B. Shaw

Overview

As noted in the previous chapter, healthcare leaders have begun to look to other industries to find effective strategies and solutions to improve the quality of healthcare. As with lean, Six Sigma is such an answer. *Six Sigma* was leveraged quite successfully and notably by Motorola and General Electric to achieve high levels of quality and cost savings as well as overall organizational performance. Since then, many healthcare organizations and providers have also implemented Six Sigma strategies with significant success. One of the most widely used Six Sigma methodologies is the five-phase problem-solving strategy—*Define, Measure, Analyze, Improve, and Control (DMAIC)*. We explore DMAIC applied to healthcare through several examples of Six Sigma tools and case studies. We also discuss organizational strategy and change management with the Six Sigma paradigm. Finally, we look to integrate Six Sigma and *Quality Improvement (QI)* more broadly with engineering methods that render it useful in the translational research framework that is now widely being used in healthcare.

7.1 SIX SIGMA PHILOSOPHY

At many organizations Six Sigma has come to represent a philosophy of striving for perfection. This philosophy stemmed from its development at Motorola in the early 1980s to improve product reliability 10-fold in a five-year period (Samuels & Adomitis (2003). As they set about achieving this goal, engineers viewed reliability as inversely proportional to errors or defects per million opportunities (DPM) in a process. In statistical terms, Six Sigma marks a high efficacy standard: 3.4 DPMO. General systems theory creates a relationship between all or nearly all of the processes in an organization. Motorola's striving for Six Sigma performance in its technical operations began to pervade every operation in the organization, and in 1988 it was awarded the Malcolm Baldrige National Quality Award for its accomplishments. The receipt of this honor prompted Motorola to release its proprietary control of Six Sigma and share the methodology with other businesses (Carrigan & Kujawa, 2006).

Subsequently, General Electric CEO Jack Welch became a very well-known proponent of Six Sigma. He integrated Six Sigma into the corporate culture of GE and used it to drive GE's strategy goal of being number 1 or 2 in every market in which it had products. Welch initiated a tremendous quality campaign in his organization and announced:

> We want to change the competitive landscape by being not just better than our competitors, but by taking quality to a whole new level. We want to make our quality so special, so valuable to our customers, so important to their success, that our products become their only real value choice. (Snee & Hoerl, 2003)

Now, Six Sigma is characterized by its customer-driven approach, emphasis on decision making based on careful analysis of quantitative data, and a priority on cost reduction. Six Sigma is deployed through the careful execution of improvement projects. Project selection should be based on a translation of the company strategy into operational goals. Six Sigma provides an organizational structure of project owners and project leaders. Members of upper management are project owners, or Champions. Project leaders are called Master Black Belts or Black Belts. Green Belts and Yellow Belts provide contextual support for the project.

The challenge for Six Sigma in healthcare is to find a way to leverage the data from Six Sigma to drive new behaviors—in other words, to combine the technical efficacy with cultural strategy. The patient has to be kept at the center—from both a satisfaction and outcomes point of view.

7.2 SIX SIGMA QUALITY

One of the primary goals of Six Sigma is to reduce the number of defects that a process generates. All processes are a combination of tools, materials, methods, and people brought together to produce a measurable output. All processes have inherent statistical variability and all processes have defects. Under Six Sigma, process defects are any factor that negatively impacts quality, customer satisfaction, profitability or cash flow—they are all termed **critical to quality** (CTQ) defects and are in fact interrelated: if customers are not satisfied by quality, they will go elsewhere and profitability will decrease. Six Sigma focuses attention on the *causes* of CTQ defects and thus prevents the escalation of defects into more costly problems at a later time. Three metrics are often used: **process capability, defects per million opportunities** (DPMO), and **sigma level** (σ).

Process capability is a measure of how well a process outputs meet customer requirements or specifications. It is typically expressed as a process capability index, such as Cp or Cpk, or as a process performance index such as Pp or Ppk. Process capability can also be assessed using control charts.

DPMO involves a specific way of looking at a process. In a healthcare context, a routine surgery might consist of dozens of opportunities for error along the way: initial visit, surgical scheduling, hand washing, room sanitation, instrument sterilization, medication administration, staff scheduling, surgeon capability, discharge processes, claims and billing, and so on.

A $\pm$ 6σ performance level translates to approximately 3.4 DPMO (or 0.034 defects per 1,000) when process drift is considered. A $\pm$ 6σ performance level translates to approximately 0.002 DPMO (or 0.000002 defects per 1,000) when process drift is not considered. Table 7.1 relates the various levels of sigma to defects per million and per thousand and both with and without the assumption of process drift. Figure 7.1 depicts the same information in Table 7.1b.

Although reaching 6σ levels may be easily achievable in some technical environments, the same may not be true for medical operations or management decisions in a healthcare environment. Nevertheless, the goal must be set and the journey undertaken. Table 7.2 shows a comparison of selected healthcare quality examples and airline quality examples in terms of DPMO. There are 3.4 crashes, or defects, every 1 million times a commercial jet takes off and lands. John S. Toussaint, MD, CEO emeritus of ThedaCare, points out that anesthesia delivery is one healthcare process that has reached the 6σ performance level (Toussaint, n.d.). He also remarks that it is also among the few processes that are actually standardized across the industry: "All anesthesia machines look similar to each other, so anesthetists don't have to learn new technology every time they

Table 7.1 Quality Levels and Corresponding Number of Defects

(a) With 1.5σ drift

Sigma Quality Level	Defects per Million Opportunities (DPMO)	Defects per 1,000 Opportunities	Percent Defective	Yield
±1σ	697700	698	69.77%	30.23%
±2σ	308500	309	30.85%	69.15%
±3σ	66810	67	6.68%	93.32%
±4σ	6210	6.2	0.62%	99.38%
±5σ	233	0.23	0.023%	99.977%
±6σ	3.4	0.034	0.00034%	99.99966%

(b) Without 1.5 σ drift

Sigma Quality Level	Defects per Million Opportunities (DPMO)	Defects per 1,000 Opportunities	Percent Defective	Yield
±1σ	317,310	317	31.73%	68.27%
±2σ	45,500	45	4.55%	95.45%
±3σ	2,700	2.7	0.27%	99.73%
±4σ	63	0.063	0.0063%	99.9937%
±5σ	0.57	0.00057	0.000057%	99.9999430%
±6σ	0.002	0.000002	0.0000002%	99.9999998%

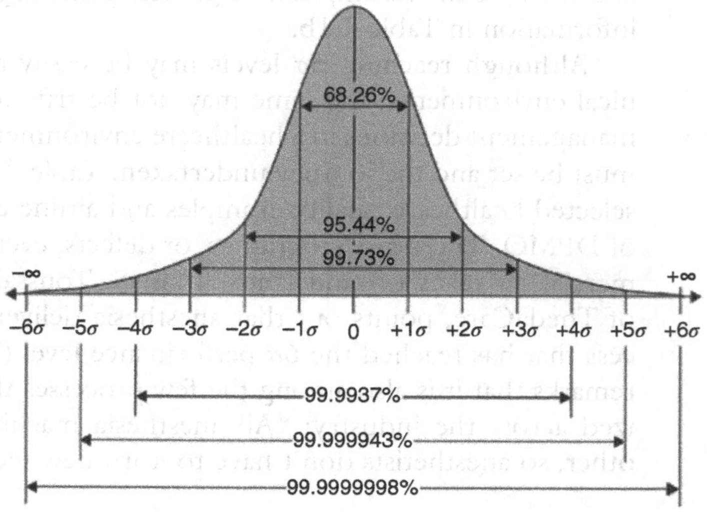

Figure 7.1 Frequency Distribution

Table 7.2 Examples of Healthcare Quality Problems Viewed as Defects per Million and Compared with Airline Examples

Sigma Quality Level	Defects per Million Opportunities (DPMO)	Healthcare Examples	Airline Examples
1σ	691,462	Eligible heart attack survivors who receive beta blockers	
2σ	308,500	Patients with depression detected and treated adequately	On-time flight departures
3σ	158,700		
4σ	6,200		Baggage handling
5σ	230		
6σ	3.4	Patient survival after general anesthesiology	Successful takeoff and landings on a commercial jet

walk into a different operating room. Error-proofing techniques like color coding and making it physically impossible to connect the wrong gas to a patient also help reduce the potential for defects." The DMPO for other processes in healthcare must be driven down as well.

Six Sigma Costs and Benefits

Six Sigma results can be very impressive, but can come at a considerable cost and effort. Another way to look at Tables 7.1 and 7.2 is that taking a process from, for example, a 2σ level to a 6σ level requires a more than 20,000 times improvement. This would be an audacious goal for any organization and would require strong commitment from its leaders. Several questions should quickly arise, not the least of which is how much such an initiative would cost and what would be the return on investment.

While it would be difficult to predict the answers for any company, Table 7.3 provides some perspective. In particular, it should be noted:

- It has been said that quality isn't free and that it takes money to make money. A substantial investment will have to be made in Six Sigma if it is to succeed.
- There are likely to be some fixed costs with a Six Sigma implementation. A large company may enjoy some benefits of scale.

Table 7.3 Estimating the Costs of a Six Sigma Initiative

Number of employees	100	500	1,000
Annual revenue ($ million)	10	50	100
Cost of poor quality at 10% of sales ($ million)	1	5	10
Cost of Implementing Six Sigma ($ million)	0.5	1	2
Savings per Six Sigma project ($ million)	0.12	0.14	0.16
Minimum number of projects to be identified for breakeven	4	7	13

- There is likely a sizable opportunity to improve quality. A figure of cost of poor quality at 10% is probably conservative.
- There is likely to be some carryover benefits of Six Sigma in subsequent years so the number of projects needed to break even in future years would decrease.
- As the culture of continuous improvement takes root in the organization there should be a multiplier effect to the Six Sigma imitative.

As a Six Sigma rollout gets under way, it is important to consider a set of projects that will be done across the organization. Another way of looking at the cost element is to select projects based on the project prioritization index (PPI), which is calculated as follows:

$$PPI = \frac{\textit{Probability of the project's success}}{\textit{Time to complete the project in years}} \times \frac{\textit{Benefits}}{\textit{Cost}}$$

The benefits of a project may include bottom-line hard dollar, cost avoidance, lost profit avoidance, productivity improvements, and profit enhancements (Breyfogle, 2003). The recommended PPI is approximately 3 × (Benefit/Cost). Management may accept lower numbers on any project if there is the sense that it is critical to the operation or will establish a window to better customer satisfaction. There are also intangible benefits such as avoiding high legal liabilities that may drive a project forward.

The leadership should expect to sustain the investment in Six Sigma over several years in order to derive the most benefit. For example, GE invested $1.6 billion in their Six Sigma efforts and realized savings of $4.4 billion over a four-year period (iSixSigma, n.d.). A key is to keep careful track of both the investment and the savings. Over time, Six Sigma will become more valuable when the results are definitely shown to be worth the cost.

Six Sigma DMAIC Methodology

Several variants of Six Sigma methodologies have been used in industry, including the *Define, Measure, Analyze, Design, Verify (DMADV)* framework for *Design for Six Sigma (DFSS)* and others. In industry and healthcare, however, the define, measure, analyze, improve, and control (DMAIC) methodology is by far the most commonly applied. De Koning, Verver, Bisgaard, and Does (2006) note that:

> Six Sigma's approach is similar to that of good medical practice used since the time of Hippocrates-relevant information is assembled followed by careful diagnosis. After a thorough diagnosis is completed, a treatment is proposed and implemented. Finally, checks are applied to see if the treatment was effective.

In the define phase, a charter is drafted that provides a clear definition of the project, its scope, key milestones, and timeline for completion. It also should show the business case for the project and the project sponsor. The team is expected to reach a common understanding of the process and identify its general steps and elements. The problem statement should be developed from the standpoint of the objective of the *Voice of the Customer (VOC)*.

In the measure phase, baseline data are collected, and the process performance is reviewed to establish the critical to quality (CTQ) elements. Central tendency measures (e.g., mean, median), variation or inconsistency measures (e.g., standard deviation, range) and the current defect rate or capability are determined. It is also vital to ensure that there is a valid quality measurements are obtained. This phase also begins to explore possible cause and effect relationships.

In the analyze phase, diagnosis continues to model and assess cause and effect relationships. The goal is to focus on the variables with a look ahead to those for which remedial actions can be taken.

In the improve phase, solutions are developed to achieve the desired process outcomes and the best solution is selected based on statistical evaluation. The goal is to develop a breakthrough solution that would lead to substantial improvement in the process.

In the control phase, steps are taken to make sure defects do not recur and that gains are sustained.

As a part of a continuous improvement cycle to optimize CTQ factors, the DMAIC methodology may be restarted from the beginning.

As a practical matter, even within the execution of one project, subcycles may be repeated. For example, after the analyze phase, it may be necessary to go back to the measure phase, or after the improve phase, there may be a need to double back to the analyze phase.

Table 7.4 lists some of the concepts and tools associated with the phases of the DMAIC methodology. Tool selection and when to use it depends upon several factors including the project scope. A number of the tools can be used in multiple phases. For example, although listed in the control phase, control charts can also be useful in the measure, analyze, and improve phases. Some of these tools are discussed in the subsequent sections of this chapter as they pertain to healthcare. For a more complete in-depth treatment of the tools, Breyfogle (2003), Pyzdek and Keller (2014) and (Montgomery (2012) are recommended.

It should be remarked upon that lean thinking and lean tools are listed in the improve phase. The concept of combining **lean and Six Sigma** is sometimes referred to as **Lean Six Sigma** and abbreviated as LSS. Black (2009) points out that since lean is the concept of reducing inefficiency, waste, and cost in any process, and Six Sigma is the concept of reducing process variation and minimizing defects when they are used together, the combination is characterized by a process improvement focused on

Table 7.4 Examples of Six Sigma DMAIC Concepts and Tools

Define		
CTQ flowdown Kano analysis	SIPOC (supplier, inputs, process, output, customer)	Process mapping Affinity diagram
Measure		
Cause-and-effect diagram Histogram Checksheets	Pareto chart Cost of poor quality Measurement system analysis	Performance measurements (Cp, Cpk, DPU, DPMO)
Analyze		
Failure mode and effects analysis (FMEA) Analysis of variance (ANOVA)	Regression Multi-vary analysis Prioritization matrix	Hypothesis testing Scatter plots
Improve		
Comparative experiments Impact-effort matrix	Design of experiments	Lean thinking and lean tools Triz
Control		
Control charts Internal audits	Scorecard/Dashboard Theory of constraints	Documentation

reducing waste by following the DMAIC methodology. De Koning et al. (2006), who also prefer the term *Lean Six Sigma,* state that the distinction between lean and Six Sigma is "artificial and often not helpful. An integration of the two approaches and a general focus on process innovation regardless of the origin of the tools and approaches would be more productive." More broadly, the field of quality improvement (QI) has benefited from the contributions of many luminaries. Joseph Juran and Amand Feigenbaum, for example, did much to advance QI through their development of quality management and total quality management from the 1950s. As lean and Six Sigma gained recognition, some early criticisms dismissed them as just a "repackaging" of previous work. What is clear now is that each dimension has made significant contributions to the whole of QI. Just as importantly, advancements are still needed in order to continue to solve problems of growing complexity.

Define Phase Concepts and Tools

At a high level, the define phase calls for concepts and tools that will help the team identify (1) the services and products—what is being offered or produced; (2) what is the customer's perception of an error-free product or service; and (3) how often and where those errors occur.

Critical-to-Quality (CTQ) Flowdown

The **critical to quality** (CTQ) flowdown is a tool used to go from a project definition to specific and measureable CTQs. It is based on hearing the voice of the customer to create a chart that shows how the higher-level concepts such as performance indicators and focal points have a significant impact on the experience of the customer. It indicates which characteristics are CTQ. This tool helps the team explicitly represent the structure and rationale underlying the project.

Example 7-1: CTQ Flowdown for Projects Aimed at Increasing Efficiency

In listening to the voice of the customer in healthcare, an often-mentioned concern is around waiting for treatment or service. Figure 7.2 shows a CTQ flowdown for projects aimed at increasing efficiency for an emergency department.

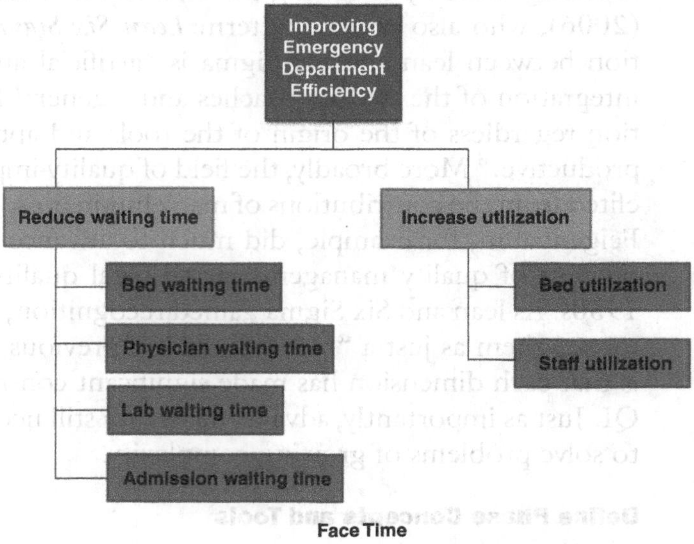

Figure 7.2 CTQ Flowdown for Emergency Department

SIPOC

A *Supplier, Inputs, Process, Outputs, and Customer (SIPOC)* diagram is designed to support the understanding of a projects major components and boundaries. This is done at a high-level to jump-start the identification of potential gaps between suppliers and input specifications and between outputs and customers' expectations. The process that bridges the inputs and outputs is typically represented in four to seven steps.

Example 7-2: SIPOC for Patient Discharge

After a hospital stay, many patients will need a period of rest and recovery at home or other care setting. However, patients are routinely ill-prepared for the transition and, as a result their care and outcomes may be compromised. The patient discharge process is considered to be central to this transition. The SIPOC diagram in

Figure 7.3 clearly shows how the discharge process is impacted by and impacts several other care points.

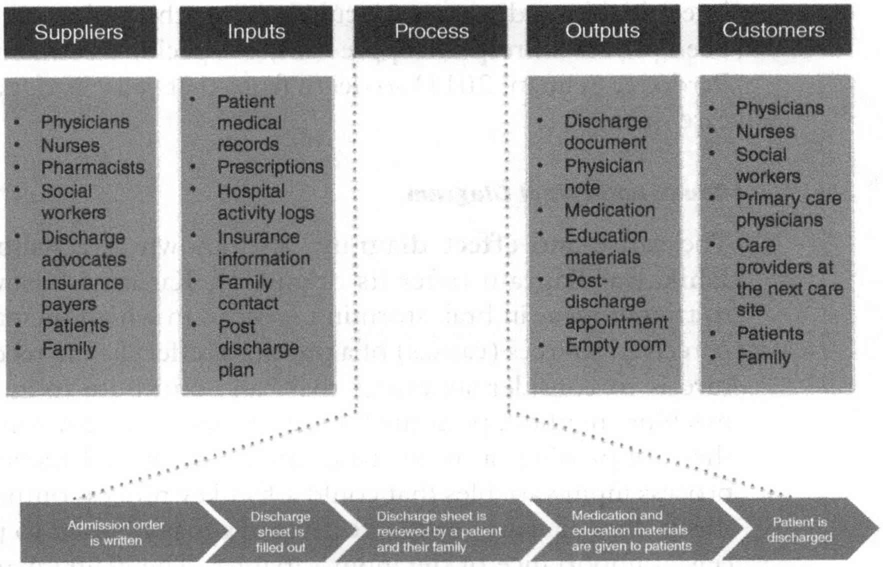

Figure 7.3 SIPOC Diagram for Patient Discharge

Measure Phase Concepts and Tools

At a high level, the measure phase calls for concepts and tools that will help the team determine how often and where errors occur that disrupt the ability to meet customer's requirements. Measurements are needed to assess either processes, such as diagnostic or therapeutic interventions, or outcomes such as the health states that patients experience. Process measures are valid quality measures when their relation to important health outcomes has been proven. The frequency with which antibiotics are administered to patients with sepsis or apparent infection is a valid quality measure because these medications improve survival. Outcome measures are valid quality measures when there is a proven relation to a process that can be modified to improve the outcome measure. Mortality due to sepsis is a valid measure of care quality because it is known that sepsis bundles (i.e., a set of protocols that combine several medical practices to promote rapid adoption of proven therapies and improve patient outcomes) dramatically reduce the chance of death (Barochia et al., 2010). On the other hand, interstitial lung disease, particularly idiopathic pulmonary fibrosis, has not been proven to respond appreciably to specific treatment regimens (King, Pardo, & Selman, 2011) so death from that cause is not a valid measure of care quality.

Cause-and-Effect Diagram

The **cause-and-effect diagram**, also known as a **fishbone diagram** or **Ishikawa diagram** (after its originator, Karoru Ishikawa), is often used to trigger ideas in brainstorming sessions in which the participants list the perceived sources (causes) of a problem (effect). The recommended structure is to consider six causes that can contribute to an effect: materials, machine, method, personnel, measurement, and environment. In essence, the tool provides a means to create an organized framework for the key process input variables that could affect key process output variables. Once this is done, a **cause-and-effect matrix** may be used to prioritize the perceived importance of the input variables. The diagram and/or matrix can be useful, for example, to determine the factors for regression analysis or design of experiments.

Example 7-3: Fishbone Diagram on the Sources of Medication/Prescription Error

Clinicians have access to more than 10,000 prescription medications. The Agency for Healthcare Research and Quality (AHRQ) defines a medication error as "an error (of commission or omission) at any step along the pathway that begins when a clinician prescribes a medication and ends when the patient actually receives the medication" ("Medication Errors," n.d.). The fishbone diagram in Figure 7.4 is a result of brainstorming possible reasons for this error.

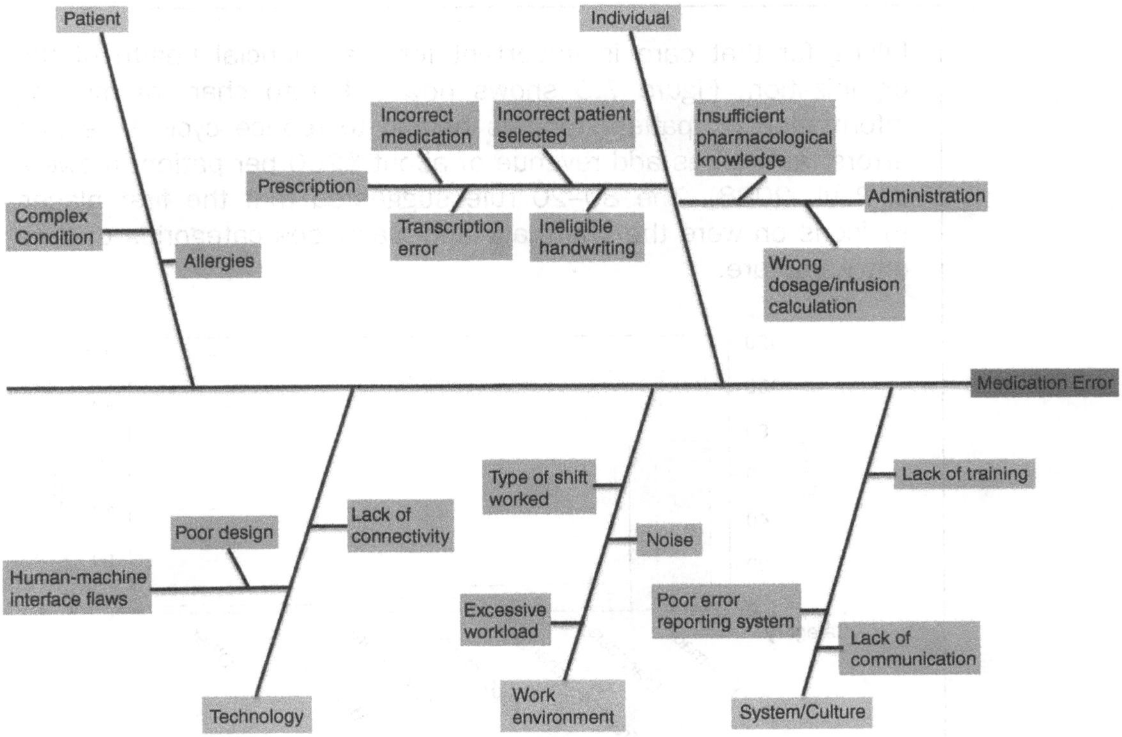

Figure 7.4 Fishbone Diagram of Medication Errors

Pareto Chart

Pareto charts are a tool that can help identify the source of problems in a process. It is rooted in the **Pareto principle** (also known as the **80–20 rule**, the **law of the vital few,** and the **principle of factor scarcity**), which states that, for many events, roughly 80% of the effects come from 20% of the causes. The chart provides a visual sort of the "vital few" causes that are responsible for "so many" of the problems.

Example 7-4: Pareto Chart of Category of Missing Information on Patient Records

Complete and accurate patient records are essential for patient care. When records have missing drugs, treatments, and visits it can create gaps in communication among the care team and between the care team and the patient. Furthermore, accurate billing for that care is important for the financial health of the organization. Figure 7.5 shows how a Pareto chart of missing information on patient records helped to reduce cycle time and errors, as well as add revenue of about $200 per patient (Powers & Paul, 2008). The 80–20 rule suggested that the first places to focus on were the pharmacy and diagnoses categories of data entry/capture.

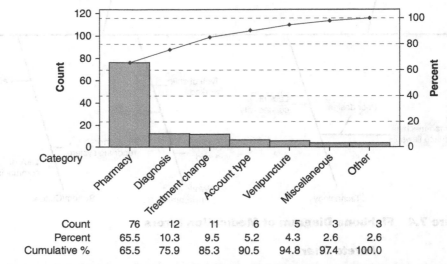

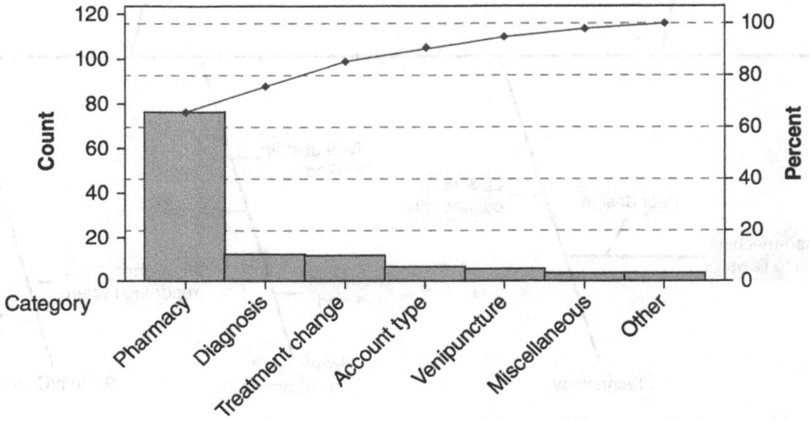

Count	76	12	11	6	5	3	3
Percent	65.5	10.3	9.5	5.2	4.3	2.6	2.6
Cumulative %	65.5	75.9	85.3	90.5	94.8	97.4	100.0

Figure 7.5 Pareto Chart of Missing Information on Patient Records

Analyze Phase Concepts and Tools

The analyze phase calls for tools that will allow the team to focus on a deeper understanding of the relation between input and output variables. This is often represented as $\mathbf{Y} = f(\mathbf{X})$ where the desired process output **Y** is a function of inputs **X.** The goal is to establish the models for this relationship and analyze their results.

Failure Modes and Effects Analysis

The **failure modes and effects analysis** (FMEA) is a systematic technique for analyzing problems that might arise from system gaps or malfunctions.

It involves a review of the unit subsystems, processes, and components to identify the failure modes, and their causes and effects.

For each unit, the failure modes and potentially resulting effects are recorded in a FMEA worksheet. Although several styles of the worksheet are used, it typically contains the following for each failure:

- Potential cause(s)/mechanism of the failure
- Local effects of failure
- Next higher level of failure
- System level or end effect of failure
- Probability estimate of the failure, *P*
- Description and rating of the severity of the failure, *S*
- Detection indicators or mechanisms if/when the failure should occur, *D*
- Risk probability number (RPN), $P*S*D$
- Actions for further investigation/evidence
- Mitigation requirements

The values of *P*, *S*, and *D* are usually determined based on the assessment opinion of the review team. The team may agree to a specific ranking of the severity, *S*; typically a 1–10 scale is used.

At its best, the FMEA is intended to be prospective in the sense that organizations can use it to identify and eliminate concerns early in the development of a process or design and provide a form of risk analysis. The quality of the system can improve when organizations work with their stakeholders to implement FMEAs.

Example 7-5: FMEA for Emergency Department Triage

Emergency Department (ED) crowding has been recognized as a major problem in hospitals world-wide. The FMEA method can be used to drill-down into this challenge. Table 7.5 shows a simulated FMEA worksheet (based on the study in Kang et al. (2014)) that could potentially result from a team effort to investigate ED crowding and provide better triage of patients to improve patient flow.

Table 7.5 FMEA worksheet for ED Triage

Task	Potential Failure Mode	Potential Failure Effects	Potential Causes	A. Probability of the Failure Rate 1–10 (10 = Highest Probability)	B. Severity Rate 1–10 (10 = Most Severe)	C. Probability of Detection (10 = Lowest Probability)	Risk Probability Number (RPN) A × B × C	Method of Control
Check-in	Long waiting time	Patients leaving without being seen	Staff shortage	2	7	2	28	Additional staff during peak time
Triage	Patient identification errors	Wrong ID bracelet	Data entry errors by triage nurse	4	9	7	252	New policy for double check
			Computer system errors	1	9	1	9	Regular check and maintenance
	Wrong assessment	Wrong severity index assignment, wrong patient route	Triage nurse's insufficient knowledge	6	4	7	168	Education and regular assessment
			Patient's complex symptom	5	4	7	140	Physician in triage
Diagnostic tests	Wrong tests	Additional costs and time	Physician's wrong assessment	4	2	8	64	Decision support system
		Adverse patient outcomes	Physician's wrong assessment	5	8	8	320	Decision support system
	Wrong labels on the specimen	Wrong medication decision	Simultaneous collection of specimen	3	9	6	162	New policy for specimen collection
			High ratio of patients to nurses	5	9	6	270	Additional staff during peak time
			Miscommunication	4	10	6	240	Implementation of computerized system
	Equipment malfunction	Adverse patient outcomes	Poor maintenance	3	10	6	180	Regular check and maintenance

Improve Phase Concepts and Tools

Lean Thinking and Lean Tools

As discussed in Chapter 6 (as well as earlier in this chapter), lean is built on the concept of redesigning organizational processes in order to reduce waste based on a set of analytical tools and coupled with creating a culture of continuous improvement. Tools that are traditionally seen as a part of the lean toolkit (see Table 6.1) can be integrated into the DMAIC improve phase.

Comparative Experiments

Comparative experiments can be used to evaluate process improvement. A two-sample experiment will have one sample as a control group representing the present process and the other sample as an experimental group representing the modified process. Both samples are assumed to come from normal populations with possibly different means and variances and compared using a two-sample *t*-test statistic. The two-sample *t*-test is basically a signal-to-noise ratio: it looks at the difference between the sample means relative to the standard deviation of the difference of the sample means.

Example 7-6: Two-Sample Experiment on Impact of a Discharge Checklist on Readmission Rate

Suppose that a hospital decides to implement a new discharge checklist with the expectation that it reduces 30-day readmission rate. We record the rate of readmissions (per 100 patient discharges) for 12 months prior to the checklist implementation and for 12 months afterward. Figure 7.6 gives a sense of the distribution of the data using a dotplot and boxplot. Figure 7.7 gives the Minitab output for the *t*-test and related statistics. The mean readmission rate has decreased slightly under the new discharge checklist. The *t*-statistic turns out to be a value of 3.28. We can look further and see that *p*-value is 0.0036—less than a 5% alpha level. We can also see that the 95% confidence interval does not contain zero, so the null hypothesis that the two means are equal can be rejected. In this case, it is fairly clear that the checklist has positively impacted the outcome.

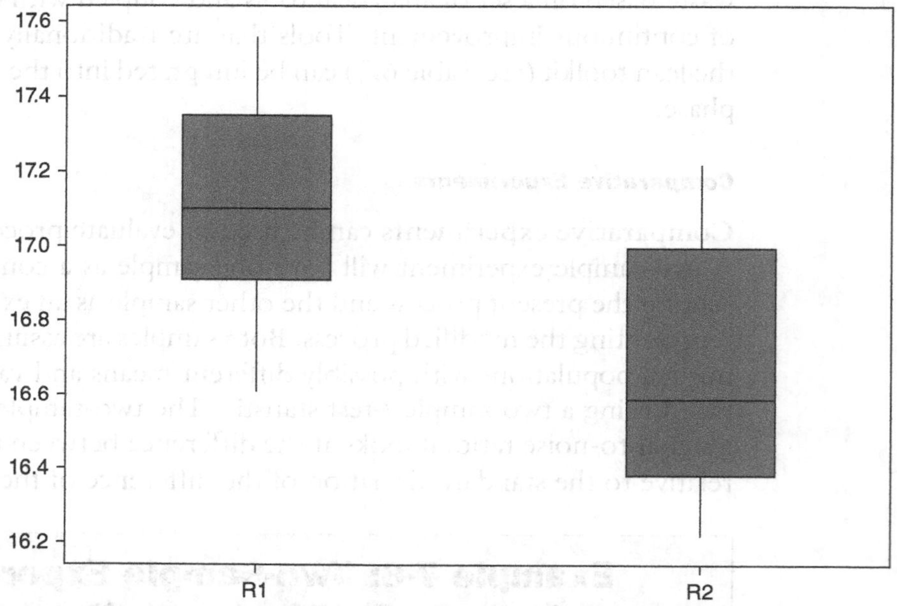

Figure 7.6 Dotplot and Boxplot of 30-Day Readmission Rates

Method

μ_1: mean of R1

μ_2: mean of R2

Difference: $\mu_1 - \mu_2$

Equal variances are not assumed for this analysis

Descriptive Statistics

Sample	N	Mean	StDev	SE Mean
R1	12	17.10917	0.28934	0.08353
R2	12	16.68750	0.33936	0.09797

Figure 7.7 Two-Sample *t*-Test and Related Statistics

Design of Experiments

Design of experiments (DOE) is a method for studying the effect of multiple variables simultaneously. In an absence of such an approach, we could be conducting experiments by changing "one factor at a time," which would certainly be time consuming and may lead to erroneous results. Many effective DOE methods have been developed including full and fractional factorials, central composite designs, Box-Behnkin designs, and others (Box, Hunter, & Hunter, 2005). With all of these methods, controlled experiments are used to understand the effect of some treatment or intervention on some output or experimental unit.

While the statistical foundation of DOE was established by R. A. Fisher in the 1920s to study agricultural effects (e.g., how much rain, fertilizer, etc. improved crop yield), (Dunn, 1997) points out that James Lind carried out systematic clinical trial to compare remedies for scurvy while serving as a surgeon on the *HMS Salisbury* in 1747.

Control Phase Concepts and Tools

The control phase aims to "lock in the gains." This is critical in order for the return on investment to be truly realized in the organization.

Statistical Process Control

Walter Shewhart was a physicist, engineer, and statistician who developed control charts in the 1920s while working at a unit that would become a part of AT&T. Shewhart framed quality in relation to **assignable-cause variability** and **natural cause variability**. Assignable-cause variability, as the name suggests, is change in the process that arise due to some special problem. On the other hand, natural cause variability is change in the process that will always be present simply because no system is perfectly constant. Shewhart designed the control chart as a tool for distinguishing between the two in order to manage process output. Control charts document the evidence of statistical control or natural process behavior.

A control chart compares data on the current process performance with the probability of the outcome based on a normal distribution. Table 7.6 shows some commonly used control charts, which are classified as **variable control charts** and **attribute control charts**, depending on the type of data sampled from the process. Variable control charts are used in pairs—one that monitors the expectation and one that monitors the variance. The control charts all have *Upper Control Limits (UCLs)* and *Lower Control Limits (LCLs)*—equidistant from

Table 7.6 Types of Control Charts

(a) Variable Control Charts

Type of chart	**Description**
X-mR	Individual values and their moving range
Xbar-R	Mean and range
Xbar-S	Mean and standard deviation
EWMA-EWMR	Exponentially weighted moving average and exponentially weighted moving range

(b) Attribute Control Charts

Type of chart	**Description**
u	Defects (errors) per unit
np	Number of defective units
p	Percent or proportion defective
c	Number of defects

a **center line** (CL)—that determine whether the process is in statistical control or out of control.

When the process deviates from natural behavior, and an out-of-control condition arises, then and only then should corrective actions be taken to adjust or correct the process. This is not to say that the process should not be changed—the process itself may be unacceptable and thus redesigned as a part of the improve phase.

Statistical process control charts can also be an effective tool in the measure and/or analyze phases. Assessing how much variability exists the process can be done using control charts in lieu of or in addition to using capability indices. They can also be helpful in analysis of variance (ANOVA) studies. The broad applicability of control charts makes them one of the most versatile tools to use across the DMAIC methodology.

Example 7-7: Wait Times in the ED

One of the important metrics around both patient satisfaction and care quality in the ED is the wait time "from door to doctor." Suppose that a Six Sigma project was launched at a particular hospital and the improve phase led to redesigned facility that yielded 35%

reduction in wait time. Figure 7.8 shows X-mR charts for ED wait times that are to be used to monitor this new improved process.

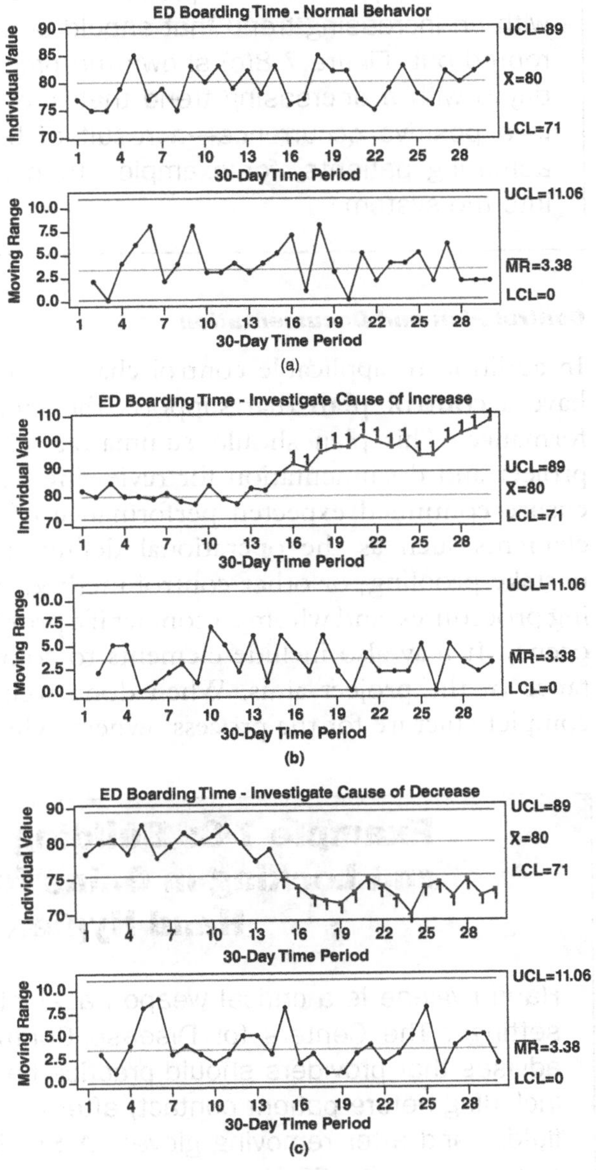

Figure 7.8 Control Charts for ED Wait Times

Figure 7.8(a) suggests that the underlying process that leads to this wait time is in control with a mean time of 25 minutes. Figure 7.8(b) shows the process is out of control from day 15 with an increasing trend that should be investigated and hopefully rooted out. Figure 7.8(c) shows the process is out of control from day 5 with a decreasing trend that should be investigated. If this is a positive occurrence—a result of better decision making on admitting patients, for example—then it should be incorporated into the system.

Control Plan and Documentation

In addition to applicable control charts, every Six Sigma project should have a **control plan** that supports the monitoring of the process performance. This plan should summarize all the CTQs involved in the project and documentation for reviewing, revising, and taking action to ensure continued expected performance of the process. It may include elements such as the operational definition of the process; checklists, mistake-proofing, or other control methods to be used; standard operating procedures; and whom to contact if a problem is detected or emergency occurs. It may also include elements to communicate the overall importance of the project aims. When done well, the control plan creates a complete picture for the process owners, who may change over time.

Example 7-8: Building Awareness and Locking in Gains for Improved Hand Hygiene

Hand hygiene is a critical weapon against infections in healthcare settings. The Centers for Disease Control and Prevention (CDC) advises that providers should practice hand hygiene at key points, including before patient contact, after contact with blood and body fluids, and after removing gloves. A Six Sigma project to improve compliance with CDC recommendations was capped off with the posters to build awareness as shown in Figure 7.9.

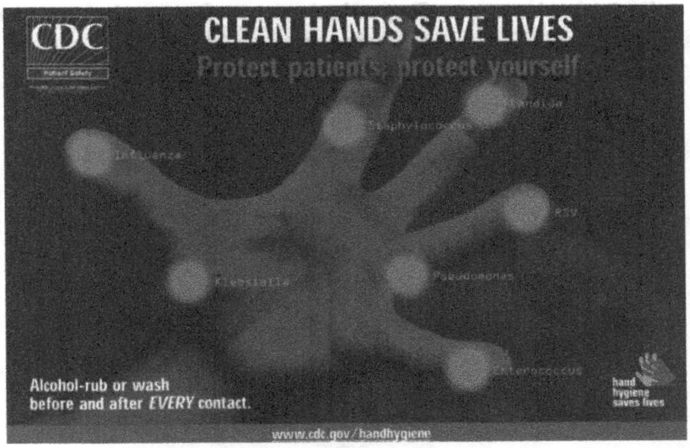

Figure 7.9 A Hand Hygiene Communication Poster ***Source:*** **CDC (2011)**

Six Sigma Implementations

As it is with lean, the published literature as well as Internet sites are replete with accounts of Six Sigma implementations in healthcare. The discussions regarding these implementations seem to largely point to mounting success and substantial cost savings. Table 7.7 shows a small subset of this literature selected primarily to show that such efforts date back many years, have been used across the United States and around the globe, have had a substantial impact on a wide variety of processes that improve patient care, and have come at a range of investments and savings.

The same table is also illustrative of several gaps:

1. The deployments seem to rarely move beyond the basic of statistical tools—and there are certainly good reasons to "keep it simple" and attack the "low-hanging fruit." However, given the scale of the quality problems in healthcare, there are undoubtedly many opportunities "left on the table" to go beyond the basics and reap the benefits from more sophisticated methods.
2. Outcomes and impact are often reported in fairly pedestrian terms and lack statistical analyses to test for significance due to the changes and improvements that have been made.
3. Many studies do not report even basic financial information on how much was invested and saved. Perhaps this is for confidentiality reasons, but it could also be a result of such information not being collected or carefully assessed.

Table 7.7 Six Sigma Implementations in Healthcare Using DMAIC Methodology

Organization, Location, and Implementation Period	Reported Investment and Savings	Concepts and Tools	Impact
Commonwealth Health Corporation Bowling Green, Kentucky 1998–2001	Investment: $1.25 million Savings: $2.9 million		A part of the effort focused on radiology, where decreased time interval between radiology report dictation and physician signature, reduced patient wait times for procedures, and improved radiology staff scheduling; all led to a 21.5% reduction in the cost per unit of service (Simmons, 2002).
Presbyterian Healthcare Services (PHS) in Albuquerque, New Mexico 2006	Investment: not reported Savings: $276,500	SIPOC CTQ matrix FMEA Control charts	Identified multifaceted reasons for hand hygiene noncompliance, and thus implemented multiple solutions that saved an estimated 2.5 lives by reducing MRSA by 51% (Carboneau, Benge, Jaco, & Robinson, 2010).
Midsized community teaching hospital, United States, 2006	Investment: not reported Savings: not reported	Histogram Control Charts Conditional statistics	Identified and quantified the impact of very prolonged patient waiting times for patients and nonproductive time between schedule procedures for surgeons. Determined four improvement pathways to reduce patient length of stay (LOS) (Sedlack, 1995).
Tata Memorial Hospital, Mumbai, India		Cause-and-effect diagram	Improved sphincter preservation rates for rectal cancer as measured by a process sigma going from 1.58 to 2.10 (Shukla, Barreto, & Nadkarni, 2008).
13 hospitals in the Netherlands and Belgium	Investment: not reported Savings: $350,000 at one site; $500,000 at another site	CTQ flowdown Pareto charts Histogram ANOVA Control charts	Operating room start time delay reduced by > 25% at one site and > 30% at another; reduced cancellations of operations at the end of the day (Does, Vermaat, Verver, Bisgaard, & Van den Heuvel, 2009).

Example 7-9: Inpatient Pediatric Service Patient Discharge

The pediatric department of an academic medical center has 36 medical/surgical beds and performs approximately 1,100 admissions per year. This example of a DMAIC implementation is based on a study to improve throughput efficiency (Beck & Gosik, 2015).

Define: Delays in patient discharge times affect hospital throughput. Advancing the discharge time creates "virtual beds" that may allow the facility to accommodate new patients.

Measure: Discharge times for 515 patients over a six-month period are shown in Table 7.8.

Table 7.8 Patient Discharge Times

Time	12 A.M.–2 A.M.	2 A.M.–4 A.M.	4 A.M.–6 A.M.	6 A.M.–8 A.M.	8 A.M.–10 A.M.	10 A.M.–NOON
Current state	0.2%	0.0%	0.0%	1.2%	4.5%	8.1%
Time	NOON–2 P.M.	2 P.M.–4 P.M.	4 P.M.–6 P.M.	6 P.M.–8 P.M.	8 P.M.–10 P.M.	10 P.M.–12 A.M.
Current state	15.9%	22.6%	24.2%	18.1%	4.0%	1.2%

Analyze: The discharge times show that 13.8% of the patients were being discharged in the morning from 6 A.M. to noon, while 62.7% of the patients were being discharged in the afternoon during noon to 6 P.M. Even though the care team may know by 9 A.M. that a patient will discharged that day, the actual discharge was not occurring until late in the day, which was impacting bed availability. Furthermore, one of the physician leaders noted that the team-rounding census, that is, the number of patients that hospitalists were expected to see, had grown to a challenging level of 15 per day.

Improve: The current state used a one-team rounding and discharge process that involved one attending, one to two residents, and two to three interns. Lean thinking was used to design an intervention that would rebalance physician workload across the day and create a standard work expectation to reduce the variations in physician work sequences. The intervention used a two-team rounding and discharge process where each team had one attending, one resident, and one to two interns. This was accomplished by persuading the attendings to agree to be "on service" for two to three more weeks per year. The new approach increased the number of patients discharged in the morning by 12.7% while having no adverse impact on 30-day readmission rates.

Control: The pediatric division voted unanimously to adopt the new rounding and discharge model despite the increase in service time. Some of the perceived added benefits included spending more time at the bedside in direct patient care and in resident education. Control charts were implemented to continue to monitor the discharge time.

Leading Change

Action plans that center around Six Sigma projects are, of course, important. But in order for Six Sigma to succeed in transforming the organization, there must be strong support among the leadership. More broadly, lean and Six Sigma facilitate the process of leading and managing for quality. Figure 7.10 shows how this process is an evolution of *kanri,* the mechanism for the exercise of routine control of work (Watson, 2015). Strategy and strategic management are accomplished through *hoshin*

Figure 7.10 The Process of Leading and Managing for Quality
***Source:* Watson (2015)**

kanri, the activities of which are to recognize both assets and challenges (e.g., through the use of the strengths-weaknesses-opportunities-threats (SWOT) framework). Change and change management are accomplished through *kaizen kanri* and *nichijo kanri,* which include the deep deployment of the DMAIC methodology. Finally, execution is accomplished through *hinshitsu kanri*, which includes those repeated cycles carried out on a daily basis to integrate and standardize the improvements. In this section, these domains are discussed with a view of their relation to healthcare and transforming the health organization.

Strategy and Strategic Management

In the best case, Six Sigma is interwoven with the strategy and strategic management of the organization.

Three principles underlying **strategy** include "defining and communicating the company's unique position, making trade-offs, and forging fit among activities" (Porter, 1996). In laying the substrate for introducing Six Sigma initiatives in the organization, leaders must articulate how the project selection and prioritization process will be helpful precisely in making such trade-offs and forging such fit. Careful attention to what the organization is striving for (e.g., being a strong community partner, academic training, world-renowned treatment) and being challenged to do from external forces (e.g., CDC, Centers for Medicare & Medicaid Services [CMS], Affordable Care Act).

Strategic management involves the major goals, initiatives, decisions, actions, and analysis undertaken in order to create and sustain competitive advantages. Six Sigma can be used to facilitate competitive advantage sustainability, since through the use of the DMAIC methodology, the company can identify the errors or defects that are responsible for rising costs and/or customer loss and then do something logical about it. The strategic management facets embraced by Six Sigma include a clear focus on achieving measurable financial returns. There is an explicit and implicit direction to base decisions on verifiable information rather than assumptions. Breyfogle (2003) stresses that Six Sigma is a business improvement methodology that focuses an organization on creating not just a snapshot of the latest results but a continuous picture that describes the key outputs over time.

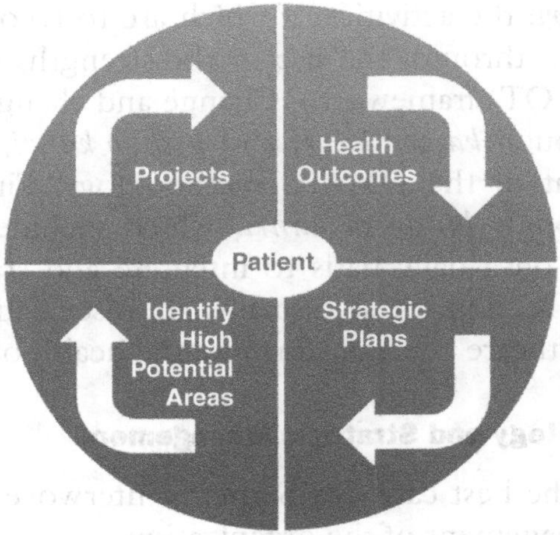

Figure 7.11 Aligning Improvement with Strategy and Health Outcomes

Each project and each undertaking should lead to improving health outcomes. These health outcomes should drive the strategic plans that are driven toward high potential areas, which drive the projects. Figure 7.11 suggests the required cycle with the patient at the center.

Change Management

It has often been said that change is hard. In *The Prince,* Machiavelli (1532) wrote, "There is nothing more difficult to take in hand, more perilous to conduct, or more uncertain in its success, than to take the lead in the introduction of a new order of things." Leaders who are implementing Six Sigma in an organization need to become change agents. They need to be able to manage change. It is helpful to recognize that change is also a process, and like any process it has stages. One view of these stages (Kotter, 1995) follows:

1. Establishing a sense of urgency
2. Forming a powerful guiding coalition
3. Creating a vision
4. Communicating the vision
5. Empowering others to act on the vision
6. Planning for and creating short-term wins

7. Consolidating improvements and producing still more change
8. Institutionalizing new approaches

The need for change may begin with some external urgent event or after some assessment of the organization's competitive/market /technology position and/or financial situation. One or two people may begin a renewal effort, then grow this circle to those who also share the commitment for change. A new image of the organization has to be developed and shared and sought by others. As victories emerge, doubters will be checked, then more progress can be made and a new face brought forth in the organization.

Some have argued that changing healthcare is perhaps more difficult than other industries, in part due to the influence of physicians who desire an environment that preserves their decision-making autonomy (e.g., Frosch, May, Rendle, Tietbohl, & Elwyn, 2012; Poon et al., 2004). Moreover, physicians and hospital staff may be skeptical about the likelihood of making improvements in hospitals, especially if they have experienced previous process improvement initiatives that failed or continuing efforts as "flavors of the month" that have little impact and quickly pass. However, caregivers are stakeholders in the hospital and depend on its smooth functioning for their own ability to deliver high-quality care. Table 7.9 provides some guidelines for gaining physician support for Six Sigma initiatives (Ettinger & Van Kooy, 2003).

Used appropriately, data obtained from the DMAIC process can be used to influence staff behavior. Control charts and other visualizations of the system performance have a visceral impact. When people can see how the steps and decision they make are impacting outcomes—for better or worse—they can become convinced about changes. This is especially true when the data comes in real time (or close). This is, in part, the reason for needing consistency with the Six Sigma effort.

Health Organization Transformation

The leaders of healthcare systems have a clear mandate: to change—dramatically and rapidly—to much better states. Gamm, Kash, and Bolin (2007) define **health organization transformation** as being "about dynamic sustainability of health systems continually promoting improvement." They also note that health organization transformation is

Table 7.9 Considerations for Gaining Physician Support for Six Sigma

Coalition building can begin with physicians by getting an upfront agreement on their role in the process and how their input will be structured into the project.	Structure focused time physician involvement so that they are assured that they will not have to attend weekly project meetings.
Focus on projects for which the processes are the responsibility of the hospital and the physician.	Maintain communication and provide regular project updates.
Recognize the contributions of physicians (e.g., formally on documentation and informally in project discussions).	Work supportively with physicians who will be impacted—both upstream and downstream—by process changes
Seek and publicize early wins.	Point out that improved financial performance may result in more resources for equipment, technology, or other favored indicatives.
Present very compelling data when asking for behavior change.	

consistent with complex adaptive systems inasmuch as it reflects relationships among many agents who must act at multiple levels both internally and with the environment. This view emphasizes the need to understand processes that can contribute to the development and transformation at the patient, population, team, organizational, network, and environment (PPTONE) system levels (Figure 2.1). Transformational change strategies such as Six Sigma call for significant reexamination and, often, realignment of values, processes, and structures characterizing a health system or organization.

In Chapter 2, we discussed the background and motivation for taking a systems approach to healthcare along with how systems thinking is a framework for understanding the complexity of the system as an interconnected whole. While Six Sigma is rooted in the concept of reducing process variation and minimizing defects, at its best and in health organization transformation, it should become a framework for innovation that leverages the interactions between the levels and stakeholders in the healthcare system—from the customer/patient to the provider and payer.

In order to fully realize this potential, system reinforcing loops and balancing loops need to be designed, tested, and implemented that integrate Six Sigma *theory*, not just methodology. However, Six Sigma theory development still has a number of important needs, including more research on user experiences reflecting the Six Sigma pros and cons (Aboelmaged, 2010), data collection methods that can offer greater insights into how information is networked on behalf of the patient, and how managing risks through Six Sigma can improve system resiliency. This agenda will go a long way toward health organization transformation.

Six Sigma in Translational Research

The foundation of knowledge in healthcare and healthcare-related sciences has been established through a tremendous allocation of resources by governmental and private agencies. Nevertheless, there is still a large gap between knowledge discovery and its implementation in clinical practice (Committee on Quality of Health Care in America & Institute of Medicine, 2001; Glasgow et al., 2012; Green, Ottoson, García, & Hiatt, 2009). To address this gap, a science emerged to systemically study how specific strategies can accelerate the rate at which new discoveries become clinical practice and impact population's health. This emerging science goes by many titles, including translational research, translation research, knowledge translation, knowledge exchange, technology transfer, implementation research, and dissemination and implementation research (Brownson, Colditz, & Proctor, 2012).

More formally, translational research has been defined as the process of integrating basic research, patient-oriented research, and population-based research, with the final aim of improving the health of the public (Rubio et al., 2010). This long process, as many other processes in healthcare, is complex, fragmented, and dynamic. Hence, comprehensive tools are needed to understand and identify better ways to translate new discoveries into clinical practice and health outcomes. In this sense, Six Sigma techniques have been argued to hold the promise of being suitable for tackling some of the challenges in translational research. Some attempts have already been made to understand the role of Six Sigma and QI in translational research. One study examined how Six Sigma principles can be applied in translational research to improve coordination, timeliness, efficiency, and value (Schweikhart & Dembe, 2009). Another study examined the DMAIC framework in relation to mapping and measuring the translational research process (Gillam, Nembhard, & Munoz, 2014). Furthermore, the integration of Six Sigma–QI and outcomes research has been examined) and seen as a promising framework for developing methodologies that support the acceleration of translational research (Baldwin et al., 2012). However, there is still a huge opportunity to scope the potential benefits of Six Sigma in a more comprehensive manner.

In order to fully take advantage of Six Sigma principles in translational research, there are two limitations that must be overcome. First, the traditional Six Sigma–QI toolkit cannot be used by its own to address the large degrees of complexity in translational research. Second, in order to provide a better understanding and integration of the translational research process, more research-oriented frameworks are needed. In response, an expansion of the Six Sigma–QI toolkit would provide a more comprehensive approach to understand, measure, and close the current gaps in translational research. This expanded toolkit could be used to generate evidence that inform healthcare stakeholders at both tactical and strategic levels. A nonexhaustive illustration of the expanded basic QI toolkit and its typical scope of action to a more comprehensive research-oriented expanded QI toolkit is shown in Figure 7.12. We recognize that this is in keeping with previous calls to expand and fortify the Six Sigma toolset in order to enhance its applicability to a more transactional environment (Tang, Goh, Lam, & Zhang, 2007).

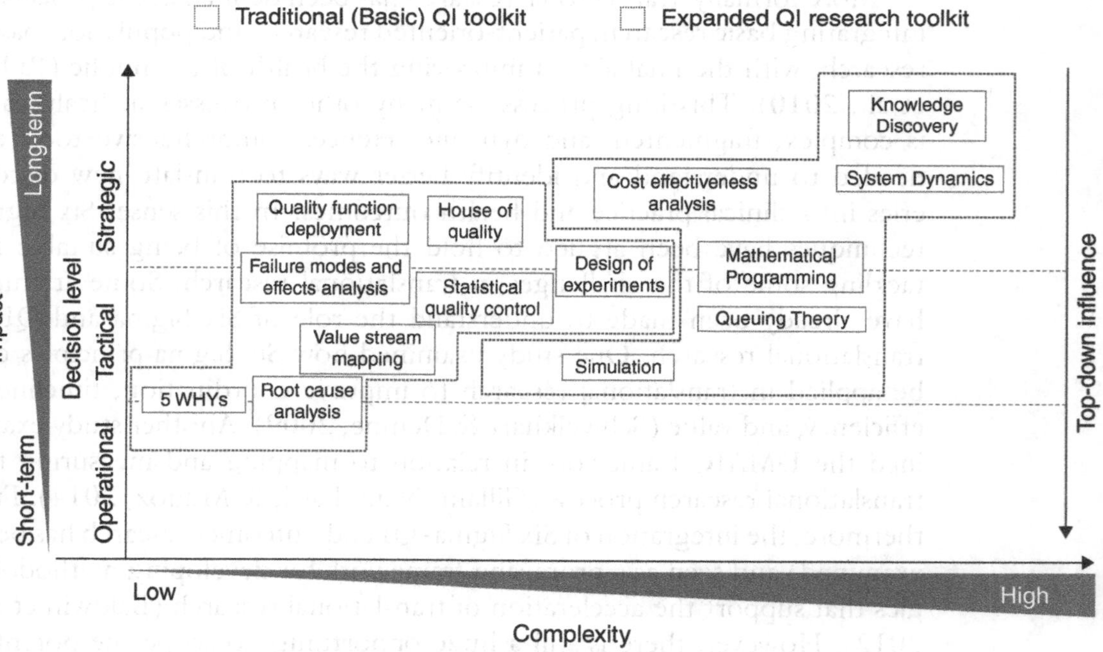

Figure 7.12 Expanded Quality Improvement Toolkit

The extended framework could be applied to a wide range of translational research efforts and its drivers. As a way to illustrate some of its applications, three case studies are presented in this section:

1. Quantification of complexity in translational research using an integrated Quality Function Deployment (QFD)–Analytical Hierarchy Process (AHP) approach.
2. Evaluating the impact of collaboration and multidisciplinarity on translational research.
3. Guiding the strategy and resource allocation of healthcare organizations based on estimated impact.

These three applications are by no means exhaustive and have the purpose of providing some initial thoughts on how a more comprehensive Six Sigma and QI toolkit may be used to meet some challenges in translational research. Full details appear in Munoz (2015).

CASE STUDY 7A: QUANTIFYING COMPLEXITY IN TRANSLATIONAL RESEARCH

OVERVIEW

Discrepancies on the meaning and scope of translational research have led to the generation of several conceptual models. The most popular models to understand the continuum of knowledge translation are based on "T" phases or "translational blocks." For instance, Sung et al. (2003) describe translational processes in two phases: T1 and T2. T1 includes the knowledge gained from laboratory testing to the development of new diagnosis and treatment tools. T2 covers the translation of those clinical studies to clinical practice. Similarly, others have proposed the use of three-phase (Dougherty & Conway, 2008; Westfall, Mold, & Fagnan, 2007) and four-phase models (Khoury, 2007). Although these models are useful to frame the scope of translational research, they fail to provide ways to quantify the translation and relative importance of each translational step. Therefore, our ability to efficiently allocate efforts and resources in the translation process has been limited. This motivates the need to investigate more comprehensive frameworks that serve to evaluate processes and quantify complexity in translational research.

CASE STUDY DESCRIPTION

This case study consists of the evaluation of complexity of a translational research intervention. More specifically, a volunteer-led intervention for weight control in primary care is assessed for complexity (Munoz, Nembhard, & Kraschnewski, 2014).

METHODOLOGY

An integrated QFD and AHP methodology was used to quantify the complexity in a translational research effort. Three main phases are used. In the first phase, a list of operational steps or markers as well as a list of technical requirements for those markers was identified. In the second phase, an AHP framework was used to quantify the relative importance or impact that the markers have on each "T" phase. In particular, a pairwise comparison was conducted to determine the importance of the markers regarding their ability to produce an impact on translational research. Finally, in the third phase, QFD with a House of Quality (HoQ) model was used to find correlations among technical requirements, and then, estimate the relative importance of those technical requirements on each marker, and overall impact on the translational research phases. The analysis was conducted using a three-phase translational research model as the reference.

MAIN RESULTS AND DISCUSSION

A list of seven, nine, and six markers were identified for translational research phases T1, T2, and T3, respectively. For instance, the list of markers for T1 was composed of $M_{1,1}$, pilot proposal; $M_{1,2}$, pilot funded, $M_{1,3}$, study proposal; $M_{1,4}$, study proposal funded; $M_{1,5}$, lab intervention; $M_{1,6}$, results analysis; and $M_{1,7}$, guidelines for clinical trial. Additionally, a list of nine general technical requirements was generated. This list included: R_1, collaboration networks; R_2, administrative support; R_3, funding availability; R_4, community engagement; R_5, information technology; R_6, regulation and standards; R_7, equipment availability; R_8, organizational leadership; and R_9, multidisciplinarity team capacity. The relative importance of the markers on each translational phase were obtained through the AHP (Figure 7.13), and the relative importance of the technical requirements were obtained using QFD (Figure 7.14).

From the analysis, a clear mapping of the required element and the dynamics of translational research was obtained. Useful guidelines regarding the drivers or technical requirements in translational research can be generated by capturing and quantifying the "voice of the customer" through the integrated QFD-AHP methodology. More frameworks like the one presented are expected to address the large-scale, complex, and dynamic translational research process, and therefore, provide guidance for a more informed resource allocation.

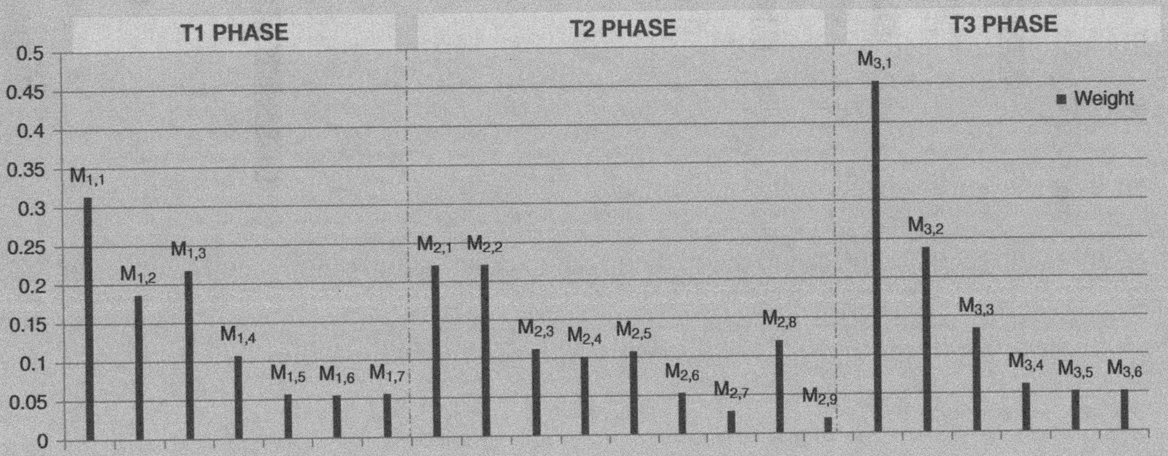

Figure 7.13 Markers' Relative Importance

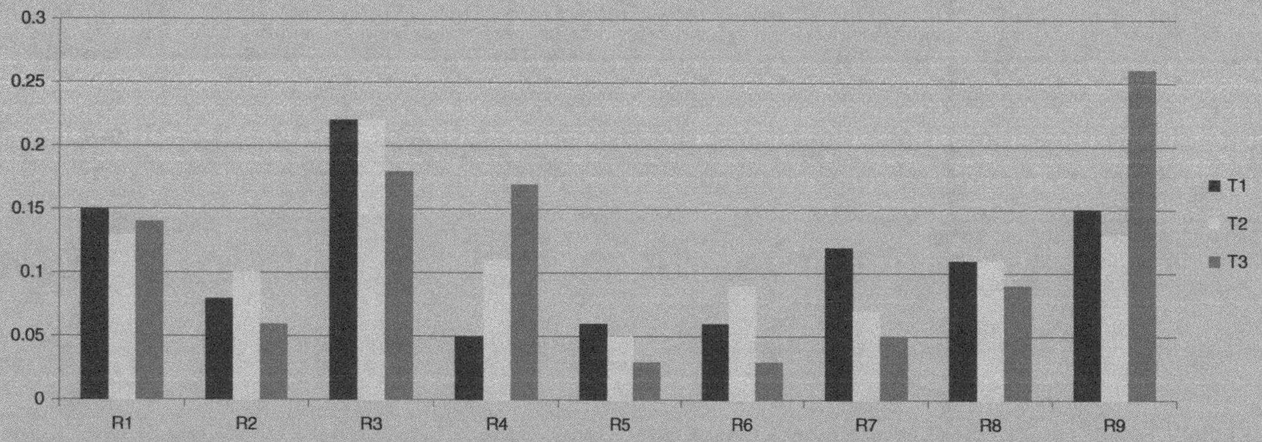

Figure 7.14 Technical Requirements' Relative Importance for Each Translational Phase

CASE STUDY 7B: EVALUATING COLLABORATION IN TRANSLATIONAL RESEARCH

OVERVIEW

Collaboration is seen as a critical element in translational research (Bennett, Gadlin, & Levine-Finley, 2010; Long, Cunningham, & Braithwaite, 2012; Marincola, 2003). In recognition of this, the National Institutes of Health (NIH) through its Clinical and Translational Sciences Award (CTSA) has focused on collaborative and multidisciplinarity research as key priorities to accelerate translational research. Certainly, translational research networks provide the conditions under which research can be efficiently initiated and spread. The ultimate goal of these efforts is to support collaborations that can strategically close the gap between basic and clinical practice research (Long et al., 2012). In consideration of these premises, quantifying evaluation seems to be a critical task to ensure that the expected collaborative patterns and outcomes are being achieved. Moreover, understanding the complexities and collaborative structure of a network may be informative on current areas of opportunity and design of network interventions.

CASE STUDY DESCRIPTION

This case study consists of the evaluation of collaboration networks for obesity research at Pennsylvania State University. Co-authorship was used as the main metric for quantifying the collaboration intensity among researchers.

METHODOLOGY

A combined quality improvement and social network analysis (SNA) approach was used to evaluate intra- and cross-institutional collaboration networks. The approach proposed was divided into five phases: (1) identification of the researchers whose interest is in obesity, (2) classification of researchers' expertise, (3) retrieval of bibliometric information, (4) conduct social network analysis, and (5) identify current collaborative gaps and design network interventions. In the first phase, a list of researchers was retrieved from an institutional profiles platform and validated against a search of keywords in a medical scientific database (PubMed). The second phase aimed to generate an initial classification of expertise based on three main researchers' attributes: expertise (number of publications), experience (number of years in the field), and influence (average number of articles citing the researchers' work). In the third phase, a list of all the researchers' publications were retrieved and relevant bibliometric measures were calculated. The intra- and cross-institutional collaboration network graphs and metrics were obtained

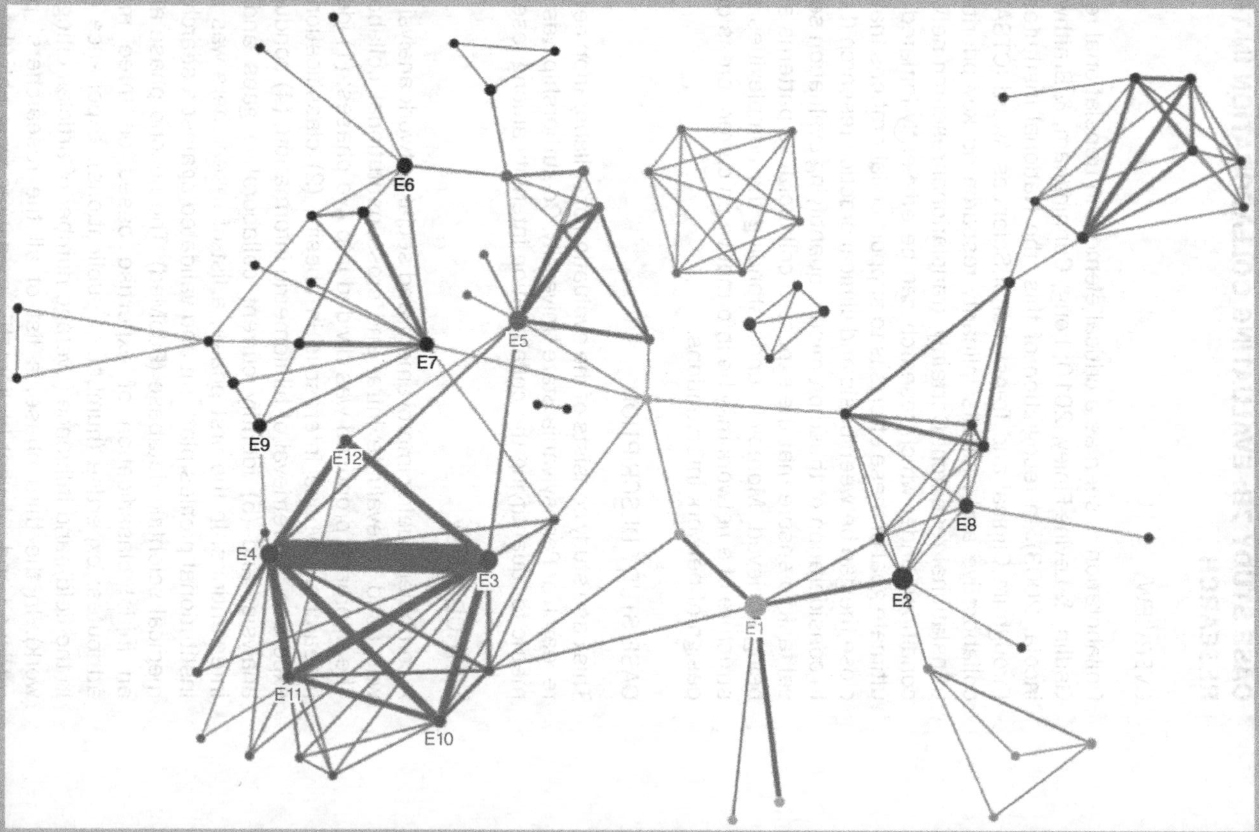

Figure 7.15 Intra-Institutional Obesity Collaboration Network

using an SNA software (NodeXL). Main areas of opportunity were discussed using Six Sigma techniques such as root cause analysis. The proposed improvements or network interventions could be evaluated with modified versions of the FMEA, for instance.

MAIN RESULTS AND DISCUSSION

A list of 114 researchers accounting for 779 publications was retrieved from PubMed. Out of this list, 12 were classified as experts (E), 22 as semi-experts (S), and 80 as new investigators (N). The network of researchers (Figure 7.15) accounted for 170 ties among them (density = 0.026).

Key network metrics including the betweenness centrality, degree centrality, closeness centrality, eigenvector centrality were used to identify the key researchers of the network and their role as leader, influencer, facilitator, and/or bridger. In addition, clustering algorithms were run to identify collaborative patterns. From this analysis, we were able to identify collaborative gaps and unconnected subnetworks (Figure 7.15). An extension of this analysis was also conducted for the cross-institutional collaboration network for those researchers that were previously classified as experts (Figure 7.16).

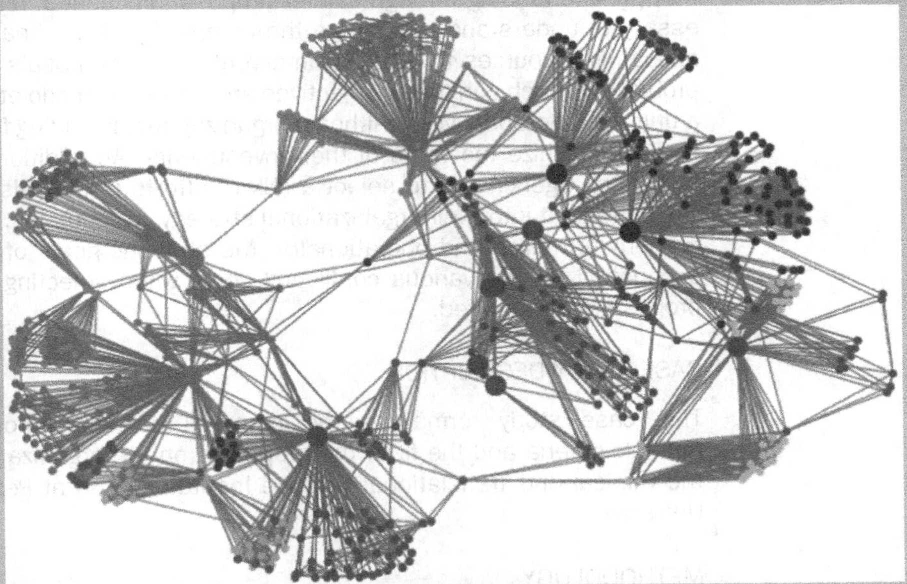

Figure 7.16 Cross-Institutional Obesity Collaboration Network

The insights obtained from the SNA were used to understand the current collaboration structure and identify strategic opportunities to better design and impact the network. In response, tools such as root cause analysis could be used to identify a detailed list of opportunities in the network (e.g., increasing connectedness, promoting a specific cluster, finding collaborative efforts between specific disciplines, matching team members) These network interventions may be then prioritized (prioritization matrix, FMEA, etc.) according to the specific objectives of the organization and its collaborative structure.

CASE STUDY 7C: RESOURCE ALLOCATION IN TRANSLATIONAL RESEARCH

OVERVIEW

One of the main challenges in translational research is to allocate resources based on evidence or anticipated estimation of the impact that certain efforts could have. This challenge is particularly pressing for healthcare organizations that provide funding through call for proposal processes. Moreover, it is necessary to understand that, given the complexity of the healthcare systems, allocating resources to different interventions or proposals is a multicriteria problem in which conflicting objectives are present (Guindo et al., 2012). From a utilitarian perspective, healthcare organizations, including funding agencies, aim to maximize the value of their investments. An additional consideration of funding agencies is to select a mix of efforts (proposals) that might provide a good fit into their organizational strategy. In response, frameworks that properly formalize and operationalize the strategic goals of an organization, accounting for the various conflicting goals when selecting the best mix of proposals are required.

CASE STUDY DESCRIPTION

This case study formalizes the proposal selection process based on multiple-criteria and the fit of the proposals on the organizational strategy of the Clinical and Translational Science Institute (CTSI) at Pennsylvania State University.

METHODOLOGY

A strategic goal programming (GP) framework was used to solve the proposal selection problem. In the model, the goal constraints were aligned to different organizational goals of the CTSI. Four main phases are needed for the SGPF: (1) understanding strategy, (2) understanding the constraints, (3) formulating the model, and (4) solving and validating. In the first phase, the key strategic goals of the organization are identified, a target for each goal is set, and the relative importance of each goal is quantified. Many techniques such as experts group, borda count, rating method, or AHP could be used for this quantification. In the second phase, the set of hard constraints is identified and the limits or boundaries for each constraint are set (the limits could be obtained, for instance by an expert group, budget analysis, or mathematical bound approach). The goal programming model is formulated in the third phase. This phase aims to clearly define the coefficients to be used, formulate goal and hard constraints, and prepare the objective functions based on the minimization of deviations from the targets. Typically, historical data can be used to infer some of the coefficients used in the formulation. Finally, in phase 4, the model is solved and validated.

MAIN RESULTS AND DISCUSSION

Four strategic goals were identified: (1) promoting multidisciplinarity, (2) training the new generation of investigators, (3) improving institute's prestige by potential benefits of the proposals, and (4) balancing the levels of risk to be taken. A rating method was used to quantify the relative importance of each goal. Goals 1 (multidisciplinarity) and 3 (potential benefits) were the two most important. In addition, six hard constraints were identified: (1) budget constraints, (2) quality constraints, (3) mutual exclusiveness constraints, (4) minimum number of proposals constraint, (5) proportion of resources allocated by department, and (6) minimum cross-campus collaboration required. For the GP model, the coefficients for the goal constraints were normalized from 0 to 1 to assure that the model was reflecting the initially calculated goals' relative importance. For instance, the risk parameter (r_i) of a given proposal i used the principal investigator's expertise (number of papers in the field) and experience (number of years in the field) as the main drivers. Hence, assuming that the investigators' expertise is fully achieved after five publications and expertise after eight years in the field, r_i was modeled as follows:

$$r_i = \frac{\left[min\left\{\left(\frac{Yexp_i-8}{8}\right),0\right\}+1\right]+\left[min\left\{\left(\frac{Npaper_i-5}{5}\right),0\right\}+1\right]}{2}$$

Here, $r_i = 0$ indicates the highest level of risk for a given proposal i. In contrast, $r_i = 1$ indicates that there is no risk associated to proposal i. This is an example of how the different coefficients could be modeled; other options are possible. Finally, the model was validated using real data obtained from a past CTSI call for proposals. The model was able to match 7 out of the 9 proposals selected from a pool of 24. Two of the goals (multidisciplinarity and training new investigators) were fully satisfied, while the other two (potential benefits and balancing risk) were partially satisfied. From a practitioner's perspective, the GP model provides an effective way to quantify the proposals' merit and provides additional guide the selection process based on organizational fit. This data-driven approach formalized and operationalized the proposal selection problem by having a better understanding of the organizational goals, system constraints, and historical data. Similarly to the previous case studies, a key component of this case was to capture the voice of the customer, in this case the CTSI's leaders. The process of "walking" through the strategy was helpful for quantifying levels of importance and generating agreements on targets that suffice for each goal.

QUESTIONS AND LEARNING ACTIVITIES

1. Identify a healthcare organization that interests you and for which you can access their mission statement. Discuss how it relates to a Six Sigma approach of customer-driven (or patient-centered) quality.
2. The Define, Measure, Analyze, Improve, Control (DMAIC) methodology can be thought of as a roadmap for process/produce improvement. Summarize and critically assess the DMAIC steps that were undertaken in the following cases:
    a. Using the Six Sigma Approach to Meet Quality Standards for Cardiac Medication (Elberfeld et al. (2004))
    b. Reducing Central Venous Catheter (CVC)-Related Bloodstream Infections (BSI) at a Florida Hospital (Trusco et al. (2007))
    c. Decreasing Turnaround Time Between General Surgery Cases: A Six Sigma Initiative (Adams et al. (2004))
3. The Malcolm Baldridge Award has long been recognized as a mark of organizational excellence and is now managed by the National Institute of Standards and Technology (NIST). Explore how organizations in the health sector have used Six Sigma to achieve results.

4. Consider a surgical operation comprised of the following steps (adapted from Trusco et al. (2007)):

   a. *Initial visit.* Visit with the doctor for problem assessment and fill out insurance paperwork.
   b. *Follow-up visit.* Schedule the surgery, determine the surgical method, and sign a liability release.
   c. *Presurgical preparation.* Give anesthesia and prepare the patient and room for surgery (the operation table, tools, scans)
   d. *Surgery.* Perform the actual surgery (mark, cut, close).
   e. *Postsurgical monitoring.* Monitor recovery and administer medications.
   f. *Discharge.* Provide instructions for recovery at home.
   g. *Follow-up visit.* Verify surgery outcome and patient's satisfaction.
   h. *Receive payment.* Collect the copayment from the patient.
   i. *File claims.* Submit the necessary paperwork to insurance companies or Medicaid/Medicare.

   Determine the overall DPMO and estimated sigma level for the surgical operation based on data collected over a one-month period as shown in the following table. Assume a 1.5σ process drift.

Operation Number	Number of Patients Treated	Errors	Estimated Opportunities
1	40	8	2
2	40	0	3
3	35	0	2
4	35	12	20
5	35	2	2
6	35	4	2
7	35	3	3
8	35	0	1
9	22	8	10
Total		37	

REFERENCES

Adams, R., Warner, P., Hummard, B., Goulding, T. (2004). Decreasing Turnaround Time Between General Surgery Cases: A Six Sigma Initative. *Journal of Nursing Administration*, *34*(3), 140–148.

Aboelmaged, M.G. (2010). Six Sigma quality: A structured review and implications for future research. *International Journal of Quality and Reliability Management*, *27*(3), 268–317.

Baldwin, L. M., Keppel, G. A., Davis, A., Guirguis-Blake, J., Force, R. W., & Berg, A. O. (2012). Developing a Practice-Based Research Network by Integrating Quality Improvement: Challenges and Ingredients for Success. *Clinical and translational science*, *5*(4), 351–355.

Barochia, A. V, Cui, X., Vitberg, D., Suffredini, A. F., O'Grady, N., Banks, S. M., … Eichacker, P. (2010). Bundled care for septic shock: An analysis of clinical trials. *Critical Care Medicine, 38*(2), 668–678. Retrieved from http://doi.org/10.1097/CCM.0b013e3181cb0ddf

Beck, M. J., & Gosik, K. (2015). Redesigning an inpatient pediatric service using lean to improve throughput efficiency. *Journal of Hospital Medicine, 10*(4), 220–227. Retrieved from http://doi.org/10.1002/jhm.2300

Bennett, L. M., Gadlin, H., & Levine-Finley, S. (2010). Collaboration & team science: A field guide. Bethesda, MD: National Institutes of Health.

Black, J. (2009). Transforming the patient care environment with Lean Six Sigma and realistic evaluation. *Journal for Healthcare Quality, 31*(3), 29–35.

Box, G. E. P., Hunter, J. S., & Hunter, W. G. (2005). Statistics for experiments (2nd ed.). Hoboken, NJ: Wiley.

Breyfogle, F. W. (2003). Implementing Six Sigma (2nd ed.). Hoboken, NJ: Wiley.

Brownson, R. C., Colditz, G. A., & Proctor, E. K. (2012). *Dissemination and implementation research in health: Translating science to practice.* Oxford, England: Oxford University Press.

Carboneau, C., Benge, E., Jaco, M. T., & Robinson, M. (2010). A Lean Six Sigma team increases hand hygiene compliance and reduces hospital-acquired MRSA infections by 51%. *Journal for Healthcare Quality, 32*(4), 61–70.

Carrigan, M. D., and Kujawa, D. (2006). Six Sigma in health care management and strategy. *Health Care Management*, *25*(2), 133–41.

Centers for Disease Control and Prevention (CDC). (2011). Clean hands save lives [Online image]. Retrieved May 30, 2015, from http://www.cdc.gov/handhygiene/Resources.html#Patients

Committee on Quality of Health Care in America & Institute of Medicine. (2001). *Crossing the quality chasm: A new health system for the 21st century*. Washington, DC: National Academies Press.

De Koning, H., Verver, J. P. S., Bisgaard, S., & Does, R. J. M. M. (2006). Lean Six Sigma in healthcare. *Journal for Healthcare Quality, 28*(2), 4–11.

Does, R. J. M. M., Vermaat, T., Verver, J., Bisgaard, S., & Van den Heuvel, J. (2009). Reducing start time delays in operating rooms. *Journal of Quality Technology, 41,* 95–109.

Dougherty, D., & Conway, P.H. (2008). The "3T's" Road Map to Transform US Health Care The "How" of High-Quality Care. *Journal of the American Medical Association, 299*(19), 2319–2321.

Dunn, P. M. (1997). James Lind (1716–94) of Edinburgh and the treatment of scurvy. *Archives of Disease in Childhood—Fetal and Neonatal Edition, 76*(1), F64–F65. Retrieved from http://doi.org/10.1136/fn.76.1.F64

Elberfeld, A., Goodman, K., and Van Kooy, M. (2004). Using the Six Sigma Approach to Meet Quality Standards for Cardiac Medication. *J Clinical Outcomes Management, 11*(8), 510–516.

Ettinger, W., & Van Kooy, M. (2003). The art and science of winning physician support for Six Sigma change. *Physician Executive, 29*(5), 34–38.

Frosch, D. L., May, S. G., Rendle, K. A. S., Tietbohl, C., & Elwyn, G. (2012). Authoritarian physicians and patients' fear of being labeled "difficult" among key obstacles to shared decision making. *Health Affairs, 31*(5), 1030–1038. http://doi.org/10.1377/hlthaff.2011.0576

Gamm, L., Kash, B., & Bolin, J. (2007). Organizational technologies for transforming care. *Journal of Ambulatory Care Management, 30*(4), 291–301.

Gillam, P. S., Nembhard, H. B., & Munoz, D. (2014). Research translation mapping and measurement: A framework for using quality improvement methods. In *Proceedings of the Industrial and Systems Engineering Research Conference,* Montreal, Canada.

Glasgow, R. E., Vinson, C., Chambers, D., Khoury, M. J., Kaplan, R. M., & Hunter, C. (2012). National institutes of health approaches to dissemination and implementation science: Current and future directions. *American Journal of Public Health, 102*(7), 1274–1281. Retrieved from http://doi.org/10.2105/AJPH.2012.300755

Green, L. W., Ottoson, J. M., García, C., & Hiatt, R. A. (2009). Diffusion theory and knowledge dissemination, utilization, and integration in public health. *Annual Review of Public Health, 30,* 151–174. Retrieved from http://doi.org/10.1146/annurev.publhealth.031308.100049

Guindo, L. A., Wagner, M., Baltussen, R., Rindress, D., van Til, J., Kind, P., & Goetghebeur, M. M. (2012). From efficacy to equity: Literature review of decision criteria for resource allocation and healthcare decision making. *Cost Effectiveness and Resource Allocation, 10*(1), 9. Retrieved from http://doi.org/10.1186/1478-7547-10-9

iSixSigma. (n.d.). Six Sigma costs and savings. Retrieved from http://www.isixsigma.com/implementation/financial-analysis/six-sigma-costs-and-savings/

Kang, H., Nembhard, H. B., Rafferty, C., and DeFlitch, C. (2014). "Patient Flow in the Emergency Department: A Classification and Analysis of Admission Process Policies," *Annals of Emergency Medicine, 64*(4), 335–342. 10.1016/j.annemergmed.2014.04.011

Khoury, M. J., Gwinn, M., Yoon, P. W., Dowling, N., Moore, C. A., & Bradley, L. (2007). The continuum of translation research in genomic medicine: how can we accelerate the appropriate integration of human genome discoveries into health care and disease prevention?. *Genetics in Medicine, 9*(10), 665–674.

King, T. E., Pardo, A., & Selman, M. (2011). Idiopathic pulmonary fibrosis. *Lancet, 378*(9807), 1949–1961. Retrieved from http://doi.org/10.1016/S0140-6736(11)60052-4

Kotter, J. P. (1995, March–April). Leading change: Why transformation efforts fail. *Harvard Business Review, 73*(2), 59–67, Reprint 95204.

Long, J. C., Cunningham, F. C., & Braithwaite, J. (2012). Network structure and the role of key players in a translational cancer research network: a study protocol. *BMJ open, 2*(3), e001434.

Marincola, F. M. (2003). Translational medicine: a two-way road. *Journal of Translational Medicine, 1*(1), 1.

Medication Errors. (n.d.). Retrieved from http://psnet.ahrq.gov/primer.aspx?primerID=23

Montgomery, D. (2012). *Statistical quality control* (7th ed.). Hoboken, NJ: Wiley.

Munoz, D. A. (2015). Quality improvement to assess and audit complexity in research translation (Dissertation). Pennsylvania State University, University Park, State College, PA.

Munoz, D. A., Nembhard, H. B., & Kraschnewski, J. L. (2014). Quantifying complexity in translational research: An integrated approach. *International Journal of Health Care Quality Assurance, 27*(8), 760–776. Retrieved from http://doi.org/10.1108/IJHCQA-01-2014-0002

Poon, E. G., Blumenthal, D., Jaggi, T., Honour, M. M., Bates, D. W., & Kaushal, R. (2004). Overcoming barriers to adopting and implementing computerized physician order entry systems in U.S. Hospitals. *Health Affairs, 23*(4), 184–190. Retrieved from http://doi.org/10.1377/hlthaff.23.4.184

Porter, M. E. (1996). What is strategy? *Harvard Business Review, 74*(6), 61–78. http://doi.org/10.1016/j.cell.2005.09.009

Powers, D., & Paul, M. (2008, February). Healthcare department reduces cycle time and errors. *ASQ Six Sigma Forum Magazine,7*(2), 30–34.

Pyzdek, T., & Keller, P. (2014). *The Six Sigma handbook* (4th ed.). New York, NY: McGraw-Hill.

Rubio, D. M., Schoenbaum, E. E., Lee, L. S., Schteingart, D. E., Marantz, P. R., Anderson, K. E., … Esposito, K. (2010). Defining translational research: Implications for training. *Academic Medicine: Journal of the Association of American Medical Colleges, 85*(3), 470–475. Retrieved from http://doi.org/10.1097/ACM.0b013e3181ccd618

Samuels, D. I., & Adomitis, F. L. (2003). Six Sigma can meet your revenue cycle needs. *Health Care Financial Management. 57* (November), 70–75.

Schweikhart, S. A., & Dembe, A. E. (2009). The applicability of lean and Six Sigma techniques to clinical and translational research. *Journal of Investigative Medicine, 57*(7), 748–755. Retrieved from http://doi.org/10.231/JIM.0b013e3181b91b3a

Sedlack, J. D. (1995). The utilization of Six Sigma and statistical process control techniques in surgical quality improvement. *Journal for Healthcare Quality, 32*(6), 18–26.

Shukla, P., Barreto, S., & Nadkarni, M. (2008). Application of Six Sigma towards improving surgical outcomes. *Hepatogastroenterology, 55,*

311–314. Retrieved from http://europepmc.org/abstract/med/18613355

Simmons, J. (2002). Using Six Sigma to make a difference in health care quality. *Quality Letter for Healthcare Leaders, 14,* 2–11.

Snee, R. D., & Hoerl, R. W. (2003). Leading Six Sigma: A step-by-step guide based on experience with GE and other Six Sigma companies. Upper Saddle River, NJ: Prentice Hall.

Sung, N. S., Crowley JR, W. F., Genel, M., Salber, P., Sandy, L., Sherwood, L. M., Johnson, S. B., Cantanese, V., Tilson, H. & Getz, K. 2003. Central challenges facing the national clinical research enterprise. *Journal of the American Medical Association, 289*(10), 1278–1287.

Tang, L.-C., Goh, T.-N., Lam, S.-W., & Zhang, C. W. (2007). Fortification of Six Sigma: Expanding the DMAIC toolset. *Quality and Reliability Engineering International, 23,* 3–18. Retrieved from http://doi.org/10.1002/qre

Toussaint, J. S. (n.d.). Deploy data and a consistent methodology to drive improvement and change your culture. Retrieved from http://createvalue.org/wp-content/uploads/2013/11/Toussaint-Article-Ver211.pdf

Trusko, B. E., Pexton, C., Harrington, J. and Gupta, P. K. (2007). *Improving Healthcare Quality and Cost with Six Sigma.* FT Press. Upper Saddle River, New Jersey.

Watson, G. H. (2015). Quality thinking for the next generation. Unpublished lecture.

Westfall, J. M., Mold, J., & Fagnan, L. (2007). Practice-based research—"Blue Highways" on the NIH roadmap. *Journal of the American Medical Association, 297*(4), 403–406.

Chapter 8

Reliability and Patient Safety

"Simplicity is prerequisite for reliability."

—Edsger Dijkstra

Overview

The quality crisis in healthcare has been well referenced in earlier chapters of this book. Gaps in efficiency and access are certainly important. However, when the focus turns to gaps connected to human reliability that affects patient safety, the picture is especially poignant. For example:

The *New York Times* reported that Linda McDougal, 46, underwent a double mastectomy after being advised that she had an aggressive form of cancer. However, two days later it was discovered that it was an error: her tissue had been mistakenly switched with another woman's tissue in the hospital's laboratory. Ms. McDougal in fact never had cancer ("Hospital Apologizes for Surgical Mistake," 2003).

This chapter covers the background and methods involved in human reliability and medical errors—especially those dealing with medication and patient falls. It also addresses the area of human factors and ergonomics for patient safety.

8.1 HUMAN RELIABILITY

Human reliability refers to the degree to which humans can be expected to perform within a system. That performance can be affected by many factors such as age, physical health, attitude and state of mind, cognitive biases, propensity for making mistakes and errors, and so forth. Human reliability is very important due to the contribution of humans to system resiliency and due to the possible adverse consequences of human errors or oversights. This is especially true when the human is a crucial part of large sociotechnical systems, as is the case in healthcare.

Human reliability analysis (HRA) refers to methods for studying human reliability for the purpose of improving human performance, especially in the face of risk or hazardous events. Two general classes of methods are those based on **probabilistic risk assessment** (PRA) and those based on **cognitive control.** In the case of PRA, a task analysis is specified at a level of detail that allows the assignment of failure or error probabilities. For example, the Technique for Human Error Rate Prediction (THERP) (Swain & Guttmann, 1983) is intended to generate human error probabilities that can be incorporated into a PRA. In the case of cognitive control, human performance is specified as a set of control modes—strategic (based on long-term planning), tactical (based on steps or procedures), opportunistic (based on present context), and scrambled (random)—and proposes a model of how transitions between these control modes occur. For example, the Cognitive Reliability and Error Analysis Method (CREAM) is an HRA method based on cognitive control (Hollnagel, 1998). Extensions to both THERP and CREAM have been that build out for detailed contexts have been proposed by several authors including Hollnagel (1998), Mosleh and Chang (2004), Parry (1995), Reason (1990), Swain (1990), and Vaughn-Cooke (2012), and Vaughn-Cooke et al. (2015).

In HRA, failures to take action to prevent risk or hazardous events are called **human errors**. This term means only that an action was omitted (or taken) by a human that adversely influenced safety and does not imply that people are necessarily personally responsible or culpable in some way. From a system viewpoint, all errors are due to design or management. In healthcare, knowing that humans are an inextricable part of the system, it is especially important to design the system for safety.

8.2 ERRORS IN HEALTHCARE

In its 2000 report, *To Err Is Human,* the Institute of Medicine (IOM, 2000) estimated that as many as 98,000 deaths occur annually in U.S. hospitals due to preventable medical errors. According to more recent reports, this figure may have significantly risen. For instance, a recent study estimated that as many as 440,000 people die each year in U.S. hospitals from preventable medical errors (James, 2013). These staggering statistics lead to the conclusion that preventable medical errors in hospitals are the third-leading cause of death in the United States (Leapfrog Group,

2013). The 2013 Global Burden of Disease Study ("Global, Regional, and National Age–Sex Specific All-Cause and Cause-Specific Mortality for 240 Causes of Death, 1990–2013: A Systematic Analysis for the Global Burden of Disease Study 2013," 2015) estimated yearly deaths for 188 countries and found that that 142,000 people died in 2013 from adverse effects of medical treatment, up from 94,000 in 1990.

In addition to deaths, medical errors place significant financial burdens on society. Measurable medical errors that harm patients were estimated to cost $17.1 billion per year (Van Den Bos et al., 2011). When including other indirect costs such as lost productivity, preventable medical errors cost the United States nearly $1 trillion every year (Andel, Davidow, Hollander, & Moreno, 2012).

Medical errors have been defined in various ways, and can be classified into two major types of definitions: **outcome-dependent** type and **process-dependent** type (Grober & Bohnen, 2005). An outcome-dependent type of definition focuses on patients who are affected by the errors. *Adverse patient outcomes* represent injury to patients that have occurred because medical errors are not identified and managed in time. James Reason (1997) explained how and why the adverse consequences of error affect patients through the "Swiss cheese" model of error causation as shown in Figure 8.1. *Adverse events* indicate unintended injury or complication that results in disability, death, or prolonged hospital stay, which are caused by medical management rather than the underlying condition of the patient (Brennan et al., 1991). Not all adverse patient outcomes

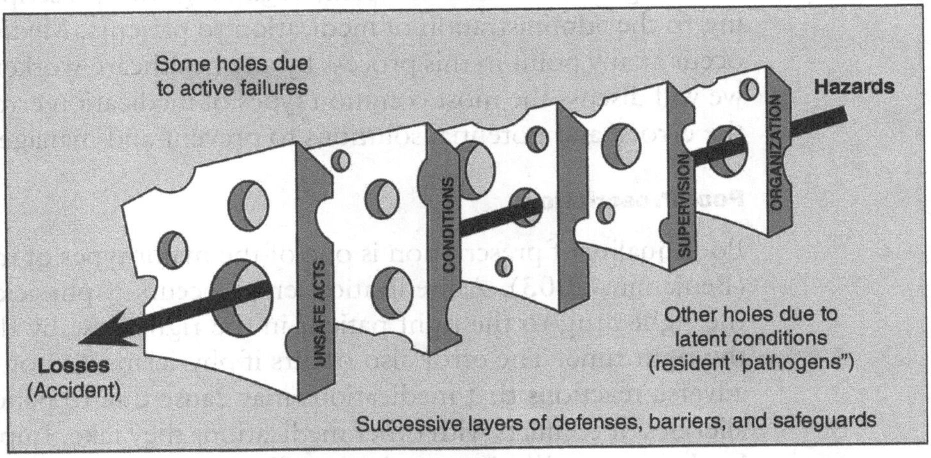

Figure 8.1 Swiss Cheese Model of Error Causation

occur as a result of a medical intervention. However, adverse events frequently occur as a result of medical errors, and they are *preventable adverse events.*

A process-dependent type of definition extends the scope of medical errors to processes and system failures regardless of the magnitude of harm resulting from such events. This broader definition makes it possible to capture not only errors that actually affect patient outcomes but also errors that have the potential to harm patients. *Medical errors* are the failure to complete a planned action as intended or deviations from the process of care, which may or may not result in adverse patient outcomes (Grober & Bohnen, 2005). *Potential adverse events* are those that expose patients to risk but do not cause them harm.

Additional definitions and classifications of medical errors can be found in Elder and Dovey (2002) and Grober and Bohnen (2005).

8.3 MEDICATION ERRORS

Medication errors, also called **adverse drug events** (ADEs), are one of the most prevalent medical errors that threaten patient safety. Annually, ADEs account for nearly 700,000 emergency department visits and 120,000 hospitalizations, along with $3.5 billion extra medical costs in the United States (Budnitz et al., 2006; IOM, 2006). These errors occur when individuals or systems fail to achieve "the five rights"—the right drug, right dose, right route, right time, and right patient. In a hospital, the process involving medication is complex, beginning with prescription and extending to the administration of medication to patients. Medication errors can occur at any point in this process by any healthcare worker. In this section, we will discuss the most common types of medication errors, the causes of the errors, and potential solutions to prevent and manage the errors.

Poor Prescription

Poor quality of prescription is one of the major types of medication errors (Benjamin, 2003). A medication error occurs if physicians fail to order the right drug to the right patient in the right dose by the right route at the right time. The error also occurs if physicians do not identify possible adverse reactions that medications may cause due to patients' medication allergies or conflicts with other medications they take. Implementing **medication reconciliation** can be an effective strategy to prevent these types of medical errors. Medication reconciliation is the process of gathering the most accurate list of all medications a patient is prescribed and taking, and comparing the list at all levels of care, care setting, or points in time. In other words, medication reconciliation can improve physicians' prescribing decisions by identifying potential medication errors. The adoption and appropriate use of technologies that support a medication reconciliation process can facilitate the identification and prevention of the medication errors occurring during the prescription stage.

Illegible Handwriting

Illegible handwriting also causes medication errors because it can be interpreted differently than intended. Pharmacists who misinterpret words or numbers may dispense a wrong medication or the wrong dosage of the medication. Nurses who misread a physician's order can administer a wrong medication at the wrong time through a wrong route. There have even been cases of such errors resulting in a patient's death. An illegible signature on a prescription makes it difficult for other medical teams to contact the prescribing physician when they cannot read the doctor's handwriting. This forces pharmacists or nurses to guess, which increases the chances of medical errors. To reduce prescribing and ordering errors caused by physicians' handwriting, hospitals have adopted a *Computerized Physician Order Entry (CPOE)*. Many studies have shown that this system is effective for decreasing medication errors, enhancing patient safety, and improving pharmacists' workflow.

Wrong Medication Administration

Many medication errors are associated with drug administration. The most common types of errors made by nurses include a wrong dosage and infusion rate (Ehsani et al., 2013). Contributory factors to these medication errors include both systems and individual issues. Using abbreviations instead of the full names of drugs and similar names is one of the system-related issues that may cause medical errors. Nurses' excessive workload and their working conditions, such as the type of shift worked, also can lead to medical errors associated with systems and management. Human factors contributing to medical errors include nurses' insufficient pharmacological knowledge and their lack of mathematical skills. Nurses' insufficient knowledge of side effects and correct dosage of the drug they administer can lead to medical errors. In addition, nurses who have low mathematical skills are more likely to commit errors in dosage calculation, intravenous regulation, and other calculations related to medication administration. Strategies that can help avoid these adverse medication errors committed by nurses include increasing the number of nurses, adjusting their workload, and providing training to improve their pharmacological knowledge and mathematical skills. Environmental and cultural changes that help nurses concentrate while preparing and administering medication can also contribute to preventing medical errors caused by distractions and interruptions.

In summary, application of both a systematic approach and individual efforts are required to reduce medication errors and improve patient safety. Information technologies can also play a significant role in preventing errors from medication prescription to administration. Continuous training to help healthcare workers to adequately use these technologies is required. Effective communication among workers and the active reporting of medical errors are also critical to reduce the occurrence of medical errors.

8.4 PATIENT FALLS

Falls represent a growing public health problem that results in injuries of various severity and even death. In particular, falls and fall-related injuries are one of the leading causes of injuries for older adults. In 2012, the overall rate of nonfatal fall injury episodes receiving medical attention was 43 per 1,000 population, and the rate for those age 75 and older was highest (121 per 1,000) among all other age groups (Adams, Kirzinger, & Martinez, 2013). The death rates from falls among the older population have continuously increased over the past decade, and the number of deaths of persons age 65 and older stood at 25,000 in 2013 (Centers for Disease Control and Prevention [CDC], 2013).

Fall rates of acute care hospitals range from 1.3 to 25 falls per 1,000 patient days depending on the care area (Agency for Healthcare Research & Quality [AHRQ], 2013; Currie, 2008). Inpatient falls cause prolonged hospital stays, and related injuries lead to rehospitalizations after discharge from the hospital (Stevens, Corso, Finkelstein, & Miller, 2006). In addition to the acute care setting, falls and injuries frequently occur in long-term care settings including nursing homes and assisted living facilities. For example, each year, a nursing home with 100 beds reports 100 to 200 falls, and about 10% to 20% of these falls result in serious injuries (CDC, 2015a). Falls in various care settings are medical errors because they occur because of the failure of performing the planned action to be completed as intended. The Centers for Medicare & Medicaid Services (CMS) includes falls on its list of *Hospital-acquired Conditions (HACs)* for which reimbursement is limited. Falls and resulting injuries are one of the major adverse events reported in hospitals. When considering that many falls go underreported and the number of older adults who are at risk of falls increases, it is imperative to identify solutions to prevent and improve falls across care settings.

Risk Factors

Falls are results of interactions between intrinsic and extrinsic factors. Intrinsic factors are associated with patients' physical and psychological conditions, while extrinsic factors are associated with environment such as room conditions, equipment, and medications. Table 8.1 shows examples of fall risk factors.

Morse (2008) classified falls into three categories: anticipated physiological falls, unanticipated physiological falls, and accidental falls. When patients who are indicated as fall-prone actually fall, this type of fall is an ***anticipated physiological fall***. For example, a patient may be flagged as fall-prone because he/she has some of the intrinsic factors shown in Table 8.1. If the fall of the patient is not prevented, it is an anticipated physiological fall. If a person with none of the risk factors falls because of other physiological factors such as a seizure and/or sudden faint, it is classified as an ***unanticipated physiological fall***. Finally, patients who are not

Table 8.1 Risk Factors for Falls

Intrinsic Factors	Extrinsic Factors
• Recent fall history	• Wet floor
• Mobility, balance function	• Low toilet seat
• Cognitive, psychological function	• High bed
• Comorbidities	• Unsafe, broken equipment
• Overall poor health status	• Medications that increase fall risk

indicated as fall-prone who happen to fall due to accidents are classified as an *accidental fall*. Extrinsic factors shown in Table 8.1 can cause accidental falls even if patients do not have any of the intrinsic risk factors.

Risk Mitigation Strategies

According to the research conducted by Morse (2008), anticipated physiological falls account for 78% of all falls in hospitals, while accidental falls make up 14% and unanticipated physiological falls make up 8%. In other words, as much as 92% of falls are preventable before the first fall. In this section, we discuss different risk mitigation strategies targeting the preventable falls.

Risk Factor Assessment and Screening

To reduce anticipated physiological falls, we first need to identify who is likely to fall. Patients who score high on risk assessment tools can be closely monitored and managed to avoid anticipated falls. There are many fall risk screening/assessment tools such as the Hendrich II Fall Risk Model™ developed by Hendrich, Bender, and Nyhuis (2003); the Morse Fall Scale (MFS) developed by Morse, Morse, and Tylko (1989); the Falls Risk Assessment Tool (FRAT) developed by Peninsula Health; and the Fall Risk Assessment & Screening Tool© (FRAST). The goal of this assessment is to ensure that each patient's risk for falls are identified and prevention strategies are incorporated into the care plan. The Joint Commission requires the risk evaluation as a National Patient Safety Goal for home care. The CDC (2015b proposed three questions that healthcare providers should ask as a routine part of their exam when they see older patients (+65 years):

- Have you fallen in the past year?
- Do you feel unsteady when standing or walking?
- Do you worry about falling?

The CDC recommends further assessment if patients answer "yes" to any of these key screening questions. The CDC also has created tools for fall reduction called Stopping Elderly Accidents, Deaths & Injuries (STEADI) that consist of screening and clinical decision support tools, tips for care providers, and educational materials for patients. The appropriate use of these assessment tools will result in the screening of more patients, and the incorporation of the risks into the patient's care plan will help prevent more falls in various care settings. This prevention will be beneficial for both patients and care providers in that unnecessary injuries and additional treatment can be avoided and resultant medical costs can be saved.

Individual Interventions

Once patients are identified as fall-prone, care providers can use various intervention strategies to monitor, manage, and improve the risk factors. Individual fall prevention interventions include:

- Colored identification bracelets;
- Fall prevention alarms;
- Physical therapy;
- Toileting regimen; and
- Medication modification.

The use of colored armband identification bracelets can help identify patients at high risk for falls, and the use of alarms notify staff when the patients who are vulnerable to falls leave their beds or wheelchairs without supervision (Figure 8.2). Physical therapy during the patient's toileting regimen can also help train patients to strengthen their general physical function and increase their awareness of the risk of falls.

System Redesign (Environmental Interventions)

It is important to ensure that the environment is optimally safe for patients because improvement of environmental risks can largely prevent both anticipated physiological and accidental falls. Commonly recommended approaches to removing environmental fall risk factors include:

- Handrails;
- Adjustment of bed height;
- Improved lighting;
- Surface contrast; and
- Clear floor space in the patient room and corridors.

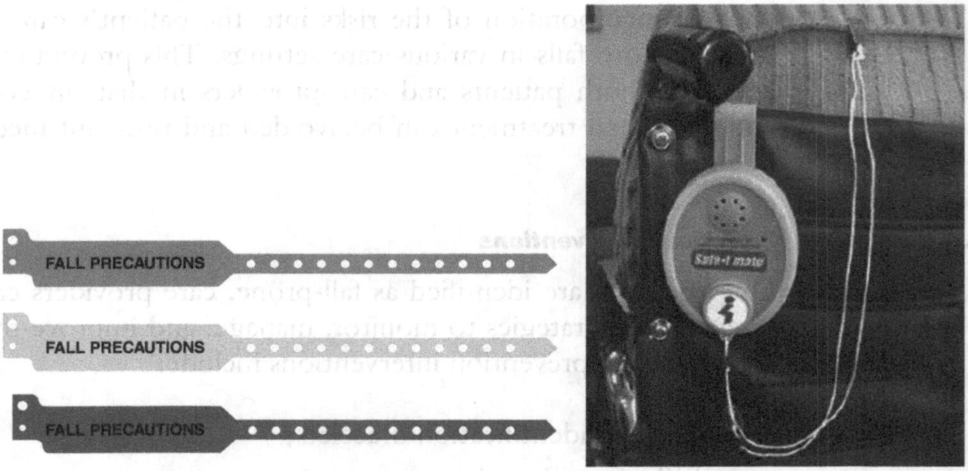

Figure 8.2 Left: Colored Identification Bracelet (http://www.pdchealthcare.com) Right: Fall Prevention Alarm (http://www.rehabmart.com/product/fall-prevention-monitor-260.html)

Handrails placed at various heights allow patients to keep their balance and provide leverage when changing positions. The use of adjustable-height and high-low beds enables patients to get in and out of beds easily. Lighting is a significant visual environmental factor. Improved lighting, including automatic motion-sensor lights, especially night lights, can reduce functional deficits associated with poor vision and help patients better navigate in their spaces. The use of contrast between floor and wall surfaces or between the toilet and the surrounding floor may help mitigate falls of patients who have reduced contrast-sensitivity function.

A universal fall prevention strategy has not been identified for patients in hospitals and long-term care facilities. Although many of these interventions have been implemented in practice, study results showed mixed outcomes. However, there is a general consensus that multifaceted and integrated strategies, along with multidisciplinary support, will be effective to avoid falls and associated injuries. Since nurses can play a crucial role in reduction in falls, nurse-led strategies should be encouraged. Also, further research is required to investigate both individual-level and system-level vulnerabilities and to identify and disseminate evidence-based best practices that can overcome barriers.

8.5 HUMAN FACTORS AND ERGONOMICS FOR PATIENT SAFETY

In the previous sections, we reviewed two major medical errors in healthcare, their main causes, and possible interventions. This section explains different components of a healthcare system that affect patient safety. Those components can be mainly classified into three categories: individual (care providers), technological, and organizational components (see Figure 8.3). We will highlight one or two components in each category and show various human factor and ergonomics approaches that can be used to improve those components.

Individuals

Healthcare delivery systems consist of interactions between healthcare workers and patients. The mental and physical status of healthcare workers influences the delivery of care, eventually affecting patient safety and outcomes. Job stress and burnout are two major common mental health problems associated with healthcare workers. Therefore, managing stress is of high importance and relevance when considering patient safety.

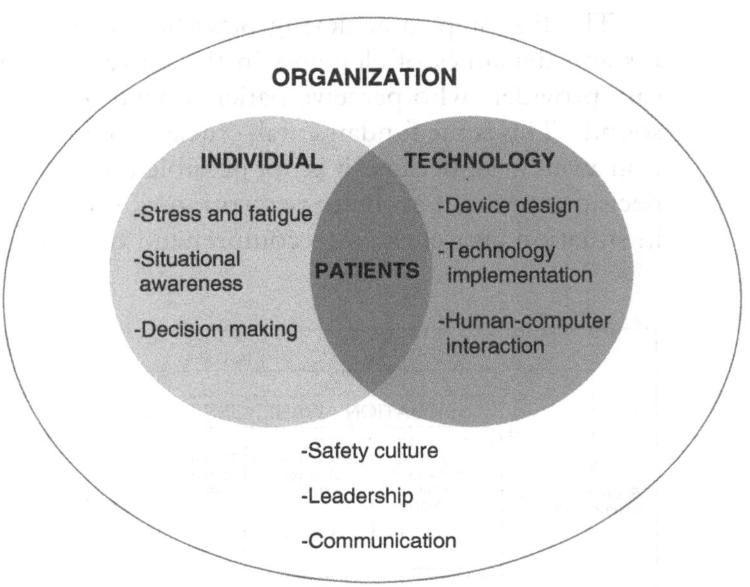

Figure 8.3 Three Categories of Healthcare System Components That Affect Patient Safety

Situation Awareness

Situation awareness refers to "an individual's perception of the elements in the environment within the volume of time and space, the comprehension of their meaning, and the projection of their status in the near future" (Endsley, 1995b). Workers' adequate situation awareness is especially important in the healthcare domain, where information flow and work processes are complex and poor decisions can lead to severe consequences for patients. Medical practitioners who are capable of performing situational assessments and who understand the complexities that could arise can predict risky situations ahead of time and prevent failures in the first place. Figure 8.4 represents a conceptual model of situation awareness.

The model shows the three hierarchical levels of situation awareness (Endsley, 1995b):

- Level 1: Perception
- Level 2: Comprehension
- Level 3: Projection

The first step in achieving *Situation awareness* is to perceive the status and dynamics of elements in the current situation. Examples include care providers who perceive patients' vital signs, symptoms, and monitor sounds. This is the fundamental, crucial step involved in formation of situation awareness, because it is not possible for care providers to make correct decisions without accurate perception of the key elements. The next step in situation awareness is to comprehend disjointed elements perceived in

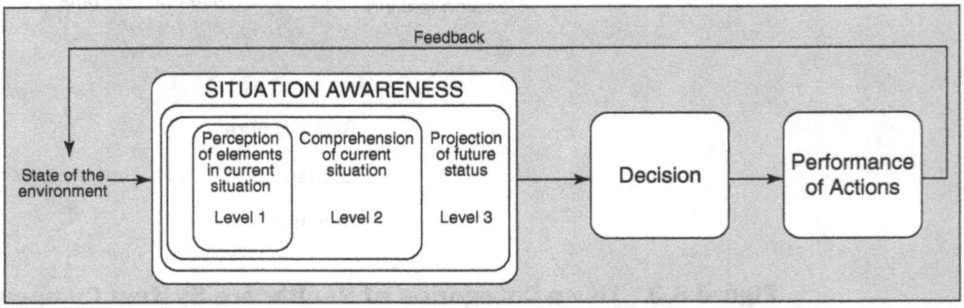

Figure 8.4 Conceptual Model of Situation Awareness
***Source:* Adapted from Wright (2004)**

Level 1 by deriving operationally relevant meaning and patterns and evaluating them. Level 2 situation awareness includes an understanding of the significance of the element pieces in light of one's goals. For example, upon seeing changes in a patient's vital signs, a physician has to integrate information from symptoms, charts, and nurses' reports, along with pertinent goals, and must quickly determine the seriousness of the problem. The final level of situation awareness involves the capability of projecting the future status of the elements in the environment. This level of situation awareness is accomplished by extrapolating information from knowledge accumulated through previous levels. For example, physicians think about various potential effects of a treatment including both positive and negative effects, given the patients' condition. Anticipating these effects allows the physicians to decide the most favorable care plan for better patient outcomes.

There are a number of techniques that can be used to measure situation awareness, and they can be classified into two major categories: subjective and objective measures. These techniques tend to be used in simulation environments.

The Situation Awareness Rating Technique (SART) developed by Taylor (1990) is one of the most well-known subjective measures. The SART, a self-rating technique, measures situation awareness of workers indirectly by using 10 dimensions, some of which capture impressions of workload, while others capture cognitive dimensions.

One of the standard objective instruments is the Situation Awareness Global Assessment Technique (SAGAT) (Endsley, 1995a; Endsley & Garland, 2000;). Here, participants are asked questions, in the middle of a dynamic simulation, about states in task processes and related circumstances. Results of SAGAT evaluations can be used to train medical/nursing students and care providers—both individual and team levels—for correct decision making. For example, the SAGAT was used to evaluate the performance of nursing students on timely appropriate recognition and response skills during an occurrence of sudden patient deterioration (Bogossian et al., 2014). The tool was also used for interprofessional obstetrical team training with high-fidelity simulation (Morgan et al., 2015). One study conducted by Crozier and colleagues (2015) developed Team SAGAT to assess situation awareness in multidisciplinary trauma teams using a human patient simulation.

Decision-Making Support

Many times, clinicians make medical decisions using informal reasoning based on their previous knowledge and experiences rather than using analytical models with specific data. That is, clinicians use cognitive skills of estimating probabilities and synthesizing information to make decisions in day-to-day practice. "Rules of thumb" or intuitive procedures to make a judgment or prediction are called *heuristics.* Using such heuristics for medical decision making is inevitable in our complex and dynamic healthcare environment, but its use can lead to medical errors due to cognitive biases and errors.

Decision making is a key skill for healthcare providers because clinical decision making has a crucial influence on patient safety in many ways. In particular, medical decisions affected by cognitive errors can lead to improper testing and missed diagnoses. For example, a quick diagnosis based on pattern recognition or disease prevalence for a specific patient population can lead to under- or overestimation of diseases. Such premature closure may result in physicians choosing to stop performing additional tests, and as a result, fail to consider other possible diagnosis. However, physicians can misdiagnose a disease by relying too much on test results that could be false positive.

Methods for Better Medical Decision Making

Medical decision making can be improved by understanding cognitive tasks that decision makers perform and by implementing systems that support decision-making processes.

Cognitive Task Analysis Cognitive task analysis (CTA) is a type of task analysis that identifies and analyzes the structure and the attributes of tasks that require cognitive activities from performers. The analysis also shows cognitive skills required for particular tasks. There are various CTA methods including applied cognitive task analysis (ACTA), the critical decision method (CDM), skill-based CTA framework, and the cognitive function model (CFM). Let's look at one of the methods.

First, the CDM is a retrospective interview method that involves an interviewee describing non-routine incidents that require expert judgment or key decision-making, and an interviewer who identifies the required decision making skills. This method has been used in healthcare to understand cognitive activities in critical care settings and to elicit knowledge from experts that novices can learn. For example, a study conducted by Crandall and Getchell-Reiter (1993) showed the effect of using the CDM on understanding the cognitive skills of nurses in a neonatal intensive care unit (NICU). The study indicated that the information, such as cues and exemplars, extracted from a CMD incident can help establish a guide to early sepsis assessment in the unit. A study conducted by Baxter, Monk, Tan, Dear, and Newell (2005) showed the use of the CTA to understand how staff make decisions about changing ventilator settings in a NICU, and explained how this information obtained from the CTA contributed to identifying factors that may aid in the successful utilization of a new medical expert system. In addition to the benefits shown in these examples, it is known that the CTA methods help examine the mental workload associated with complex controls, the development of mental models, and information requirements for decision making (Clark, Feldon, Merriënboer, Yates, & Early, 2007).

Clinical Decision Support System Care providers are often required to make healthcare decisions when there is not an obvious optimal treatment choice. As the amounts of health data increase while time for decision making is limited, it becomes more complex and challenging for care providers to make the optimal medical decisions. Clinical decision support systems (CDSSs) can help care providers address such challenges and make informed medical decisions by assisting in the cognitive process. In particular, CDSSs based on health information technology (HIT) play a significant role in facilitating the application of evidence-based medicine to patient care by allowing users to access data and clinical evidence more efficiently and effectively. CDSSs are not designed to replace care providers and make decisions regarding medical care. Rather, CDSSs reinforce their decision-making processes by supplementing the lack of information or cognitive biases.

HIT-based CDSSs can be classified into three categories (Musen, Middleton, & Greenes, 2014; Yu, 2015) as follows:

- Information management
- Situational awareness
- Patient specific

The first type of CDSS allows users to access medical literature or clinical evidence information as needed. For example, the CDSS presents information associated with various therapeutic choices that care providers can consider in the management of disease. The second type of CDSS enhances decision makers' situational awareness by alarming possible adverse events such as drug interactions. Such CDSSs also help promote care providers' clinical insights by effectively visualizing complex, high-volume data. This visualization tool can also help facilitate shared decision making by helping patients understand their health status more easily. The third type of CDSS provides patient-specific recommendations or interventions from diagnosis to treatment. Using data analytics and artificial intelligence algorithms, the CDSS integrates patient data into medical evidence and knowledge. Therefore, this kind of CDSS enables care providers to make the best individualized care decisions by taking into account the best scientific evidence available and estimating the potential harms and benefits of a screening, diagnosis, and treatment for specific patients.

Multiple-Criteria Decision Making Care providers encounter situations that require decision making in the presence of multiple, often conflicting criteria, such as maximizing patient satisfaction while minimizing costs. A multiple-criteria decision-making (MCDM) approach can be an effective method to support decision makers facing such problems. There are many different modeling methods that can be used based on the various problems and objectives. MCDM is discussed in more detail in Chapter 9.

Technology

In healthcare delivery systems, both care providers and patients consistently use technologies including medical devices, information technologies, and communication technologies. These technologies affect both patient flow and providers' workflow. To deliver safer care for patients, organizations should work to measure and control the potential problems associated with technologies and associated work environment.

Medical Device Design

The design of a medical device is critical because it affects users' efficiency, effectiveness, and safety, which eventually impact patient safety and outcomes. Systems engineering, especially human factors and ergonomics, seeks approaches to good design by applying knowledge about human capabilities.

Figure 8.5 shows a model using a systems engineering approach for good design that addresses the intended use of the device and provides safe medical care (Ward & Clarkson, 2011). Establishing a knowledge base is the first stage of the design process. This is a crucial stage for identifying patient safety problems and developing effective design requirements. A good understanding of the problem can be accomplished by identifying components of the system and their relationships, analyzing clinical tasks, and discerning the needs of users such as physicians. In the second stage, the knowledge base gained from the previous stage is translated into more specific requirements with respect to the managerial, technical, and regulatory needs for the device. To capture the device requirements, Shefelbine, Clarkson, Farmer, and Eason (2002) proposed an approach: "who,

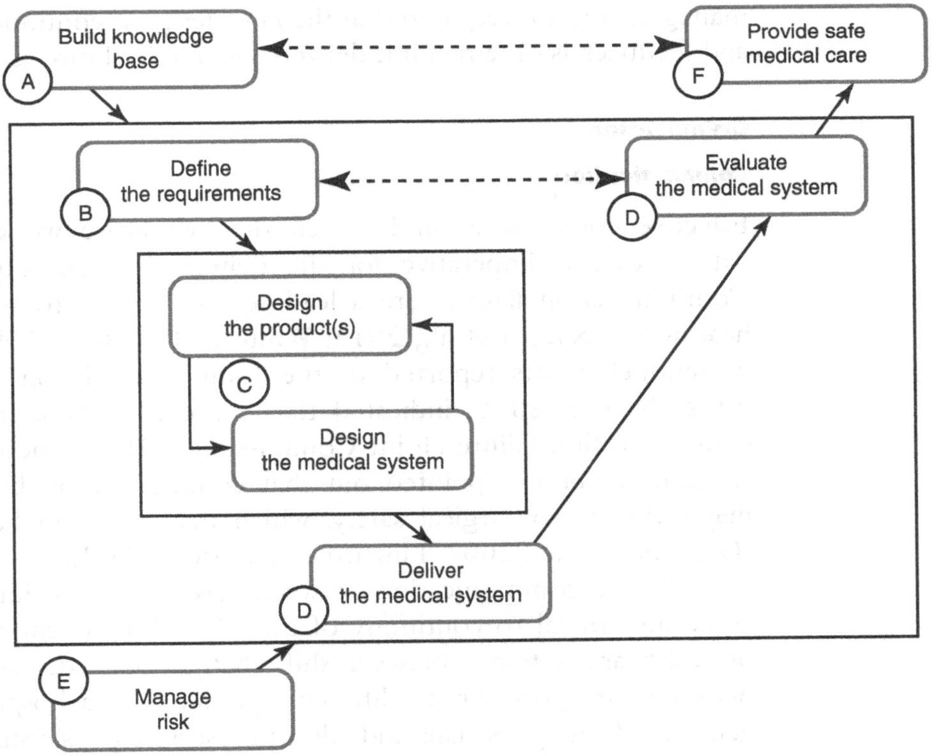

Figure 8.5 Framework for Device Design Based on Intended Use and Safety ***Source:*** **Ward and Clarkson (2011)**

what, where, when, and why of device use." In the third phase, the product and the system surrounding the device are designed. Since products are interrelated to the environment where the products are used, it is essential to look at them together, rather than designing products and systems independently. Effective and timely evaluation is critical for medical device design. Depending on the evaluation methods, there is a trade-off between accuracy and expenses (time and cost). Early evaluation can assist in saving the resources required for redesigning the products when errors are detected, while also increasing the spending of unnecessary costs in case of false alarms. Throughout the device design process, risk management should be conducted to identify potential sources of harm. Commonly used risk management techniques include *Fault Tree Analysis (FTA), Failure Mode and Effects Analysis (FEMA), and Probabilistic Risk Assessment (PRA)*. Those methods, separate or combined, can help identify potential risks and hazards that can treat patient safety. After evaluation and risk management, it is required that the new design is continuously monitored and feedback is obtained if it delivers safe medical care as intended.

Organization

Communication

Effective communication between the healthcare workers involved in patient care is imperative for the delivery of safe, high-quality care. Communication failures are a leading contributor to adverse events in healthcare (Nagpal et al., 2012; White & Del Rey, 2009). The analysis of sentinel events reported to the Joint Commission for three years, 2012 through 2014, indicated that more than 60% are the result of communication failure (Joint Commission, 2015). Studies that analyzed surgical errors also pointed out that communication breakdowns were major threats to surgical safety, which may result in harm to patients (Greenberg et al., 2007; Lingard et al., 2005, 2008).

Effective communication between care providers during handoffs is especially critical to continuity of care. Handoffs occur at various levels of healthcare systems—between shift changes (e.g., day and night shifts), between care providers in different specialties in a hospital (e.g., emergency medicine physicians and admitting service physicians), and between care settings (e.g., hospital and primary care). Improper interactions and communication between these transitions can lead to adverse patient outcomes, hospital readmissions, and high cost.

Communication between care providers and their patients is also important for safe, patient-centered care. If care providers do not completely or effectively communicate information to patients in a timely fashion, patients are likely to misunderstand care guidelines such as prescriptions, follow-up activities, and treatment options, which eventually can cause adverse events. The possibility of miscommunication increases if patients have language barriers or complex activity limitations (Chevarley, 2011).

Table 8.2 shows a number of individual and organizational factors contributing to communication failures in healthcare.

These approaches for overcoming individual barriers are important. However, when considering that communication is highly affected by organizational process, structure, and culture, it is necessary to make organizational changes for effective communication. First, care providers need communication training. Although face-to-face oral presentation is part of a care providers' vital skills, communication training for healthcare professionals has received less attention throughout the training process compared to other clinical tasks (Rajashree, 2011). For better communication capability of care professionals, it is necessary to incorporate communication skills into the academic curriculum and training programs.

Second, a standardized, structured communication method should be developed and implemented. This is especially important for communication during handoffs. Variability in the preparation and performance of patient handoffs causes fragmented communication between care providers

Table 8.2 Factors Associated with Communication Barriers

Individual Patient	Individual Care Provider	Organization
• Physical status • Linguistic competence • Limited English proficiency • Cultural background	• Cognitive workload • Communication skills • Personality differences	• Organizational culture (hierarchy, teamwork) • Information technology • Differences in language and jargon • Differences in accountability, payment, and rewards

and between different care settings. The use of standardized checklists or questions that guide care providers to include required information can reduce the variability and improve communication processes (Malekzadeh, Mazluom, Etezadi, & Tasseri, 2013). The SBAR (situation, background, assessment, and recommendation) is one of the best practices for standardized communication in healthcare. Many studies have shown the effectiveness of adopting the SBAR technique in practice such as decreased time to treatment, increased care providers' satisfaction with communication, higher rates of resolution of patient issues, and rapid decision making (Dingley, Daugherty, Derieg, & Persing, 2008; Randmaa, Mårtensson, Leo Swenne, & Engström, 2014; Vardaman et al., 2012).

Finally, for effective communication between care providers, it is fundamental to create an organizational culture in which individuals can speak up and address problems with their colleagues or with persons at higher levels. Also, it is required to develop an environment where healthcare professionals use standard terminology when sharing patient information and alerting team members to unsafe situations.

In summary, effective communication is the foundation of any healthcare team for safe, high-quality care. Good communication involves appropriate documenting and sharing patient information between care transitions. This is critical to prevent medical errors for patients who have complex medical conditions and who go through a series of care transitions between different care providers and care settings. To promote effective communication between patient and care providers and between care professionals, organizational changes are essential. Appropriate use of information technology can also facilitate communication.

QUESTIONS AND LEARNING ACTIVITIES

1. Human reliability analysis (HRA) has primarily been applied to intensive cognitive applications such as surgery. The Observational Clinical Human Reliability Assessment (OCHRA) has been applied to assess technical skill in laparoscopic cholecystectomy and other general surgical procedures. HRA was subsequently established as a validated system of objective surgical skill assessment (Joice, Hanna, & Cuschieri, 1998; Tang, Hanna, Bax, & Cuschieri, 2004; Tang, Hanna, Joice, & Cuschieri, 2004). Review the example of the Human Reliability Analysis of Cataract Surgery tool (Gauba, et al., 2008). How is the tool used to

identify the frequency and pattern of technical errors observed during phacoemulsification cataract extraction by surgeons with varying levels of experience?

2. Determine a fault tree for a patient monitoring system in case the wrong medication is administered.

REFERENCES

Adams, P., Kirzinger, W., & Martinez, M. (2013). Summary health statistics for the U.S. population: National Health Interview Survey. *Vital and Health Statistics*, *10*(259). Retrieved from http://www.cdc.gov/mmwr/preview/mmwrhtml/mm6329a8.htm

Agency for Healthcare Research & Quality (AHRQ). (2013). Acute care prevention of falls: Rate of inpatient falls per 1,000 patient days. Retrieved from http://www.qualitymeasures.ahrq.gov/content.aspx?id=36944

Andel, C., Davidow, S. L., Hollander, M., & Moreno, D. A. (2012). The economics of health care quality and medical errors. *Journal of Health Care Finance*, *39*(1), 39–50. Retrieved from http://www.ncbi.nlm.nih.gov/pubmed/23155743

Baxter, G. D., Monk, A. F., Tan, K., Dear, P. R. F., & Newell, S. J. (2005). Using cognitive task analysis to facilitate the integration of decision support systems into the neonatal intensive care unit. *Artificial Intelligence in Medicine*, *35*(3), 243–257. doi:10.1016/j.artmed.2005.01.004

Benjamin, D. M. (2003). Reducing medication errors and increasing patient safety: Case studies in clinical pharmacology. *Journal of Clinical Pharmacology*, *43*(7), 768–783. Retrieved from http://www.ncbi.nlm.nih.gov/pubmed/12856392

Bogossian, F., Cooper, S., Cant, R., Beauchamp, A., Porter, J., Kain, V., … Phillips, N. M. (2014). Undergraduate nursing students' performance in recognising and responding to sudden patient deterioration in high psychological fidelity simulated environments: An Australian multi-centre study. *Nurse Education Today*, *34*(5), 691–696. doi:10.1016/j.nedt.2013.09.015

Brennan, T. A., Leape, L. L., Laird, N. M., Hebert, L., Localio, A. R., Lawthers, A. G., … Hiatt, H. H. (1991). Incidence of adverse events and negligence in hospitalized patients. Results of the Harvard Medical Practice Study I. *The New England Journal of Medicine*, *324*(6), 370–376. doi:10.1056/NEJM199102073240604

Budnitz, D. S., Pollock, D. A., Weidenbach, K. N., Mendelsohn, A. B., Schroeder, T. J., & Annest, J. L. (2006). National surveillance of emergency department visits for outpatient adverse drug events. *Journal of the American Medical Association*, *296*, 1858–1866.

Centers for Disease Control and Prevention (CDC). (2013). National Center for Injury Prevention and Control. Web–based Injury Statistics Query and Reporting System (WISQARS). Retrieved August 15, 2013, from http://www.cdc.gov/HomeandRecreationalSafety/Falls/adultfalls.html

Centers for Disease Control and Prevention (CDC). (2015a). Falls in nursing homes. Retrieved April 22, 2015, from http://www.cdc.gov/HomeandRecreationalSafety/Falls/nursing.html

Centers for Disease Control and Prevention (CDC). (2015b). Make STEADI part of your medical practice. Retrieved April 23, 2015, from http://www.cdc.gov/homeandrecreationalsafety/Falls/steadi/index.html

Chevarley, F. M. (2011). Patient-provider communication by race/ethnicity and disability status: United States, 2007. *Statistical Brief #312*. Rockville, MD. Retrieved from http://www.meps.ahrq.gov/mepsweb/data_files/publications/st312/stat312.shtml

Clark, R. E., Feldon, D. F., Van Merriënboer, J., Yates, K. A., & Early, S. (2007). Cognitive task analysis. In J. M. Spector, M. D. Merrill, J. Van Merrienboer, & M. P. Driscoll (Eds.), *Handbook of research on educational communications and technology* (pp. 577–594). New York, NY: Lawrence Erlbaum.

Crandall, B., & Getchell-Reiter, K. (1993). Critical decision method: A technique for eliciting concrete assessment indicators from the intuition of NICU nurses. *Advances in Nursing Science*, *16*(1), 42–51. Retrieved from http://www.ncbi.nlm.nih.gov/pubmed/8311424

Crozier, M. S., Ting, H. Y., Boone, D. C., O'Regan, N. B., Bandrauk, N., Furey, A., … Hogan, M. P. (2015). Use of human patient simulation and validation of the Team Situation Awareness Global Assessment Technique (TSAGAT): A multidisciplinary team assessment tool in trauma education. *Journal of Surgical Education*, *72*(1), 156–163. doi:10.1016/j.jsurg.2014.07.009

Currie, L. (2008). Fall and injury prevention. In R. G. Hughes (Ed.), *Patient safety and quality: An evidence-based handbook for nurses, Chapter 10*. Rockville, MD: Agency for Healthcare Research and Quality. Retrieved from http://www.ncbi.nlm.nih.gov/books/NBK2653/

Dingley, C., Daugherty, K., Derieg, M. K., & Persing, R. (2008). *Improving patient safety through provider communication strategy enhancements*. Rockville, MD: Agency for Healthcare Research and Quality. Retrieved from http://www.ncbi.nlm.nih.gov/books/NBK43663/

Ehsani, S. R., Cheraghi, M. A., Nejati, A., Salari, A., Esmaeilpoor, A. H., & Nejad, E. M. (2013). Medication errors of nurses in the emergency department. *Journal of Medical Ethics and History of Medicine*, *6*, 11. Retrieved from http://www.pubmedcentral.nih.gov/articlerender.fcgi?artid=3885144&tool=pmcentrez&rendertype=abstract

Elder, N. C., & Dovey, S. M. (2002). Classification of medical errors and preventable adverse events in primary care: A synthesis of the literature. *Journal of Family Practice*, *51*(11), 927–932.

Endsley, M. R. (1995a). Measurement of situation awareness in dynamic systems. *Human Factors*, *37*, 65–84.

Endsley, M. R. (1995b). Toward a theory of situation awareness in dynamic systems. *Human Factors*, *37*(1), 32–64.

Endsley, M. R., & Garland, D. J. (Eds.). (2000). *Situation awareness analysis and measurement*. Boca Raton, FL: CRC Press.

Gauba, V., Tsangaris, P., Tossounis, C., Mitra, A., McLean, C., & Saleh, G. M. (2008). Human reliability analysis of cataract surgery. *Archives of ophthalmology*, *126*(2), 173–177.

Global, regional, and national age–sex specific all-cause and cause-specific mortality for 240 causes of death, 1990–2013: A systematic analysis for the Global Burden of Disease Study 2013. (2015). *The Lancet*, *385*(9963), 117–171. doi:10.1016/S0140-6736(14)61682-2

Greenberg, C. C., Regenbogen, S. E., Studdert, D. M., Lipsitz, S. R., Rogers, S. O., Zinner, M. J., & Gawande, A. A. (2007). Patterns of communication breakdowns resulting in injury to surgical patients. *Journal of the American College of Surgeons*, *204*(4), 533–540. doi:10.1016/j.jamcollsurg.2007.01.010

Grober, E. D., & Bohnen, J. M. A. (2005). Defining medical error. *Canadian Journal of Surgery*, *48*(1), 39–44. Retrieved from http://www.pubmedcentral.nih.gov/articlerender.fcgi?artid=3211566&tool=pmcentrez&rendertype=abstract

Hendrich, A. L., Bender, P. S., & Nyhuis, A. (2003). Validation of the Hendrich II Fall Risk Model: A large concurrent case/control study of hospitalized patients. *Applied Nursing Research*, *16*(1), 9–21. doi:10.1053/apnr.2003.YAPNR2

Hollnagel, E. (1998). *Cognitive reliability and error analysis method: CREAM*. Oxford, UK: Elsevier Science, Inc. ISBN 0-08042-848-7

The Associated Press (2003, January 19). Hospital apologizes for surgical mistake. *The New York TImes.* Retrieved from http://www.nytimes.com/2003/01/19/national/19SURG.html

Institute of Medicine (IOM). (2000). *To err is human: Building a safer health system*. L. T. Kohn, J. M. Corrigan, & M. S. Donaldson (Eds.). Washington, DC: National Academies Press.

Institute of Medicine. (2006). *Committee on identifying and preventing medication errors: Preventing medication errors*. Washington, DC: National Academies Press.

James, J. T. (2013). A new, evidence-based estimate of patient harms associated with hospital care. *Journal of Patient Safety*, *9*(3), 122–128. Retrieved from http://journals.lww.com/journalpatientsafety/Fulltext/2013/09000/A_New,_Evidence_based_Estimate_of_Patient_Harms.2.aspx?TB_iframe=true&width=288&height=432

Joice, P., Hanna, G. B., & Cuschieri, A. (1998). Errors enacted during endoscopic surgery—a human reliability analysis. *Applied ergonomics*, *29*(6), 409–414.

Joint Commission. (2015). *Sentinel event data: Root causes by event type*. Retrieved from http://www.jointcommission.org/Sentinel_Event_Statistics/

Leapfrog Group. (2013). Hospital errors are the third leading cause of death in U.S., and new hospital safety scores show improvements are too slow. *Hospital Safety Score*. Retrieved April 20, 2015, from http://www.hospitalsafetyscore.org/newsroom/display/hospitalerrors-thirdleading-causeofdeathinus-improvementstooslow

Lingard, L., Espin, S., Rubin, B., Whyte, S., Colmenares, M., Baker, G. R., … Reznick, R. (2005). Getting teams to talk: Development and pilot implementation of a checklist to promote interprofessional communication in the OR. *Quality & Safety in Healthcare*, *14*(5), 340–346. doi:10.1136/qshc.2004.012377

Lingard, L., Regehr, G., Orser, B., Reznick, R., Baker, G. R., Doran, D., ... Whyte, S. (2008). Evaluation of a preoperative checklist and team briefing among surgeons, nurses, and anesthesiologists to reduce failures in communication. *Archives of Surgery*, *143*(1), 12–17, 18. doi:10.1001/archsurg.2007.21

Malekzadeh, J., Mazluom, S. R., Etezadi, T., & Tasseri, A. (2013). A standardized shift handover protocol: Improving nurses' safe practice in intensive care units. *Journal of Caring Sciences*, *2*(3), 177–185. doi:10.5681/jcs.2013.022

Morgan, P., Tregunno, D., Brydges, R., Pittini, R., Tarshis, J., Kurrek, M., ... Ryzynski, A. (2015). Using a situational awareness global assessment technique for interprofessional obstetrical team training with high fidelity simulation. *Journal of Interprofessional Care*, *29*(1), 13–19. doi:10.3109/13561820.2014.936371

Morse, J. (2008). *Preventing patient falls: Establishing a fall intervention program (Vol 2.)*. New York, NY: Springer.

Morse, J., Morse, R., & Tylko, S. (1989). Development of a scale toidentify the fall-prone patient. *Canadian Journal on Aging*, *8*(4), 366–377. Retrieved from http://onlinelibrary.wiley.com/doi/10.1111/j.1440-172X.2006.00573.x/epdf

Mosleh, A., & Chang, Y. H. (2004). Model-based human reliability analysis: prospects and requirements. *Reliability Engineering & System Safety*, *83*(2), 241–253. doi:10.1016/j.ress.2003.09.014

Musen, M. A., Middleton, B., & Greenes, R. A. (2014). Clinical decision-support systems. In *Biomedical informatics* (pp. 643–674). London, England: Springer.

Nagpal, K., Arora, S., Vats, A., Wong, H. W., Sevdalis, N., Vincent, C., & Moorthy, K. (2012). Failures in communication and information transfer across the surgical care pathway: interview study. *BMJ Quality & Safety*, *21*(10), 843–849. doi:10.1136/bmjqs-2012-000886

Parry, G. W. (1995). Suggestions for an improved HRA method for use in probabilistic safety assessment. *Reliability Engineering & System Safety*, *49*(1), 1–12. doi:10.1016/0951-8320(95)00034-Y

Rajashree, K. (2011). Training programs in communication skills for health care professionals and volunteers. *Indian Journal of Palliative Care*, *17*(Suppl), S12–S13. doi:10.4103/0973-1075.76232

Randmaa, M., Mårtensson, G., Leo Swenne, C., & Engström, M. (2014). SBAR improves communication and safety climate and decreases incident reports due to communication errors in an anaesthetic clinic: a prospective intervention study. *BMJ Open*, *4*(1), e004268. doi:10.1136/bmjopen-2013-004268

Reason, J. T. (1990). *Human error*. Cambridge, England: Cambridge University Press.

Reason, J. T. (1997). *Managing the risks of organizational accidents*. Burlington, VT: Ashgate.

Shefelbine, S., Clarkson, J., Farmer, R., & Eason, S. (2002). *Good design practice for medical devices and equipment—requirements capture*. Cambridge Engineering Design Centre.

Stevens, J. A., Corso, P. S., Finkelstein, E. A., & Miller, T. R. (2006). The costs of fatal and non-fatal falls among older adults. *Injury Prevention*, *12*(5), 290–295. doi:10.1136/ip.2005.011015

Swain, A. D., & Guttmann, H. E. (1983). NUREG/CR-1278. *Handbook of Human Reliability Analysis with Emphasis on Nuclear Power Plant Applications, US Nuclear Regulatory Commission.*

Swain, A. D. (1990). Human reliability analysis: Need, status, trends and limitations. *Reliability Engineering & System Safety*, *29*(3), 301–313. doi:10.1016/0951-8320(90)90013-D

Tang, B., Hanna, G. B., Bax, N. M. A., & Cuschieri, A. (2004). Analysis of technical surgical errors during initial experience of laparoscopic pyloromyotomy by a group of Dutch pediatric surgeons. *Surgical Endoscopy And Other Interventional Techniques*, *18*(12), 1716–1720.

Tang, B., Hanna, G. B., Joice, P., & Cuschieri, A. (2004). Identification and categorization of technical errors by Observational Clinical Human Reliability Assessment (OCHRA) during laparoscopic cholecystectomy. *Archives of Surgery*, *139*(11), 1215–1220.

Taylor, R. M. (1990). Situation awareness rating technique (SART): The development of a tool for aircrew systems design. In *Situational Awareness in Aerospace Operations.* Chapter 3). [Neuilly sur-Seine is the name of the place (commune)] Neuilly sur-Seine ,France: NATO-AGARD-CP-478.

Van Den Bos, J., Rustagi, K., Gray, T., Halford, M., Ziemkiewicz, E., & Shreve, J. (2011). The $17.1 billion problem: The annual cost of measurable medical errors. *Health Affairs*, *30*(4), 596–603. doi:10.1377/hlthaff.2011.0084

Vardaman, J. M., Cornell, P., Gondo, M. B., Amis, J. M., Townsend-Gervis, M., & Thetford, C. (2012) Beyond communication:

the role of standardized protocols in a changing health care environment. *Health Care Management Review*, *37*(1), 88–97. doi:10.1097/HMR.0b013e31821fa503 [its official journal name is Health Care Management Review—space between Health and Care]

Vaughn-Cooke, M. (2012). *A multidimensional information system for human reliability assessment: Applied to patient adherence*. University Park: Pennsylvania State University Press.

Vaughn-Cooke, M., Nembhard, H. B., Ulbrect, J., and Gabbay, R. (2015). Informing Patient Self-Management Technology Design Using a Patient Adherence Error Classification. *Engineering Management Journal*, *27*(3), 124–130. DOI: 10.1080/10429247.2015.1061889

Ward, J., & Clarkson, J. (2011). Human factors engineering and the design of medical devices. In P. Carayon (Ed.), *Handbook of human factors and ergonomics in healthcare and patient safety* (2nd ed.). Boca Raton, FL: CRS Press.

White, C., & Del Rey, J. G. (2009). Decreasing adverse events through night talks: An interdisciplinary, hospital-based quality improvement project. *Permanente Journal*, *13*(4), 16–22. Retrieved from http://www.pubmedcentral.nih.gov/articlerender.fcgi?artid=2911826&tool=pmcentrez&rendertype=abstract

Wright, M. C. (2004). Objective measures of situation awareness in a simulated medical environment. *Quality and Safety in Health Care*, *13*(Suppl 1), i65–i71. doi:10.1136/qshc.2004.009951

Yu, P. P. (2015). Knowledge bases, clinical decision support systems, and rapid learning in oncology. *Journal of Oncology Practice/American Society of Clinical Oncology*, *11*(2), e206–11. doi:10.1200/JOP.2014.000620

Chapter 9

Health Analytics

"If we have data, let's look at data. If all we have are opinions, let's go with mine."

—Jim Barksdale, former Netscape CEO[1]

Overview

Clinical data hold the potential to help transform the U.S. healthcare system. By providing greater insight to patients, providers, and policy makers into the appropriate application of interventions, and quality and costs of care, these data offer the opportunity to accelerate progress on the six dimensions of quality care—safe, effective, patient centered, timely, efficient, and equitable (Chaudhry, 2006). Rapid advancements in digitizing, integrating, and exchanging health information have changed today's healthcare environment, giving workers access to a wealth of important data, thereby assisting them in the practice of making more informed decisions. This chapter explores the mining, visualization, and analysis of healthcare data, and examines ways such data can be managed efficiently in order to positively affect an organization's ability to generate revenue, control costs, and mitigate risks.

9.1 DATA MINING

Organizations have been rapidly advancing in their ability to collect and store vast amounts of data and have used *Data mining* extensively to establish a range of capabilities. Manufacturers have leveraged data mining for quality control, financial institutions for credit scoring, and marketers for direct marketing. In healthcare, data mining has become increasingly

[1] Found at http://blog.datamarket.com/2012/07/08/the-11-best-data-quotes/

popular and useful. Data mining can help patients get tailored health alerts, help physicians detect disease earlier, and help insurers reduce fraud and abuse.

Definition and Methodologies

Data mining is the process of automatically extracting useful information from large data repositories. The field of data mining has come together as a result of researchers from disciplines focusing on the development of efficient and scalable tools that could handle large and diverse types of data. As such, it builds on ideas from many fields including statistics, computer science, database management, artificial intelligence, machine learning, signal processing, and information retrieval. **Knowledge Discovery in Databases** (KDD) was one of the early methodologies to encompass data mining and connect it with other elements (Fayyad, Piatetsky-Shapiro, & Smyth, 1996). As shown in Figure 9.1, the KDD methodology is defined with data mining as a stage between data transformation and interpretation/evaluation. Now there are many variations on this theme, including the Cross Industry Standard Process for Data Mining (CRISP-DM), which defines six stages: business understanding, data understanding, data preparation, modeling, evaluation, and deployment (Shearer, 2000). Here, we briefly describe the basic concepts

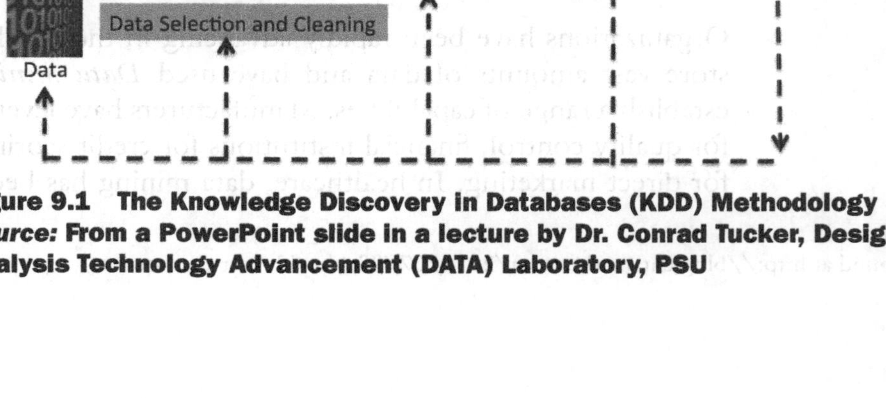

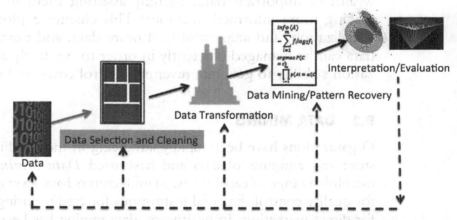

Figure 9.1 The Knowledge Discovery in Databases (KDD) Methodology
Source: **From a PowerPoint slide in a lecture by Dr. Conrad Tucker, Design Analysis Technology Advancement (DATA) Laboratory, PSU**

involved in the pre-processing stages, the actual data-mining stage, and the interpretation/evaluation stage.

Preprocessing

Data collection, preparation, selection and cleaning are not a part of the data-mining stage, but do belong to the KDD methodology as additional steps. The target data set must be large enough to contain patterns that can be discovered through data mining while being small enough to be mined within an acceptable time limit. Data sets differ in important ways, driving what types of data-mining tools and techniques can be used to analyze the data. For example, in addition to being qualitative or quantitative attributes, data objects may have special characteristics or relationships such as time dependency that need to be incorporated into the analysis.

Most data have some level of imperfection, and improving data quality typically improves the resulting analysis. Data cleaning addresses quality issues such as the presence of noise and outliers; missing, inconsistent, or duplicate data; and data that are biased or unrepresentative of the population or phenomenon that the data are supposed to describe (Tan et al., 2006, p. 19).

Data Mining

The actual data-mining stage commonly involves the following types of tasks (Tan et al., 2006):

- **Classification:** The task of assigning data objects to one of several predefined categories. Examples include categorizing email messages as "legitimate" or "spam" based on the message header and content, or categorizing cells as malignant or benign based on a medical imaging scan. The aim of a classification model is to build a model of a discrete target variable as a function of the explanatory variable, by mapping each attribute set x to one of the predefined class labels y. The goal of the classification task is to learn a model that minimizes the error between the predicted and true values of the target variable.
- **Regression:** Has the same goal of model building with error minimization, but is used for continuous variables. For example, forecasting the future price of a stock is a regression task because price is a

continuous variable. Both the classification and regression tasks are carried as a component of predictive modeling, which seeks to predict the value of a particular attribute based on the values of other attributes.

- **Association analysis:** The task of discovering patterns of relationships between variables. The discovered patterns are typically represented in the form of rules, hence some users refer to this task as association rule learning or dependency modeling. For example, analysis of point-of-sale data of a grocery story may reveal the rule {Diapers} → {Milk}, which suggests that customers who buy diapers also tend to buy milk. This rule may help identify other opportunities to cross-sell related items.
- **Cluster analysis:** The task of finding groups of closely related observations, without using known structures in the data, so that observations that belong to the same cluster are more similar to each other than observations that belong to other clusters. For example, analysis of a collection of news articles can identify groups based on the words contained in the article and the number of times the word appears in the article as shown in Table 9.1. A good clustering algorithm should be able to identify two clusters based on the similarity of the words that appear in the articles: the first cluster of the first three articles provides news about the economy, while the second cluster of the last three articles provides news about healthcare.
- **Anomaly detection:** The task of identifying unusual data records whose characteristics are significantly different from the rest of the data. These records, referred to as anomalies or outliers, might be interesting in a positive way or might indicate data errors.

Table 9.1 Word-Frequency Analysis of News Articles

Article	Words
1	(dollar: 1), (country: 2), (loan: 3), (deal: 2),
2	(machinery: 2), (labor: 3), (market: 4)
3	(job: 5), (inflation: 3), (rise: 2), (jobless: 2)
4	(patient: 4), (symptom: 2) (drug: 3), (health: 2)
5	(pharmaceutical: 2), (vaccine: 1), (flu:3)
6	(death: 2), (cancer: 4), (drug: 3)

Interpretation/Evaluation

A sufficiently exhaustive search of a large set of data will result in patterns of some kind. The trouble, however, is that many of these "patterns" may represent just a fleeting temporary or local effect. The final step of the KDD methodology is to verify that the patterns produced by the data-mining algorithms actually model the underlying structure, which is responsible for consistent and replicable patterns.

Not all patterns found by the data-mining algorithms are necessarily valid. It is common for the data-mining algorithms to find patterns in the training set that are not present in the general data set. This is called *overfitting*. To overcome this, the evaluation uses a test set of data on which the data-mining algorithm was not trained. The learned patterns are applied to this test set, and the resulting output is compared to the desired output. For example, a data-mining algorithm trying to distinguish "spam" from "legitimate" e-mails would be trained on a training set of sample e-mails. Once trained, the learned patterns would be applied to the test set of e-mails on which it had *not* been trained. The accuracy of the patterns can then be measured from how many e-mails they correctly classify. A number of statistical methods may be used to evaluate the algorithm, such as *Receiver Operating Characteristic (ROC)* curves.

If the learned patterns do not meet the desired standards, subsequently it is necessary to reevaluate and change the preprocessing and data-mining steps. If the learned patterns do meet the desired standards, then the final step is to interpret the learned patterns and turn them into knowledge.

Medical Data Mining

The quantity of data collected for healthcare purposes continues to expand rapidly, due to the electronic health record (EHR) and increasing number of studies seeking internal review board (IRB) approval.

Medical data mining can help identify healthcare-related behaviors, discover patient patterns and trends, improve the quality of provider service, achieve better patient satisfaction, enhance medical equipment utility, and reduce the cost of providing healthcare.

A few examples of data mining in the medical and health field are outlined as follows.

Mining Claims Data

An insurer could examine cases of allergies and asthma to find the most effective treatment regime. The data involved in the analysis could involve lists of drugs (e.g., antihistamines and corticosteroids), holistic treatments (such as taking a spoonful of honey every day), and pollution levels. Data mining could potentially indicate which courses of action prove effective (Milley, 2000).

Mining Clinical Trial Data The process of looking for data that can estimate the feasibility of a clinical trial (Zhu and Davidson, 2007).

Mining Adverse Drug Reaction Incidents "In adverse drug reaction surveillance, the Uppsala Monitoring Centre has, since 1998, used data-mining methods to routinely screen for reporting patterns indicative of emerging drug safety issues in the WHO global database of 4.6 million suspected adverse drug reaction incidents." (Bate et al., 1998).

Mining Healthcare Market Segments Healthcare market segmentation offers insights into healthcare consumers' behaviors and attitudes, which is critical information in an environment where healthcare is moving rapidly toward patient-centered care that is premised upon individuals becoming more active participants in managing their health. Personalized healthcare considers patient data from the electronic medical record to help diagnose diseases, predict their onset, and suggest models for innovative healthcare delivery systems that better utilize resources to treat patients while improving health promotion in the community. Although every patient is unique, there are commonalities among patient characteristics (clinical, diagnostic, demographic, etc.) that can be leveraged through data-mining methods to improve health promotion.

For example, one can use data-mining methods (e.g., clustering) to explore patient data from electronic medical records at a hospital or other healthcare organizations, to identify different healthcare market segments in a community so that tailored marketing strategies can be designed and deployed to target these unique patient groups to improve health promotion. Further data-mining methods can be employed to build predictive models (e.g., classification/regression trees) using these segments to effectively reach each type of patient in the population and propel them forward into greater engagement and self-management.

CASE STUDY 9A: PREDICTING PARKINSON'S DISEASE USING DATA MINING[2]

Parkinson's disease (PD) is the second most common neurological disorder after Alzheimer's disease. Key clinical features of PD are motor related, especially during the early stages. The movement dysfunctions are typically assessed by healthcare providers based on qualitative visual inspection of movement/gait/postural and clinical testing procedures that can vary among human raters. More advanced diagnosis techniques such as those using computed tomography scans to measure brain function can be cost prohibitive and expose patients to radiation and other harmful effects.

This case looks at how data mining can be used to identify novel gait patterns that can be used to model and predict the presence (or lack therefore) of PD movement abnormalities. Tucker et al. (2013) provide a full methodology, driven by data mining, to implement the approach in nonclinical environments.

DATA ACQUISITION AND PREPROCESSING

A Microsoft Kinect system is used capture human gait data. While multimodal sensors such as the Kinect are capable of capturing multiple data streams (VGA, depth using infrared sensors), only the 3D coordinate skeletal data (XYZ coordinate data of each node) is required for this study. Figure 9.2 shows the 20 nodes that are collected. As a patient walks through space, the coordinate data for each of these nodes is captured and stored in a structured table similar to the one in the figure.

Once these raw data are captured, irrelevant/noisy data from are removed, leaving only the XYZ data needed for the data-mining stage. The velocity and acceleration of each single node based on the original XYZ position data of each of the 20 nodes is also calculated. Furthermore, the ratio in each dimension from each pair of nodes in the position, velocity, and acceleration data are also generated in order to normalize variations in human characteristics (size, height, weight) that may cause biases in the resulting data-mining predictive models.

DATA MINING AND KNOWLEDGE DISCOVERY

The aim of the data-mining stage is to develop a model that quantifies novel feature combinations that are relevant to predicting the class variable. A function $f(X) = Y$ is needed that would map the selected features $X = \{x_1, x_2, \ldots, x_k\}$ to the class variable (Y), where k is the total number of features relevant to the class variable Y. Here, the class variable is binary,

[2]Based on Tucker, Han et al. (2015).

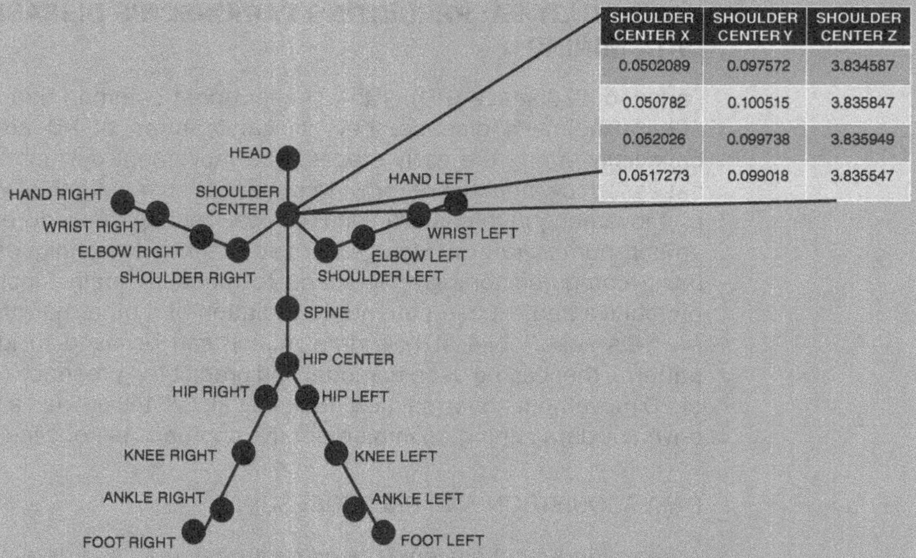

Figure 9.2 Skeletal Image of a Person with 20 Nodes and Example Data Table for the Shoulder Center Node

that is, separating PD patients from controls. There are several data-mining algorithms that have been employed to solve binary classification problems:

Binary logistic regression is a common classification model applied to the likelihood of one observation falling into one of the two categories. That is, if a case is selected randomly, then we could know the probability of which group or category it will belong. In this case, the category would be if someone would be classified as PD patient or control, based on a combination of PD features.

Support vector machines (SVMs) are a relatively new branch of data-mining classification technique that aim to maximize the linear boundary in logistic regression modeling. SVMs attempt to construct a decision boundary based on a kernel function, which could optimally separate two classes by maximizing the geometric margin between data points in two classes. In practice, since data cannot be linearly separated in some cases, SVMs are also able to transform the current data to a higher dimensional space and construct the decision boundary.

The ***C4.5*** decision tree classifier is a supervised machine learning algorithm that iteratively tests each feature for its ability to reduce uncertainty (randomness) in the *class/output variable* (Witten et al., 2011). The methodology comprises three steps: best feature evaluation, splitting point selection, and model training and evaluation. The feature evaluation step of the methodology attempts to select the most informative node in each subset of the training data set (the whole training data set for the root node selection) based on the maximum value of gain. The splitting point selection of the methodology attempts to decide the best numerical split point, which has the minimum misclassification error. The last step of the methodology attempts to train the resulting tree model, based on predefined stopping criteria, and evaluate its predictive performance.

The **Random Forest (RF)** model is an effective classifier that capitalizes on flexible fitting procedures, thereby potentially improving on accuracies, compared to traditional tree-based models. The general procedure of RF is as follows. First, it samples M cases randomly (M is the predetermined sample size) with replacement, which is the training data set for each tree. Second, an N variable sample would also be created (N is the variable sample size) to help split in each node, which is similar in the single tree construction. Finally, each tree is grown to the largest extent and the maximum voted class in the forest is accepted as the final decision (Witten et al., 2011).

The classifier **IBK** is another type of the instance-based learner. In the IBK classifier, a normalized distance is applied so that each feature has the same impact on the distance measure, and each observation would be classified in the most similar class assigned to the majority of its K neighbors (K is the number of nearest neighbors in prediction model). For example, if $K = 1$, then each new observation would be classified in the class by its nearest neighbors.

Once the data acquisition and preprocessing steps are complete, the Waikato Environment for Knowledge Analysis (WEKA) software (Witten et al., 2011) is used to discover novel, previously unknown knowledge pertaining to patient gait by employing the aforementioned data-mining/machine learning algorithms. In doing so, the proposed methodology is able to model and predict PD gait patterns based on the selected features presented in the previous subsection. Furthermore, the researchers are also able to evaluate performances of different algorithms in order to select the best classifier/model for healthcare decision support and obtain more insights in PD diagnosis. In addition, the 10-fold cross validation is applied here to help measure the model performances. In this case study, the PD-OFF data set and CONTROL data set are used. The details about evaluation results are shown as follows.

Results

Figure 9.3 presents two frames showing the experiment in progress. As the patient walks toward the sensor hardware (left side of Figure 9.3), depth sensor data are collected, with the corresponding skeletal data (right side of

Figure 9.3) used for the actual data-mining pattern discovery. After a four- to six-second run, the data-mining algorithms classify a subject as either exhibiting symptoms of PD (determined by the relevant features in the data-mining model) or a control group. Table 9.2 shows the performance of the data mining classification algorithms. This work is extended in Tucker, Behoora et al. (2015).

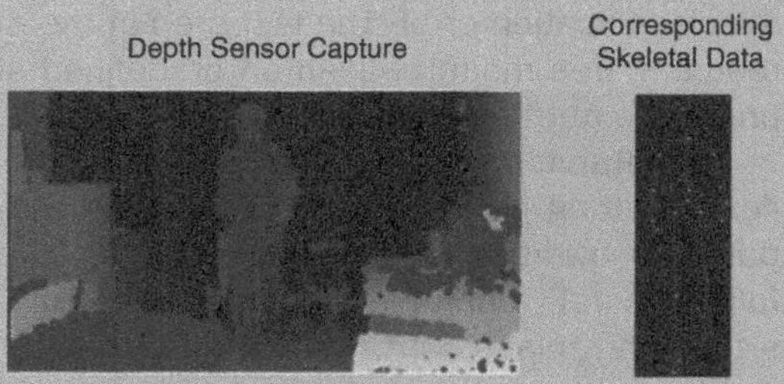

Figure 9.3 Dynamic PD Prediction

Table 9.2 Performance of the Data-Mining Classification Algorithms

Algorithm	**Confusion Matrix**				**Accuracy**
IBK		PPD	CControl	SSum	98.8%
	PPD	11498	225	11523	
	CContro	116	11349	11365	
	SSum	11514	11374	22888	
Binary Logistic Regression		PPD	CControl	SSum	64.3%
	PPD	11079	4444	11523	
	CContro	5556	7779	11365	
	SSum	11665	11223	22888	
C4.5 (J48 Application)		PPD	CControl	SSum	92.1%
	PPD	11414	1109	11523	
	CControl	1120	11245	11365	
	SSum	11534	11354	22888	
SVM		PPD	CControl	SSum	64.5%
	PPD	11055	4468	11523	
	CControl	5556	8809	11365	
	SSum	11611	11277	22888	
Random Forest		PPD	CControl	SSum	95.4%
	PPD	11472	551	11523	
	CControl	882	11283	11365	
	SSum	11554	11334	22888	

Source: Tucker et al. (2015)

9.2 DATA VISUALIZATION

Data visualization (DV) has long been used in statistics and mathematics. Indeed, the love of data visualization is not new:

> There is a magic in graphs. The profile of a curve reveals in a flash a whole situation—the life history of an epidemic, a panic, or an era of prosperity. The curve informs the mind, awakens the imagination, convinces.
>
> **Henry D. Hubbard, 1939**[3]

More recently, DV has been applied to aspects we may encounter daily including the results of a financial analysis, a weather forecasting graph presenting the movement of the storm, or a distribution of certain diseases around the world (Ward, 2010).

In every application, the purpose of DV is to present the data and information in a way so that the users can grasp the meaning it conveys both accurately and quickly. Friedman (2008) defined the main goal of data visualization as "*to communicate information clearly and effectively through graphical means.*" According to Friedman, DV should combine the form and function together and try to find balance between these two attributes.

In order to achieve this goal, many techniques have been developed to present and display data in a user friendly way. Based on different data types, from spatial data to multivariate data, visualization techniques can range from the simple bar chart to a complex tree graph, from the interaction to a glyph. This wide application has proved the effectiveness of DV to help users achieve clear and integrated information. DV may be used on a series of data in a graphical manner, helping to communicate the information in a direct way (Dzemyda et al., 2013), or it can be used to integrate multidimensional data and present them in a lower dimensional space (one dimension or two dimensions). These techniques and uses are discussed further below.

In particular, we are interested in how DV can help the healthcare system become more efficient and effective. The implementation and intervention of DV in the healthcare system is no longer a novelty. In fact, many techniques have been applied in healthcare systems for years, such as the display of blood pressure, heart rate, and blood tests. These techniques help

---

[3] Found at http://blog.datamarket.com/2012/07/08/the-11-best-data-quotes/

display the index in an expanded timeline to show the results in a dynamic manner, which can help the patients and the physicians to easily identify the levels as well as the trends in each index and thus get a more reliable understanding and self-awareness.

Definition and Function of Data Visualization

DV is a valuable tool in helping users to discover what otherwise would have remained hidden. It can also facilitate better understanding of massive and complex data sets and communicate the meaning to others. Visualization can be defined as a collection of parts of graphical speech and should consist of a series of components such as data, coordinates, elements, statistics, aesthetics, faceting, guides, interactivity and styles.

> The ability to take data—to be able to understand it, to process it, to extract value from it, to visualize it, to communicate it—that's going to be a hugely important skill in the next decades,... because now we really do have essentially free and ubiquitous data. So the complimentary scarce factor is the ability to understand that data and extract value from it. (Varian, 2009)

Visualization can explore data, summarize data, and confirm hypothesis due to its ability in providing an overview of complex and massive data sets and identifying regions of the parameters of interest for further analysis (Grinstein, 2002).

Brief History of Data Visualization

In this section, the topic of DV is organized in a temporal way to show the development of the visualization techniques. Some milestones in this field are highlighted to emphasize their significance in promoting the development of new ideas in visualization.

The history of visualization goes back to second-century Egypt, where some astronomical information was organized as a tool for better navigation (Few, 2007). But the common form of visualization didn't arise until the seventeenth century, when Rene Descartes designed the innovative method to present data based on a system of coordinates. By the late eighteenth and early nineteenth centuries, more techniques such as bar charts emerged and improved. In 1983, Edward Tufte, a pioneer in the field of DV, published his book *The Visual Display of Quantitative Information*, visually presenting many effective ways of displaying data (Few,

2007). With the spread of affordable computers, the twentieth century witnessed the emergence of information explosion, along with a new research area called *information visualization*, which is what we now experience and study (Card, 1999).

There are so many advanced DV designs that only a few of them with great significance are presented here as the milestones in the field of DV. Friendly (2006) from York University conducted an expansive study, called the Milestone Project, focusing on the important developments in the history of DV.

Data Visualization Techniques Today

The organization and classification of previous visualization techniques are essential for the design and application of new techniques. As Friendly (2006) stated, "*The past is often the fountain of ideas, as rich as the future.*" There are a number of ways to classify and organize the visualization techniques. They can be classified based on their dimensions, based on whether they are static or dynamic, or based on whether they are geometric or symbolic (Grinstein, 2002). Ward (2010) summarized and synthesized these techniques and developed eight ways to increase efficiency and effectiveness when designing or improving new techniques. The eight visual variables are position, shape, size, brightness, color, orientation, texture, and motion (Ward et al., 2010).

According to Fry (2004), there are two main categories of data visualization in healthcare practice. One category focuses on primarily numeric or symbolic data. Another category focuses on the display of the physical nature of the subjects or patients.

The Data Visualization Process

Several approaches to visualization techniques have been proposed. Although there are some differences among approaches, the general logic remains the same. It starts with the analysis of the types of data we want to display, followed by the organization of data and visualization of data. For example, Bui et al. (2009) designed a new information clinical system to help improve the presentation of knowledge and simplify the structure, which incorporate three steps (i.e., information extraction, information organization, and information visualization). Ward (2010) developed the visualization pipeline, consisting of data modeling, data selection, data to

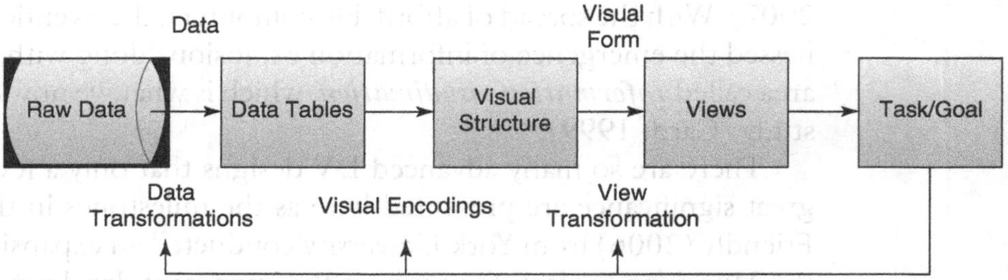

Figure 9.4 Visualization Pipeline *Source:* Ward (2010)

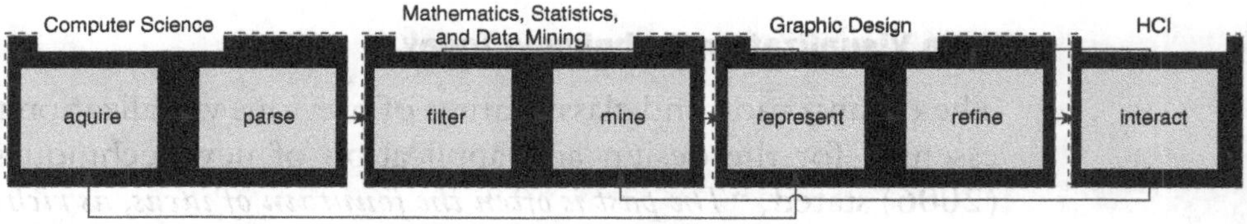

Figure 9.5 DV Process *Source:* Fry (2004)

visual mappings, and scene parameter setting. Figure 9.4 is an example of the visualization pipeline proposed by Ward. Another process by Fry (2004) has a series of more detailed steps that are also nicely classified according to their subjects. Figure 9.5 shows the visualization process developed by Fry (2004).

Perception

A good visualization not only requires knowledge in statistics and mathematics, but also calls for careful studies of human perception, and the information process mechanism (Ward, 2010). According to Ward, perception means "*the process of recognizing (being aware of), organizing (gathering and storing), and interpreting (binding to knowledge) sensory information.*" Without the external aids, the capacity of our cognition and mental power would be decreased dramatically. Ware (2000) pointed out that the human visual system is particularly good at seeking and recognizing patterns, and has a strong ability to process what is seen.

An example provided by Fry (2004) in Figure 9.6 is presented below. The two images present a set of data from regions of France using two different methods.

The left-hand image contains quantity information which requires the viewer to search the numbers, compare the numbers and turn them into a

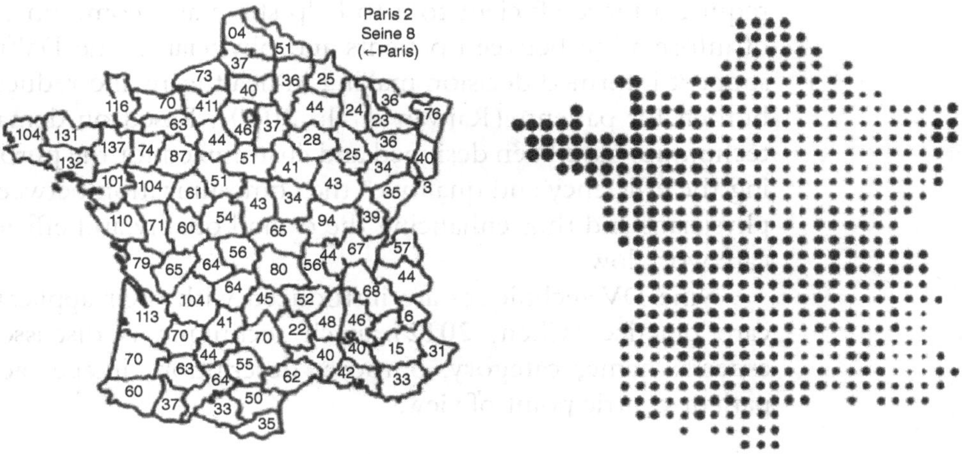

Figure 9.6 Comparison of Images with and without DV ***Source:*** **Fry (2004)**

more understandable meaning, while the right-hand image provides a qualitative understanding, where the meaning of the graph can be conveyed to the viewer immediately and clearly (Fry, 2004). Therefore, the right-hand image is much better than the left-hand one in terms of visualization.

Human Factors in Data Visualization

In recent years, more emphasis has been put into the customized design, delivering the concept of designing for targeted users rather than designing for all. This concept is extremely important in the practice of DV since the purpose of visualization is to help viewers to better grasp and understand information based on their demand. In order to incorporate human factors into the design of visualization, a general approach is recommended by many designers in the process of the DV. This approach starts by classifying users, followed by the clarification of function and objectives. Then, a clear description is required to show further details. A good design should also provide effective guidance for viewers, create meaningful graphics and images when necessary, and choose proper color for each technique. Last but not least, the integration, organization, and testing are always essential in determining whether it is a good customized design or not.

Data Visualization in Healthcare Practice

The prevalence of DV also found its way into healthcare practice. The high volume of medical data and complex nature of the medical knowledge require a more efficient tool to help share and communicate these types of information between patients and physicians. The DV techniques can support informed decision making in healthcare and reduce the cognitive burden for patients (Rajwan et al., 2010). Based on demand, many DV techniques have been designed and customized for the purpose of improving the efficiency and quality of the communication between patients and physicians and thus enhancing the overall quality and efficiency in healthcare workflow.

Nine DV techniques are introduced with their application in healthcare practice (Chen, 2012). Each technique is discussed from seven aspects: name, category, purpose, description, image, advantages, and patient-centric point of view.

Technique 1: Icon Arrays

Category: Risk analysis in healthcare

Purpose: To convey both the "numerator" and the "denominator" simultaneously using discrete level of measurement (Ancker et al., 2006).

Description: The icon arrays consist of a part-to-whole relationship with sequential arrangement. Proportions are easy to judge in this icon array because the part-to-whole information is available visually (Ancker et al., 2006). Figure 9.7 shows a visualization of risk.

Advantages: According to the research conducted by Ancker, patients tend to perform better when the data are displayed at the discrete level rather than as a number in proportion (Ancker et al., 2006).

Patient oriented: This visualization technique addresses patients' understanding ability and the different education levels among them. Patients may better understand risks with the help of this technique.

Technique 2: Magnifier Risk Scale

Category: Risk analysis in healthcare

Purpose: To elicit risk perceptions when the scale is at the low end (Ancker et al., 2006).

Description: A value *x* can be placed in the scale to allow the magnifier to describe the chance of each event that happens in the lower end (Ancker et al, 2006). Figure 9.8 depicts a visualization of scale.

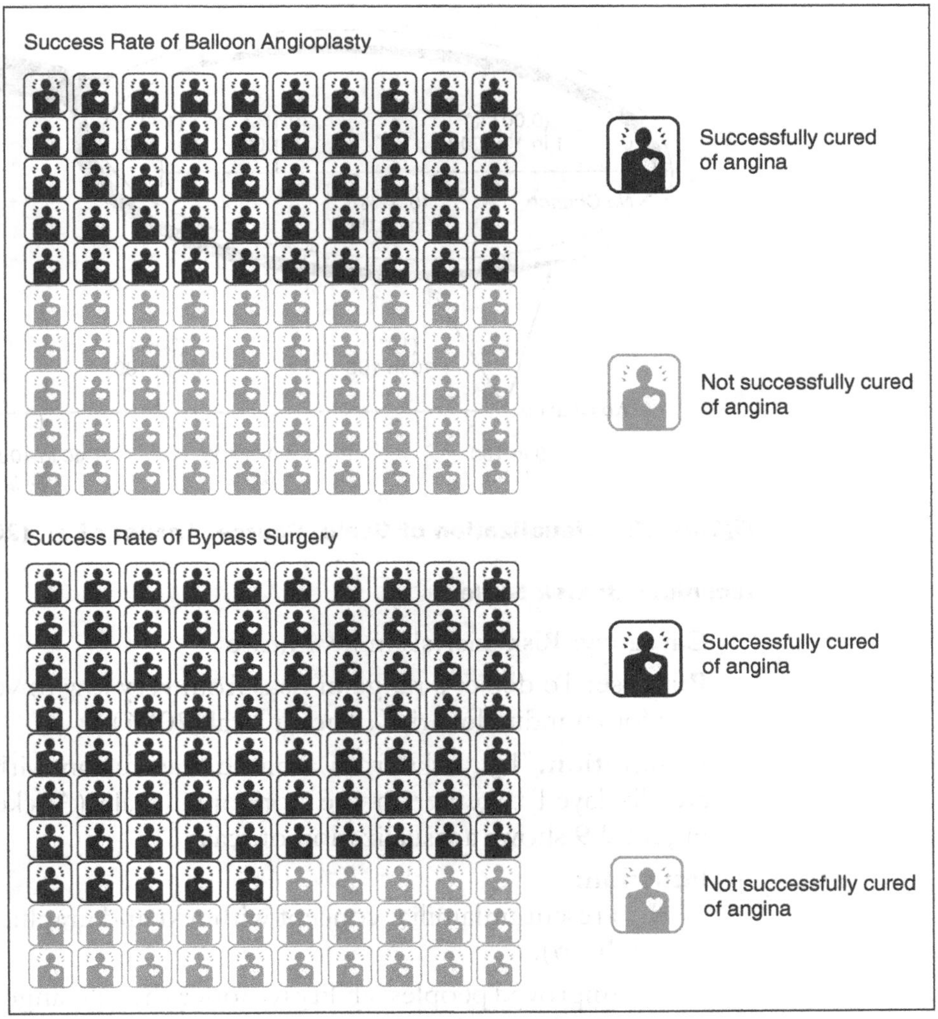

Figure 9.7 Visualization of Risk *Source:* Ancker et al. (2006)

Advantages: The magnifying lens at the low end allowed users to perceive and understand smaller values for very low risks and reduces the magnitude of higher risks (Ancker et al., 2006).

Disadvantages: May not be applied to large data sets.

Patient oriented: Allows the patient to identify his or her test results if the values are very small.

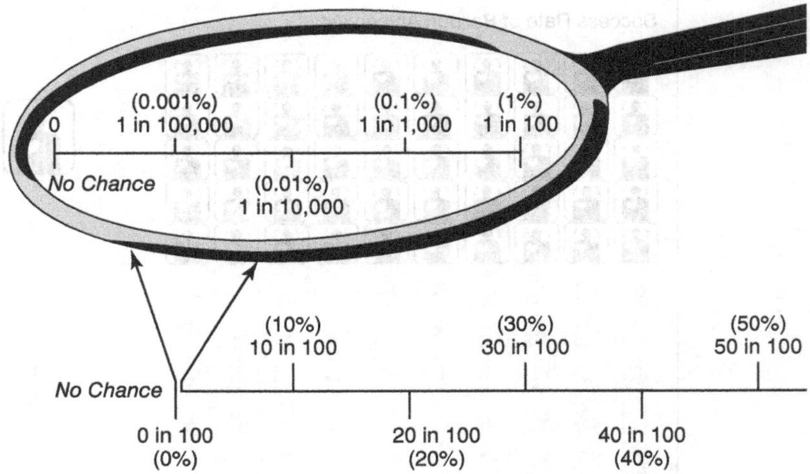

Figure 9.8 Visualization of Scale *Source:* Ancker et al. (2006)

Technique 3: Risk Scale

Category: Risk analysis in healthcare

Purpose: To depict a range of risks from very low to very high as context for an individual risk (Ancker et al, 2006).

Description: The risks from a blood transfusion with other hazards are displayed together in the horizontal scale (Ancker et al, 2006). Figure 9.9 shows a visualization of scale.

Function:

1. Present unfamiliar concept with text and graphics (Ancker et al., 2006).
2. Improved peoples' ability to notice the full range of possible risks (Ancker et al., 2006).

Advantages: This visualization technique is effective in increasing knowledge and reducing dread about rare hazards of transfusion. It can be also used for presenting unfamiliar concepts together with other familiar tasks (Ancker et al., 2006).

Disadvantages: The logarithmic scale may be unexpected for many people.

Patient oriented: Patient can compare their unfamiliar risks with familiar ones to get a better understanding of the risks of their deceases.

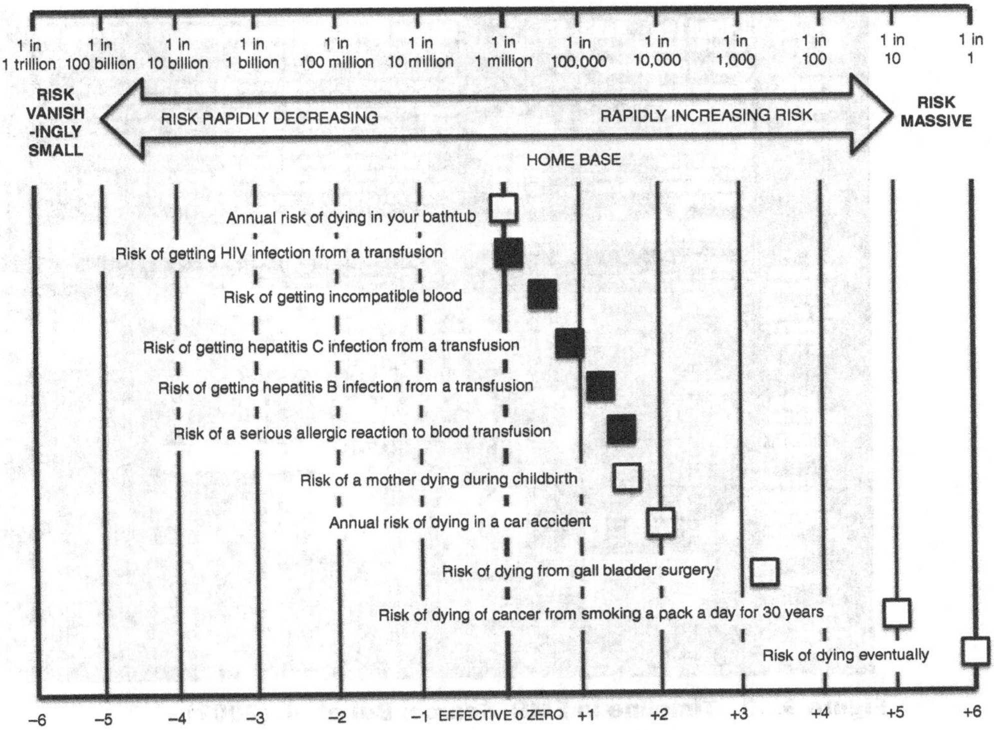

Figure 9.9 Visualization of Scale *Source:* Ancker et al. (2006)

Technique 4A: Timeline

An example of the timeline technique, shown in Figure 9.10, is one of those innovations. This visualization technique uses the integration concept to summarize all the information in the same display and thus can help users to better grasp information, combine related information, and make quick and high quality decisions in the end. According to Bui et al. (2007), this visualization is able to display data in a problem-oriented way, which is conducive to patient management and medical research. Figure 9.10 depicts a timeline in EMR.

Technique 4B: Timeline Belt

Category: Workflow evaluation

Purpose: To exhibit the fragmentation of workflow in healthcare before and after the computerized provider order entry (CPOE) implementation (Zheng et al., 2010).

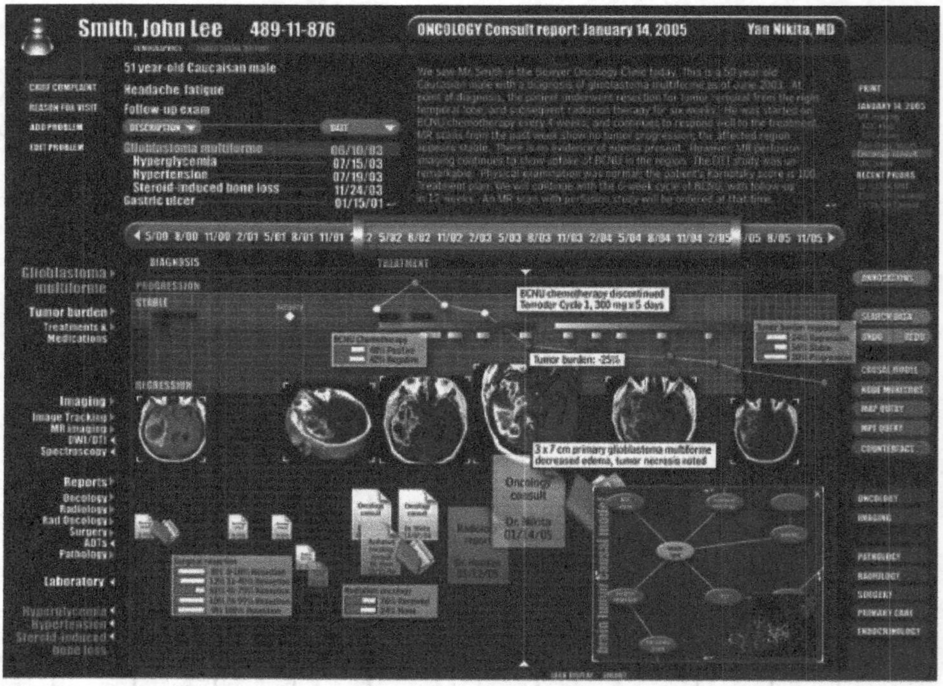

Figure 9.10 Timeline in EMR *Source:* Bui et al. (2007)

Description: Each row represents a time and motion (T&M) observation session (Zheng et al., 2010). The first 20 stripes indicate the preimplementation T&M session, while the following 22 stripes represent the postimplementation session. Different colors designate different clinical tasks execution. Length of a colored stripe is proportional to how long the task lasted (Zheng et al., 2010). Figure 9.11 provides a visualization of frequency.

Advantages: This visualization technique displays the difference between the post-CPOE representation and the pre-CPOE representation clearly and obviously. Thus it can help researchers to identify the problem within the workflow and then figure out a way to deal with it to decrease the frequencies of switching the tasks.

Technique 5: Heatmap

Category: Workflow evaluation

Purpose: To exhibit task transition probabilities (Zheng et al., 2010).

Description: Different density of color represents the transition probabilities estimated from empirical data, with higher density of

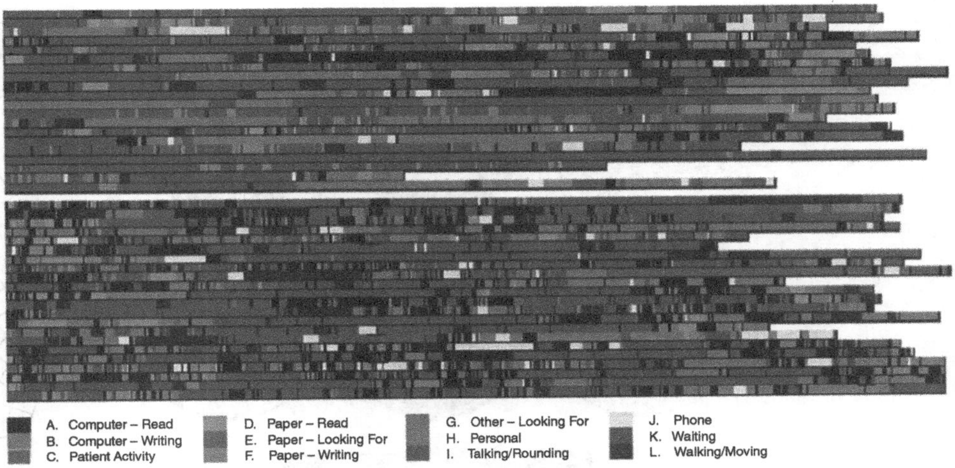

Figure 9.11 Visualization of Frequency ***Source:*** **Zheng et al. (2010)**

	A	*B*	*C*	*D*	*E*	*F*
A. Computer - Read	–	0.2472	0	0.0562	0.0112	0.0899
B. Computer – Writing	0.1181	–	0.0069	0.0139	0.0347	0.0764
C. Patient Activity	0	0.0909	–	0	0	0.0909
D. Paper – Read	0.0455	0.0114	0	–	0.0114	0.3182
E. Paper – Looking For	0.0606	0.0606	0	0.1515	–	0.2727
F. Paper – Writing	0.0498	0.0498	0.005	0.1294	0.0249	–
G. Other – Looking For	0	0.1667	0	0.1667	0	0.1667
H. Personal	0.1414	0.202	0	0.0404	0.0101	0.0505
I. Talking/Rounding	0.0773	0.1646	0.015	0.1047	0.0224	0.3092
J. Phone	0.1364	0.2727	0	0	0.0227	0.0227
K. Waiting	0	0.1667	0	0.1667	0	0.1667
L. Walking/Moving	0.0204	0.0748	0.0136	0.034	0.068	0.0748

Figure 9.12 Visualization of Workflow ***Source:*** **Zheng et al. (2010)**

color meaning higher transition probabilities (Zheng et al., 2010). Figure 9.12 shows a visualization of workflow.

Advantages: It can be easily observed which tasks are with the highest transitional probability from all the possibilities between different task pairs (Zheng et al., 2010).

Technique 6: Kaleidomaps by Bale et al. (2007)

Category: Time-dependent

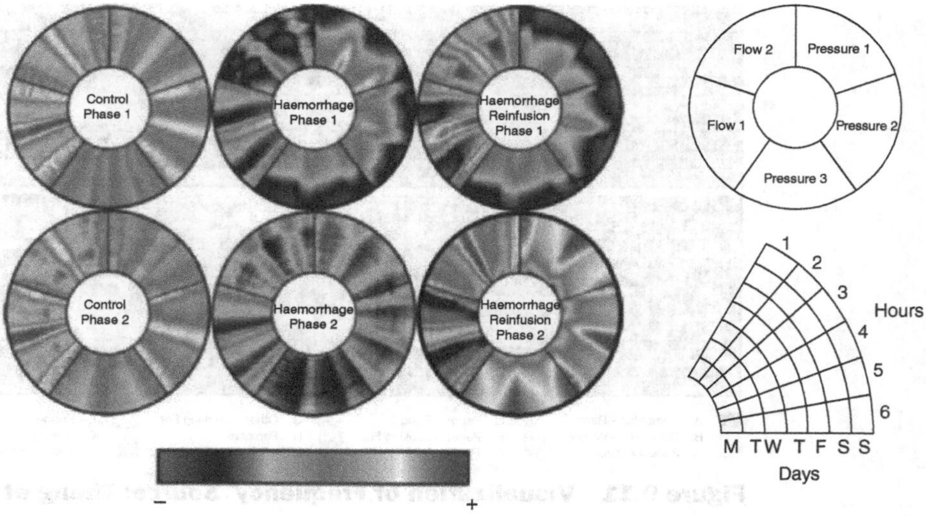

Figure 9.13 Visualization of Period *Source:* Aigner et al. (2011)

Purpose: To visualize multivariate time-series data using the curvature of a line to alter the detection of possible periodic patterns (Aigner et al., 2011).

Description: The six kaleidomaps in Figure 9.13, show the morphology of blood pressure and flow waves over two experimental phases (Aigner et al., 2011). A base circle is broken into segments of equal angles for different variables. Each circle segment has two axes representing time, one along the radius and one along the arc of the segment. The data values and categories are represented using color. Figure 9.13 provides a visualization of period.

Advantages: The kaleidomaps provide an interactive way for viewers to explore both in time and frequency to have a better understanding of the relationships between time and waveform morphologies (Aigner et al., 2011).

Disadvantages: Since the kaleidomaps have the format of a circle, it can only have a maximum of six to eight variables within one circle (Aigner et al., 2011).

Patient oriented: Kaleidomaps can help patients follow their prescription accurately and on time (Aigner et al., 2011).

Technique 7: Timeline Browser by Cousins and Kahn (1991)

Category: Time-dependent

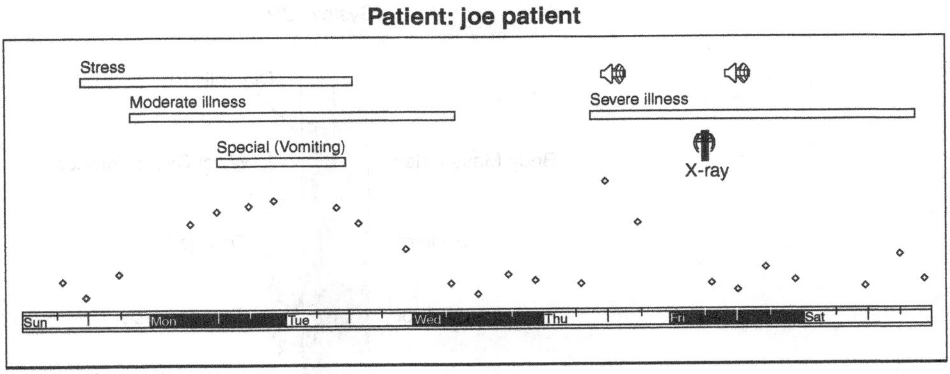

Figure 9.14 Visualization of Time ***Source:*** **Aigner et al. (2011)**

Purpose: To develop the timeline browser for visualizing heterogeneous time-oriented data.

Description: Intervals are displayed as labeled bars and events are displayed as icons. The small circles form a point plot that shows the patient's blood glucose over time. Figure 9.14 depicts a visualization of time.

Advantages: The timeline browser integrates qualitative and quantitative data as well as instant and interval data into a single coherent view.

Patient oriented: Distinguish simple events, complex events, and intervals. Simple events are represented as small circles; complex events are shown as icons. Bars are used to indicate location and duration of intervals.

Technique 8: Glyph

Category: Integration

Purpose: To help patients better understand their disease and track their health status for a chronic illnesses.

Description: Uses gestalt principles to create a "unified whole" for visual perception of multiple factors; the body shape forms a loci for recognizing and remembering important information. Figure 9.15 displays a glyph.

Advantages: Help patients to follow the dietary approaches to stop hypertension and have a better understanding of complications of hypertension.

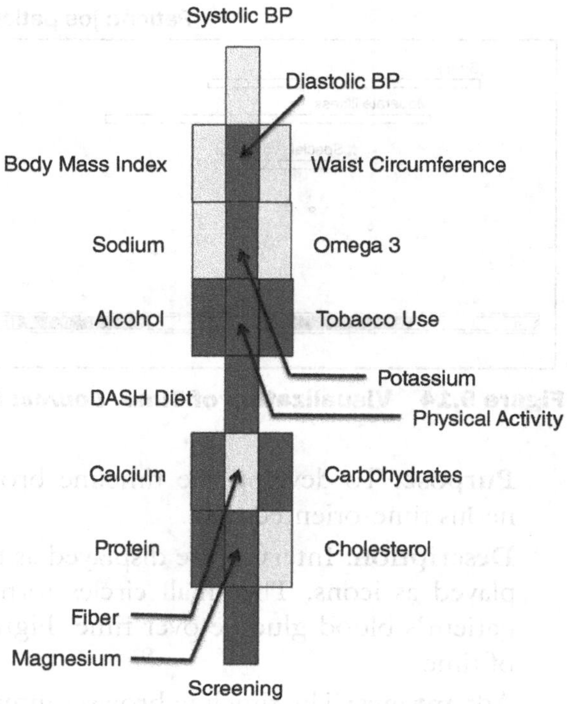

Figure 9.15 Glyph

Patient oriented: Help patients have an integrated view of their current health status and assist their understanding and comprehension of the important information for their health issues.

Technique 9: Overlaps and Interconnections

Category: Overlap

Purpose: To present the proportion of three diseases (Alzheimer's disease, high blood pressure, and heart disease), which have certain probability of overlapping.

Description: Different colors and numbers of icons present different overlapping groups of patients and the different proportions of the diseases respectively. Figure 9.16 displays a visualization of overlap.

Advantages: The ability to visually present the individual proportion of the disease and the overlapping of the disease respectively.

Patient oriented: Since the number of patients who are diagnosed with one of these three diseases is expected to increase continuously,

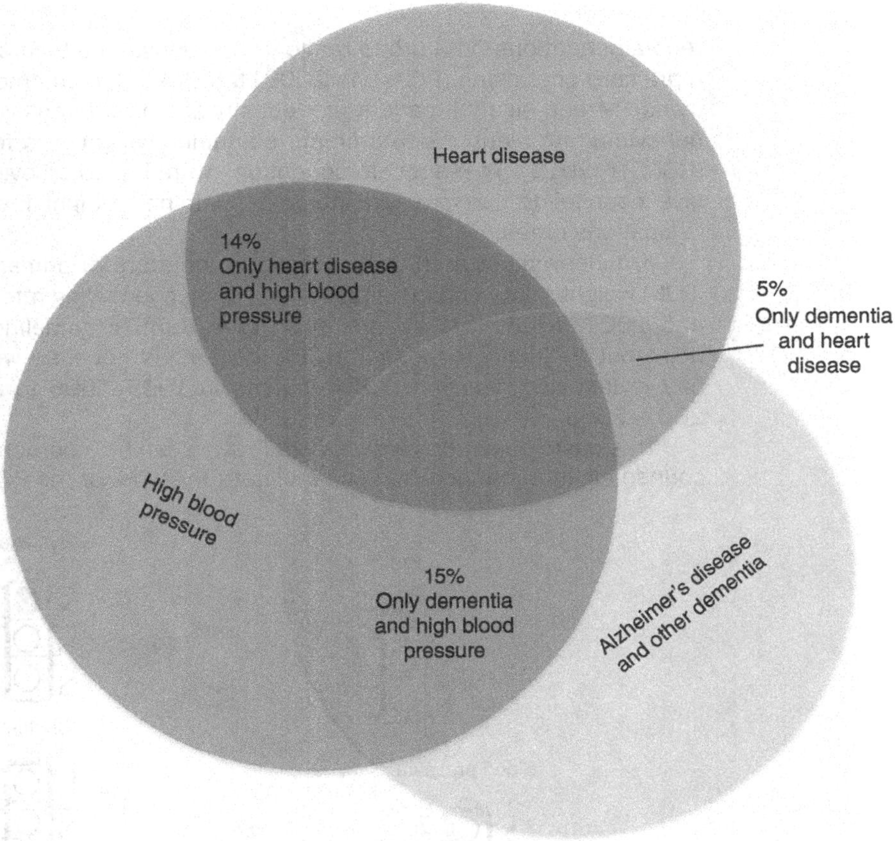

Figure 9.16 Visualization of Overlap ***Source:*** **Fairfield et al. (2013)**

it is crucial to make people aware of the overlap of the three diseases so that they can take proper measurements to prevent them (Fairfield et al., 2013).

CASE STUDY 9B: DATA VISUALIZATION IN OBESITY COUNSELING[4]

Approximately two-thirds of U.S. adults—145 million people—are overweight or obese. This epidemic has had profound economic and public health consequences. The United States Preventive Services Task Force (USPSTF) develops

[4]Based on an example developed by Gabby Pfeifer (BS, IE '12) and Hyojung Kang (doctoral student in IE), as advised by H. B. Nembhard and Jennifer Kraschnewski, MD, MPH.

recommendations on a broad range of preventive healthcare services for primary care physicians (PCPs). In 2003, the USPSTF recommended that physicians "screen all adult patients for obesity and offer intensive counseling and behavioral interventions to promote sustained weight loss for obese adults" (USPSTF, 2011). These recommendations were based on evidence that intensive counseling can promote modest sustained weight loss and improved clinical outcomes.

Krashnewski et al. (2013) described the state of primary care physician (PCP) weight-related counseling by comparing counseling rates in 1995–1996 and 2007–2008, that is, pre- and post- USPSTF guidelines. The authors point out a "silent response" to the guidelines because weight counseling had in fact declined from 7.8% of visits in 1995–1996 to 6.2% of visits in 2007–2008.

This case looks at how data visualization can be incorporated into obesity counseling in a manner than can facilitate the work of the PCP to meet these

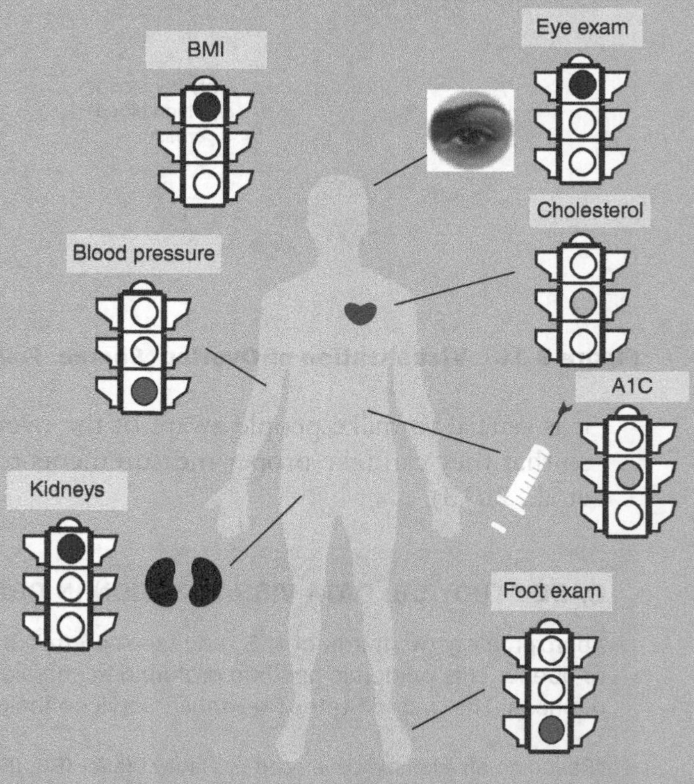

Figure 9.17 Example of a Custom Glyph for Obesity Counseling

guidelines as well as help to communicate the health status to the patient during the counseling session.

Methods

- We identified 20 key health factors in the CDC National Health and Nutrition Examination Survey (NHANES) that are appropriate to both the primary care setting and patients with obesity.
- We used Tableau software (www.TableauSoftware.com) to develop a glyph structure for primary care visits.

Results

- Data visualization with glyphs provides a highly visual and quickly comprehendible assessment of patient information (see Figure 9.17 below).
- This approach may be used as a patient-doctor communications tool, and improve the return on physician contact time with patients.
- The visual representation of health data may help patients understand their health status and improve patient compliance.

9.3 SOCIAL NETWORK ANALYSIS

Social network analysis (SNA) is a theoretical perspective and a set of techniques used to understand relationships among people and how they affect behaviors. SNA has its foundation in the fields of mathematics, social science, and statistical physics. Within the healthcare context, SNA models are useful for understanding human behavior, particularly health behaviors. In the area of public health, for example, SNA can be used to investigate the role of peer networks and their influence on smoking, substance abuse, and obesity in order to understand how to support better health outcomes.

A network is a set of entities (people, organizations, communities, nations, regions, locations, etc.) connected to each other through relationships. Types of relationships include whom you like, with whom you collaborate, with whom you communicate, whom you go to for advice, and with whom you share information. Relationships influence a person's behavior above and beyond the influence of his or her individual attributes. Attributes such as age, gender, educational level, income, occupation, and ethnicity are very important and influence a person's attitude, beliefs, and behaviors. These attributes also influence whom people know and spend time with (i.e., social networks). SNA models are constructed to show how these relations influence attitudes, beliefs, and behaviors.

SNA has been one of the primary tools that provide a rich support to make those relationships visible and systematically provide insights about individuals' behavior, patterns, and change. In general, SNA can be used to:

1. Visualize relationships within and outside the organization;
2. Facilitate identification of leaders and facilitators of a network;
3. Identify structural holes in an organization that could be affecting efficiency and effectiveness;
4. Identify isolated teams or individuals and knowledge bottlenecks;
5. Accelerate the flow of knowledge and information across functional and organizational boundaries;
6. Evaluate the effectiveness of programs and network change over time; and
7. Raise awareness of the importance of informal networks.

Networks are critical to professional successes, including success of initiatives to improve healthcare. Further, networks are important for obtaining information, solving problems, learning "how," understanding social support and service coordination, conducting collaborative research work, and understanding the underlying "true" structure of an organization. SNA can be used to improve healthcare through improving information flow, supporting collaborations, promoting innovation, developing communities of practice, and examining change in relationships over time.

The mathematical scaffold behind SNA is graph theory. Its roots go back to 1736 to Koningsberg, a thriving merchant city in Eastern Prussia (Figure 9.18). Its busy fleet of ships and the trade they brought allowed city officials to build seven bridges across the river Pregel that surrounded the town. Five of these connected the elegant island Kneiphof, caught between the two branches of the Pregel, to the mainland; two crossed the two branches of the river.

This peculiar arrangement gave birth to a contemporary puzzle: can one walk across all seven bridges and never cross the same one twice? Despite many attempts, no one could find such path. The problem remained unsolved until 1736, when Leonard Euler, a Swiss born

Figure 9.18 Map of Koningsberg ***Source:*** **http://www-history.mcs.stand.ac.uk/Extras/Konigsberg.html**

mathematician, offered a rigorous mathematical proof that such path does not exist.

Euler's insight was to represent each of the four land areas separated by the river as nodes, distinguishing them with letters A, B, C, and D. Next, he connected with lines each piece of land that had a bridge between them. He thus built a graph, whose nodes were the pieces of land and links were the bridges.

Then Euler made a simple observation: if there is a path crossing all bridges, but never the same bridge twice, then nodes with odd number of links must be either the starting or the end point of this path. Indeed, with an odd number of links you can arrive to a node and have no unused link for you to leave it. Yet, a continuous path that goes through all bridges can have only one starting and one end point. Thus, such a path cannot exist on a graph that has more than two nodes with an odd number of links. As the Koningsberg graph had three such nodes—B, C, and D—each with three links, no path could satisfy the problem.

Today, we remember Euler's proof because it was the first time someone solved a mathematical problem by turning it into a graph. In hindsight the proof has two important messages: The first is that some problems become simpler and more treatable if they are represented as a graph. The second is that the existence of the path does not depend on our ingenuity to find it. Rather, it is a property of the graph. Indeed, given the layout of the Koningsberg bridges, no matter how smart we are, we will never find the desired path.

Network Representation

Formally, a network often refers to real systems, while a graph is the mathematical representation of a network. As seen in Figure 9.19 below, the nodes (vertices) in a graph are the components (entities) in the system, and the edges (links) are the interactions (relationships) in the system. Networks can be undirected or directed (also called a digraph).

The mathematical representation of a network, *G*, can be written as follows:

$$G = (V, E)$$

where *V* is a set of entities or nodes, and *E* is a set of edges (pair of nodes).

A graph can be either directed or undirected depending on the characteristics of connection between the nodes. A graph is said to be directed if the relationship among nodes has some sense of directionality. For instance, leadership networks are usually treated as directed as some nodes influence other nodes in the network but in some cases the opposite direction is inexistent. In contrast, undirected networks only denote relationship but no directionality. A graphical representation of

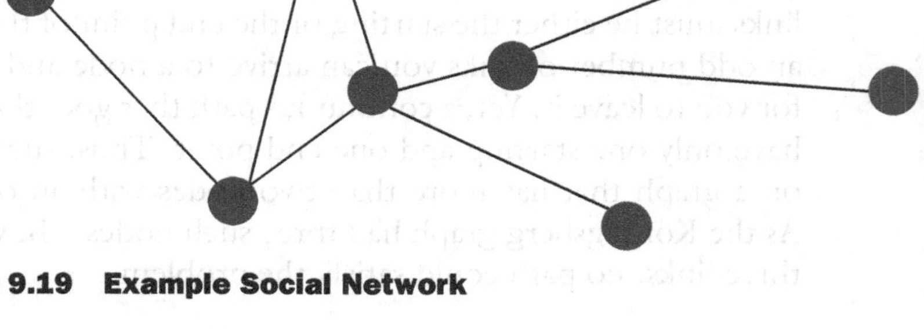

Figure 9.19 Example Social Network

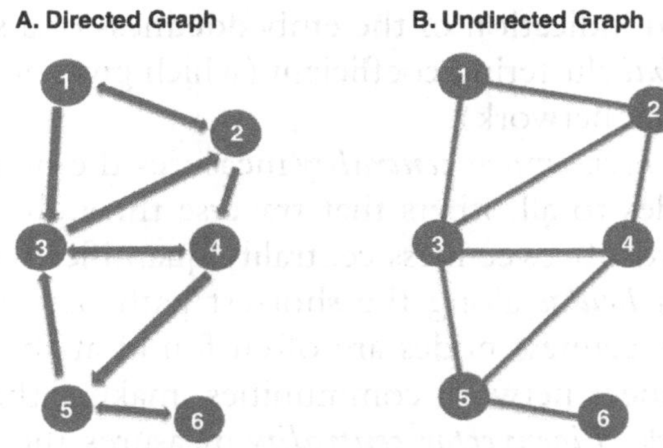

Figure 9.20 Directed and Undirected Graphs

graph and its directionality is shown in Figure 9.20. The sets of the directed graph can be written as $V = \{1, 2, 3, 4, 5, 6\}$ and $E = \{(1, 2), (1, 3), (2, 1), (2, 4), (3, 2), (3, 4), (4, 3), (4, 5), (5, 3), (5, 6), (6, 5)\}$. The undirected graph can be written as $V = \{1, 2, 3, 4, 5, 6\}$ and $E = \{(1, 2), (1, 3), (2, 3), (2, 4), (3, 4), (3, 5), (4, 5), (5, 6)\}$.

Another useful representation of these sets of nodes and edges is called *adjacency matrix*. This matrix is a square matrix of size *nxn* in which *n* represents the number of nodes in the network. The cells of the matrix represent the presence of an edge between two nodes. For instance, a nonzero entry at position (*i,j*) represents that there is an edge between nodes *i* and *j*. Additionally, this entry can represent the strength of the corresponding edge. For undirected graphs, the adjacency matrix is symmetric.

Descriptive Network Properties

We now define several foundational descriptive properties of social networks. The *diameter* of a network is the maximum *shortest path* between any two nodes in the network. *Network density* describes the ratio of the number of edges to the number of potential edges in a network. The *average path length* is the number of steps (on average) it takes to get from one node of the network to another node in the network. The *degree* of a node refers to the number of edges incident upon a node, which is also known as *degree centrality*. The *clustering coefficient* is the proportion of a node's neighbors that can be reached by other neighbors. In other words, the clustering coefficient of a node is the ratio of existing edges connecting a node's neighbors to each other to the maximum possible number of such edges. Here, we are referring to the *local* clustering coefficient (which is an indication of the embeddedness of a single node) as opposed to the *global* clustering coefficient (which gives an overall indication of clustering in the network).

Betweenness centrality measures the number of shortest paths from all nodes to all others that traverse through that node of interest. In other words, betweenness centrality quantifies the number of times a node acts as a *bridge* along the shortest path between two different nodes. High betweenness nodes are often found at the intersections of more densely connect network communities, making them very influential in the network. *Eigenvector centrality* measures the influence of each node in the network by assessing how well connected a node is to the parts of the network with the greatest connectivity. In other words, highly connected nodes within highly interconnected clusters have high eigenvector centrality. As a result, eigenvector centrality can be large either because a node has many neighboring nodes and/or because it has important neighbors. *Closeness centrality* refers to how connected a node is to its neighboring nodes. This metric represents the inverse of the average distance that each node

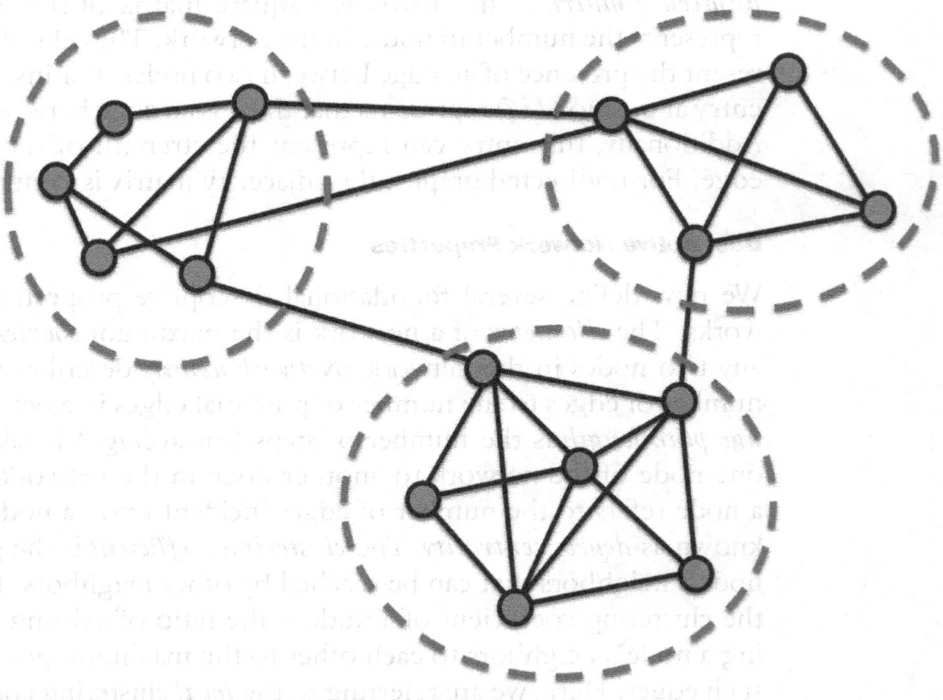

Figure 9.21 Communities in a Network

is from all other nodes in the network. Thus, the more central a node is, the closer it is to all other nodes in the network. In a typical social network setting, such nodes might have better access to information at other nodes or more direct influence on other nodes. Finally, ***modularity*** is a measure of the structure of the network, as it measures the strength of division of a network into clusters or communities (see Figure 9.21). Networks with high modularity have dense connections between the nodes within clusters but sparse connection between nodes in different clusters.

There is a vast literature of SNA theory and methods, as we only scraped the surface in this discussion. Thus, we refer the reader to Jackson (2008); Newman (2010); and Borgatti, Everett, and Johnson (2013) for additional details.

CASE STUDY 9C: SNA IN A PEDIATRIC INTENSIVE CARE UNIT (PICU)[5]

SNA was used in conjunction with quality improvement (QI) tools to provide a better understanding of the workflow issues in a Pediatric Intensive Care Unit (PICU). In particular, SNA was used to assess disruptions in PICU workflow and identify the tasks that affect efficiency and increase likelihood of errors. SNA was also useful in visualizing workflow based on the interconnectedness among the various tasks conducted in a PICU. From the results of a time-motion study, results indicated that nurses spent on average 21% of their time in Patient Monitoring, 24% in Collaboration, 7% in Medication, 27% in Documentation, 10% in Transit, and 11% in Miscellaneous activities.

In order to explore in greater detail the complexity of the workflow and interconnectedness among the several tasks, network graphs were developed. In these types of graphs, nodes represent the entity under analysis (i.e., tasks or categories) and the edges represent whether two tasks or task categories have been found sequentially in the workflow. The size of the circle is proportional to the time spent on the corresponding task and the width of the edges is proportional to the frequency or intensity in which two tasks are sequentially arranged. In Figure 9.22, a general network graph is shown to illustrate how category tasks are interrelated. It can be said that among the categories identified, Documentation is the group of tasks that is more time consuming, followed by Collaboration and Patient Monitoring. By visualizing the edges of the network graph, it can be said, for instance, that Documentation and Collaboration sequences are much more frequent than Miscellaneous and Medication sequences.

[5]From Muñoz, Alonso, and Nembhard (2014).

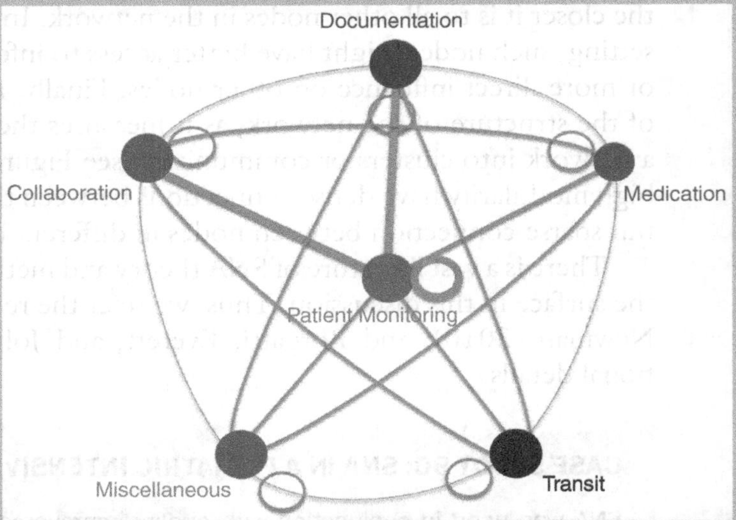

Figure 9.22 Network Graph for Task Categories ***Source:*** **Muñoz et al. (2014)**

A more detailed network graph is presented in Figure 9.23. This graph provides an expanded visualization that could serve to explore inefficiencies in a more detailed manner. The interconnectedness and sequence of any two tasks for the various categories can be visualized. One of the inferences that can be made from the graph is the high centrality of the use of HIT systems. This task is connected to almost every other task. It indicates that HIT systems play a major role not only for indirect care activities, but also as a driver for coordination and communication. A high frequency can be appreciated for the sequences Electronic–Nurse-Nurse Communication and Electronic–General Care. In addition, the density of the graph is 0.366, which indicates that 36.6% of the total possible connections among every task are present in the network. Although this number can appear low, considering the relatively large size of the network, this number is impactful. It indicates that the workflow is relatively highly fragmented and complexly interconnected.

One of the main advantages of SNA is that the levels of abstraction or details can be adjusted as required to investigate specific workflow patterns. For instance, let us assume that Coordination is required to be explored in more detail, maintaining aggrupation for all the other task categories. This new abstracted network graph is shown in Figure 9.24. Details such as how coordination occurs, what specific communication interactions appear in the workflow, and how different clinician staff supports other task categories can be visualized from this type of graph.

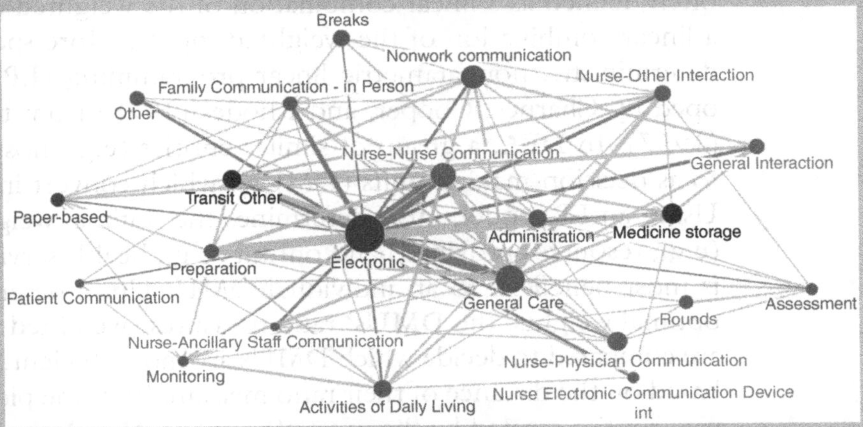

Figure 9.23 Network Graph of Detailed Tasks
***Source:* Muñoz et al. (2014)**

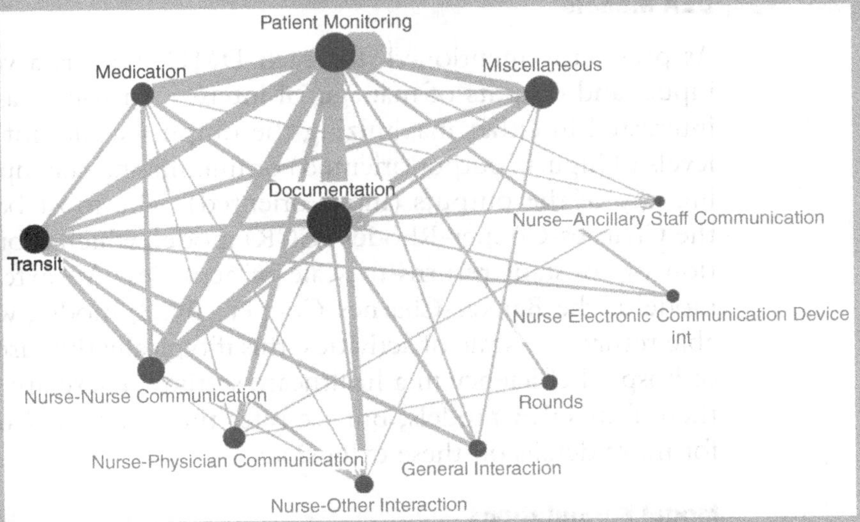

Figure 9.24 Network Graph of Coordinated Tasks
***Source:* Muñoz et al. (2014)**

9.4 DATA ENVELOPMENT ANALYSIS

Data envelopment analysis (DEA) is a set of flexible, mathematical programming approaches for the assessment of efficiency, where efficiency is often defined as a linear combination of the weighted outputs divided by a linear combination of the weighted inputs. More specifically, DEA is a deterministic, nonparametric linear programming (LP) technique developed by Charnes, Cooper, and Rhodes (1978) from the work of Farrell (1957). In DEA, a group of similar entities (e.g., hospitals) are referred to as decision-making units (DMUs), which convert inputs into outputs. Using an LP model, DEA determines the optimal weights for each input (e.g., resources provided) and output (e.g., health services delivered) that is most beneficial to an individual DMU. Upon determination of these optimal weights, the DMUs' ratio measures (weighted outputs to inputs) are compared to decide which DMUs are most efficient, where efficiency is based on the distance of their ratio measure from the piecewise-linear convex frontier created by the most efficient DMUs. Note that DEA requires no explicit identification of underlying relations between inputs and outputs, and weights are not assigned *a priori*.

DEA Models

As previously mentioned, for each DMU there is a vector of associated inputs and outputs of managerial interest. In some cases, the manager is interested in either maximizing the outputs while not exceeding current levels of inputs (output oriented) or minimizing the inputs without reducing any of the outputs (input oriented). The most basic DEA model is the Charnes-Cooper-Rhodes (CCR) model, which is built on the assumption of constant returns-to-scale of activities. An extension of the CCR model is the Banker-Charnes-Cooper (BCC) model, which assumes variable returns-to-scale of activities; this if typically the case for the evaluation of hospital efficiency in a healthcare setting. There are many variations to these four DEA models, but we refer the reader to Cooper et al. (2007) for more details on these extensions.

Model Formulations

The following mathematical programming formulation is known as the input-oriented CCR model:

$$\text{Max } \theta = \frac{\vec{u}^{T}\vec{y}_o}{\vec{v}^{T}\vec{x}_o}$$

subject to:

$$\frac{\vec{u}^T \vec{y}_z}{\vec{v}^T \vec{x}_z} \leq 1, \ \forall z$$

$$\vec{u} \geq \vec{0}$$

$$\vec{v} \geq \vec{0}$$

In this formulation, there is a vector of outputs ($\vec{y}$), a vector of inputs ($\vec{x}$), and z DMUs. Efficiency is designated as θ. The index o identifies the selected DMU for which an efficiency score will be generated. This mathematical program is run z times (the total number of DMUs), once to determine the efficiency of each DMU. The components of the vectors $\vec{u}$ and $\vec{v}$ are the weights to be determined for the outputs and inputs, respectively. This model defines efficiency for the selected DMU as the weighted linear combination of its outputs divided by the weighted linear combination of its inputs, subject to the constraint that, for each DMU (including the one whose index z is o), the efficiency cannot exceed one. All weights are restricted to be nonnegative. This formulation is nonlinear; however, it is trivial to manipulate the formulation so that we get the following:

$$\text{Max } \theta = \vec{u}^T \vec{y}_o$$

subject to:

$$\vec{u}^T \vec{y}_z - \vec{v}^T \vec{x}_z \leq \vec{0}, \ \forall z$$

$$\vec{v}^T \vec{x}_o = 1$$

$$\vec{u} \geq \vec{0}$$

$$\vec{v} \geq \vec{0}$$

For consistency with much (but not all) of the literature, this formulation is considered the primal, so taking the "dual" provides the following formulation:

$$\text{Min}\, \theta - \varepsilon(\vec{1}^T \vec{s}^+ + \vec{1}^T \vec{s}^-)$$

subject to:

$$X\vec{\lambda} - \theta \vec{x}_o + \vec{s}^- = 0$$

$$\Upsilon\vec{\lambda} - \vec{y}_o - \vec{s}^+ = 0$$

$$\vec{\lambda}, \vec{s}^-, \vec{s}^+ \geq \vec{0}, \varepsilon > 0$$

Here, the ε in the objective function is called the non-Archimedean element. This allows a minimization over efficiency score θ to preempt the optimization of slacks ($\vec{s}^{\,-}, \vec{s}^{\,+}$), which reflect output shortages and input excesses. A DMU that has an efficiency score of one and a zero-slack solution (for all slacks) is considered (technically) efficient or Pareto-Koopmans efficient. As defined in Cooper et al. (2007), efficiency is attained only if it is impossible to improve any input or output without worsening some other input or output. In all other cases, it is possible to improve one or more of the inputs or outputs without worsening any other input or output.

Next, we provide the mathematical programming representation of the dual version of the input-oriented BCC model, which minimizes the inputs without reducing any of the outputs and assumes that the relationship between inputs and outputs involve nonconstant returns-to-scale. Further, we consider the fact that there exist nondiscretionary inputs (e.g., number of patient encounters representing the population) that cannot be adjusted by the decision maker. The optimization model formulation is as follows:

$$\text{Min } \theta - \eta(es_{ND}^{-} + es^{+}) \tag{9-1}$$

subject to:

$$Y\lambda - s^{+} = y_o \tag{9-2}$$

$$X\lambda + s_D^{-} = \theta x_o \tag{9-3}$$

$$X\lambda + s_{ND}^{-} = x_o \tag{9-4}$$

$$e\lambda = 1 \tag{9-5}$$

$$x \geq 0, y \geq 0, \lambda \geq 0, s_D^{-} \geq 0, s_{ND}^{-} \geq 0 s^{+} \geq 0 \tag{9-6}$$

In this LP formulation, there are m outputs, n inputs, and z DMUs, where efficiency is designated as θ. This model is run z times, once to determine the efficiency of each DMU. The index 0 identifies the selected DMU for which an efficiency score is generated, and η is a very small positive value (also known as the non-Archimedean element). Further, λ is the vector of dual multipliers, y_o is the column vector of outputs for DMU_o, x_o is the column vector of inputs for DMU_o, Y and X are matrices of outputs and inputs, respectively, e is a row vector with all elements unity, and s_D^{-}, s_{ND}^{-} and s^{+} are the column vectors for input (discretionary and nondiscretionary) and output slack variables, respectively. As noticed, this formulation partitions the inputs and input slacks into two mutually exclusive and categorically exhaustive sets, discretionary (D) and nondiscretionary (ND). One can readily see that the nondiscretionary input slacks are not included in the objective function (1) and do not enter the measure of efficiency evaluation that is being obtained. Further, they are not multiplied by θ in the constraint set (4), so the nondiscretionary input may not be reduced.

The objective function (1) seeks to minimize the difference between the global efficiency and the product of the non-Archimedean element times the sum of the input excesses minus the output shortages. The constraints in (2) enforce the product of the dual multipliers and output data minus the dual output slack to equal the output data for the selected DMU. The constraints in (3) enforce the product of the dual multipliers and input data plus the dual input slack (discretionary) to equal the product of the efficiency and input data for the selected DMU. The constraints in (4) enforce the product of the dual multipliers and input data plus the dual input slack (nondiscretionary) to equal the input data for the selected DMU. The convexity constraint (5) forces the sum of the dual multipliers to equal one, which is required for a variable returns-to-scale optimization model. Finally, the constraints in (6) are the nonnegative constraints for the model.

Next, in Figure 9.25 we show a quick example from a hospital setting of input-oriented efficiency models to illustrate the difference between CCR (constant returns-to-scale, or CRS) and BCC (variable returns-to-scale, or VRS) DEA models.

Illustrated in Figure 9.25 is the ordinary least squares (OLS) solution for one input (say full time equivalents, or FTEs) and one output (say workload), such that we have a cost model, Cost = f(Workload). In this case, we see the OLS line represented in red and the individual hospitals (DMUs). Our hospital of interest in this case is Hospital Z.

Assume that we wish to reduce the single input while maintaining outputs at the current level (input-oriented). We wish to move along the input axis as far as possible. Ideally, we would produce the same outputs with zero inputs; of course this is not feasible. Nevertheless, it is possible to measure the distance from the location of point Z to a frontier line that we assume is feasible. The distance measurement begins with the understanding of the location of Z on the input field as depicted by the black bracket. We turn now to the issue of evaluating which input fields are possible.

In the simplest case, we might assume CRS under the assumption that the production possibility set is a cone. That is, if x,y is feasible, then tx,ty is feasible and the feasible frontier stems from the origin, where t is a scalar.

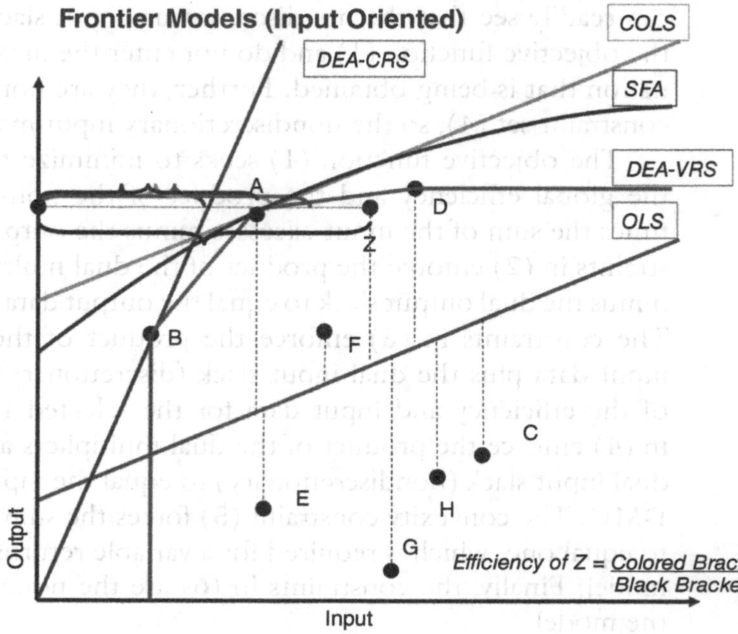

Figure 9.25 Example of Input-Oriented Efficiency Models

In this case, we have the input-oriented CCR DEA model. The efficiency of hospital Z is nothing more than the distance from zero input to the feasible frontier at output level Z divided by the distance from the zero input to point Z at output level Z: The purple bracket divided by the black bracket.

Alternatively, we might assume that the production possibility set is still CRS; however, instead of the origin, we now use a line parallel to the OLS regression line. Specifically, we shift the frontier up by the maximum of the residuals. This approach is nothing more than corrected OLS (COLS). In this case, we measure efficiency as the ratio of distances as before; however, the frontier line is different and hence the efficiency scores are automatically different.

Next, we could use a modification of DEA, which immediately allows for VRS. By adding a convexity constraint, we assume that the production possibility set is part of the convex hull of the polyhedron in a typical linear program instead of a cone. We note that the efficiency is then measured from this frontier.

The last method depicted is stochastic frontier analysis (SFA). SFA models normally (not always) assume that the error term from a traditional linear model may be split into two pieces (composed error), each with its own distribution. For simple depiction, we see that the curvilinear line represents the production frontier for SFA (which is nonlinear) and the efficiency is measured from this frontier.

CASE STUDY 9D: USING DEA FOR FINDING BEST-IN-CLASS HOSPITALS IN A NETWORK

As the growth in healthcare costs continues, large, centrally funded and operated healthcare systems, such as the Department of Defense's Military Health System (MHS), must become more efficient. The MHS operates a $52 billion system that provides health services to over 4.5 million enrolled beneficiaries such as uniformed service members, their family members, survivors, and retirees (Bastian, Fulton, Shah, & Ekin, 2014). Due to these rising healthcare costs, the MHS must optimize health system performance by efficiently allocating resources and balancing costs, quality of care, and access to care across the MHS. To do so requires careful management of major health system components.

The MHS is centrally funded, which means that dollars are sent to hospitals for expenditure. The amount of money provided to each hospital varies based largely on historical expenditures. Recently, the efficiency of hospitals within the MHS has received quite a bit of attention from leaders (Bastian et al., 2014), and the concern for improving health system performance motivates this case study.

The input-oriented BCC DEA model is used to examine 16 hospitals in a military hospital network with inputs and outputs that were deemed important to decision makers in evaluating efficiency. The data are from 2003 (as to be non-sensitive in nature), and the hospitals were chosen from 24 facilities because they are largely homogenous. The inputs that could be manipulated included the funding stream (COST)—expenditures (in 1,000s) less graduate medical education and readiness costs, and inflated in two parts to 2003 dollars—and the FTEs (FTE)—number of assigned full-time equivalents (in 1,000s) in 2003.

A nondiscretionary input was the enrollment population supported (ENROLL)—enrollment population supported (in 1,000s) in 2003. The outputs of interest included: inpatient aggregated hospital relative weighted product (in 1,000s) in 2003 (RWP), outpatient aggregated hospital relative value units (in 1,000s) in 2003 (RVU), a prevention/quality composite score found in the MHS survey scaled between [0, 100] in 2003 (PREV),

Table 9.3 Data from Real-World Example of Military Hospital Network

	ENROLL	FTE	COST	RWP	RVU	PREV	ACCESS	SAT
H1	14.81	7.13	56.66	7.05	112.21	83.28	70.55	73.19
H2	23.09	9.86	72.67	6.51	182.38	83.40	66.24	71.55
H3	68.40	17.66	163.99	21.74	372.06	78.89	57.29	63.02
H4	80.62	17.20	169.14	14.14	476.48	89.14	67.39	73.63
H5	49.84	15.25	125.44	16.87	314.98	85.65	65.72	72.02
H6	38.13	13.04	130.23	10.41	229.08	84.82	65.61	69.87
H7	32.87	8.68	67.25	10.74	187.00	79.70	67.86	70.83
H8	12.74	6.34	53.16	7.07	85.10	84.60	67.49	73.67
H9	23.95	11.73	95.60	14.31	253.72	83.15	70.59	74.81
H10	14.93	6.42	52.37	0.96	76.53	89.44	65.40	69.85
H11	47.87	16.91	129.16	21.93	339.66	85.73	69.30	74.35
H12	31.50	8.81	71.98	3.01	153.56	82.32	60.92	69.81
H13	22.99	11.13	99.60	6.71	252.20	85.63	74.52	80.99
H14	31.39	12.73	92.53	14.87	298.59	83.97	70.48	75.62
H15	10.70	6.22	38.08	3.03	60.06	80.83	64.76	72.84
H16	63.40	14.71	114.29	14.86	327.31	80.24	62.89	68.53

a satisfaction composite score found in MHS survey scaled between [0, 100] in 2003 (SAT), and an ease of access composite score found in MHS survey scaled between [0, 100] in 2003 (ACCESS). The original data are shown in Table 9.3 (Bastian et al., 2014).

Assuming that enrollment is a nondiscretionary input, hospitals with efficiency scores less than 1.0 included: H1 (.851), H2 (.928), H5 (.948), H7 (.779), H8 (.951), H9 (.850), H11 (.998), H13 (.959), H14 (.842), and H16 (.974).

The dual version, input-oriented BCC DEA model is also used as part of the case study to measure the effect of a financial incentive model (pay-for-performance, or P4P) on hospital efficiency and outcomes (Bastian et al., 2015). The data set of analysis included measures for 23 Army hospitals, 12 Air Force hospitals, and 19 Navy hospitals during the period of 2001–2012, resulting in a total sample of 648 annual observations for different variables of interest.

DEA analysis is conducted to compute the efficiency of each of the 54 military hospitals. Given the panel data, an extended DEA approach, time window analysis, is employed to measure hospital efficiency over time. The window analysis uses a moving average approach where each DMU is treated as if it were a different DMU in each of a specified window width. This approach enables catching changes in a DMU's performance from one window to another. It also has an effect of increasing the sample size, which leads to improving the discrimination power of the model. This case study uses a three-year time window.

For example, in the first window (2001, 2002, 2003), input-oriented BCC DEA is performed for 162 DMUs, where 54 DMUs are evaluated in such a way that they are different in each of the three successive years. This procedure is repeated for 12 times from 2001 to 2012 by shifting the window forward one period each time. The number of efficiency scores assigned to each DMU varied by time. For example, between 2003 and 2010, each DMU had three different efficiency scores for each year. On the other hand, DMUs had one or two efficiency scores for the first and last two years due to the moving average approach. Based on the results, the efficiency score for each DMU at each period was computed by taking an average of the efficiency scores obtained from different windows over time. The discretionary input variables included staffing, outpatient expenditures, and inpatient expenditures; number of patient encounters was used as the nondiscretionary variable. The output variables include outpatient and inpatient workload measures. The DEA time window analysis optimization model is solved to compute the efficiency score for each of the Army, Navy, and Air Force hospitals in the panel data set.

Subsequent analyses in the case study uses the average DEA time window analysis hospital efficiency scores as a basis of a response variable. Difference-in-difference estimation with OLS regression is then used to evaluate the impact of the P4P program on hospital efficiency of two groups. Reduction of hospital efficiency is found to be associated with implementation of the monetary incentive program. A decrease in hospital efficiency for the sake of patient quality, access and satisfaction improvements is financially unsustainable and not cost-effective in the long run.

9.5 MULTICRITERIA DECISION MAKING

Decision problems in healthcare often exhibit the presence of multiple, conflicting criteria for judging alternatives, as well as the need for making compromises or trade-offs regarding the outcomes of alternate courses of action. Multiple-criteria decision making (MCDM) is a practical approach for helping make better, more informed decisions for problems in the healthcare context.

MCDM can be broadly classified as mathematical programming problems or selection problems. The focus of multiple criteria mathematical programming (MCMP) problems is to fashion or create an alternative when the possible number of alternatives is high (or infinite) and all alternatives are not known in advance. MCMP problems are usually modeled using explicit mathematical relationships, involving decision variables incorporated within constraints and objectives. Although there are numerous solution methods for MCMP problems, we focus our attention on goal programming (GP); this method uses completely prespecified preferences of the decision maker. On the other hand, the focus of multiple-criteria selection problems (MCSP) is on selecting the best or preferred alternative(s) from a finite set of alternatives. Given the large variety of MCSP solution methods (max-min, min-max regret, compromise programming, TOPSIS, ELECTRE, analytical hierarchy process, PROMETHEE, etc.), we refer the reader to Yu (1985), Zeleny (1982), and Roy (1996) for more details.

MCDM Terms and Concepts

To provide some common understanding of the MCDM problem and its solution methods, we define some critical terms and concepts:

- **Alternatives:** The possible courses of action in a decision problem. These alternatives can be pre-specified by the decision maker(s), which is the case for MCSP, or there are times (such as in MCMP) where prespecification is not possible, so alternatives are defined implicitly through mathematical relationships between decision variables.
- **Attributes:** The traits, characteristics, qualities or performance parameters of the alternatives (i.e., descriptors of the alternatives from the decision making point of view). For example, if the decision situation is one of choosing the "best" hospital to use, then the attributes could be satisfaction ratings, quality scores, proximity, cost, performance measures, and the like.
- **Objectives:** The directions of improvement as perceived by the decision maker. Using a healthcare example, an objective may be to

"maximize the operating room utilization" or "minimize the patient waiting time in the emergency department." Within MCMP, the objectives form the evaluation criteria.

- **Goals:** Desired status of attributes or objectives. Typically, goals are target values set by the decision maker, which are expected to be attained by the best alternative. For example, a goal may be for the "operating room utilization to be 80%."
- **Criteria:** Encompass the attributes, goals, and objectives; these are rules of acceptability for the alternatives determined by the decision maker.
- **Best compromise solution:** In MCDM problems, there is typically no "optimum solution" due to the conflicting nature of the objectives. Thus, the best compromise solution is a feasible solution that meets or exceeds the decision maker's minimum expected level of achievement of criteria values.
- **Nondominated (efficient) solution:** A feasible solution that is not dominated by any other feasible solution. For an efficient solution, an increase in the value of any one criterion is not possible without some decrease in the value of at least one other criterion. In other words, an efficient solution has the property that an improvement in any one objective is possible only at the expense of at least one other objective.

Goal Programming

For MCMP problems, one way to treat multiple criteria is to select one objective as primary and the others as secondary. The primary criterion is then used as the optimization objective function, while the secondary criteria are assigned acceptable minimum or maximum values depending on whether the criterion is maximum or minimum and are treated as problem constraints. However, if careful consideration is not given while selecting the acceptable levels, a feasible design that satisfies all the constraints may not exist. Fortunately, this problem is overcome using GP.

GP is a practical method for handling multiple criteria. It can be used in order to achieve the best compromise solution, while considering many desires that the decision maker would like to achieve. In GP, all of the objectives are assigned target levels for achievement and relative priority on achieving these levels. The aim is to minimize deviations between the specified targets of the decision maker and what can actually be achieved for the multiple objective functions within the given constraints. In other words, GP attempts to find an optimal solution that comes as close as possible to the targets in the order of specified weights or priorities. Note that deviations can be either positive or negative, depending on whether we overachieve or underachieve a specific goal, respectively. In preemptive GP, priorities are provided by the decision maker, whereas in nonpreemptive GP, weights are provided by the decision maker for each goal. For goals that are very important, high values are used as weights and for the other goals, relatively less weights are used. In that way, there is a chance that high priority goals will be satisfied first. In other words, numeric weights are assigned to each deviation variable in the objective function, where the magnitude of the weight assigned to a specific deviation variable depends on the relative importance of that goal. Hence, the results of GP models provide a feasible, best compromise solution for the decision maker.

The following nonpreemptive (weighted) GP formulation is provided to reinforce the concept. For GP formulation, the decision maker specifies an accept level of achievement (b_i) for each criterion f_i and specifies a weight w_i to be associated with the deviation between f_i and b_i. The weighted GP model looks as follows.

Minimize

$$Z = \sum_{i=1}^{k} (w_i^+ d_i^+ + w_i^- d_i^-) \tag{9-7}$$

subject to:

$$f_i(x) + d_i^- - d_i^+ = b_i \forall i = 1, 2, \ldots, k \tag{9-8}$$

$$g_i(x) \leq 0 \forall j = 1, 2, \ldots, m \tag{9-9}$$

$$x_j, d_i^-, d_i^+ \geq 0 \forall i, j \tag{9-10}$$

The objective function in (9-7) seeks to minimize the sum of weighted goal deviations. The constraints in (9-8) represent that goal constraints relating the multiple criteria to the goal/targets for those criteria. It should be noted that d_i^- and d_i^+ are the deviational variables, representing the underachievement and overachievement, respectively.

CASE STUDY 9E: USE OF GP TO OPTIMIZE HELICOPTER EMPLACEMENT AT MEDICAL TREATMENT FACILITIES[6]

The United States Army is making remarkable strides in its systematic approach to delivering healthcare across a continuum of combat operations. According to current force health protection policy, the U.S. Army's Health Service Support (HSS) system is designed to maintain a healthy force and to conserve combat strength of deployed soldiers. Specifically, the HSS system remains particularly effective by providing prompt medical treatment to prepare patients for evacuation, employing standardized air and ground medical evacuation assets, providing a responsive field-sited medical treatment facility for the wounded soldiers evacuated from the battlefield, and providing various other health and preventive medicine services.

Although more wounded soldiers survive compared to any other war because of the HSS system, the U.S. Army can still greatly improve its systematic approach to treat and evacuate casualties from combat zones. As a pillar of military medical doctrine, optimizing the emplacement of medical treatment and aeromedical evacuation (MEDEVAC) assets can increase casualty survivability given a set of resource constraints. Therefore, thorough investigation and development of improved analytical solutions derived from objectives concerning casualty coverage, resource utilization and vulnerability measures directly supports the military medical mission.

Military commanders have faced a significant combinatorial challenge integrating these lifesaving yet limited air evacuation assets into a fully functional, comprehensive system for the entire theater. MCDM is useful for tackling such an immense combinatorial problem given the number of potential MEDEVAC helicopter locations, the number of different aircraft models for employment and its associated constraints, the potential sites for casualty sustainment, and the number of supporting medical treatment facility locations.

In this case study, a weighted GP model is used to optimize over a set of expected scenarios with stochastically determined casualty locations to emplace the minimum number of helicopters at each medical treatment facility necessary to maximize the coverage of the theatre-wide casualty demand and the probability of meeting that demand, while minimizing the maximal medical treatment facility evacuation site total vulnerability to enemy attack (Bastian, 2010). The first goal seeks to maximize the aggregate expected casualty demand covered. The second goal seeks to minimize spare capacities of MEDEVAC helicopters for each type emplaced at each mobile hospital, ensure a sufficient level of predetermined reliability that an aircraft will be available when casualties occur. The third goal seeks to minimize the value of

[6]Case study taken from Bastian (2010).

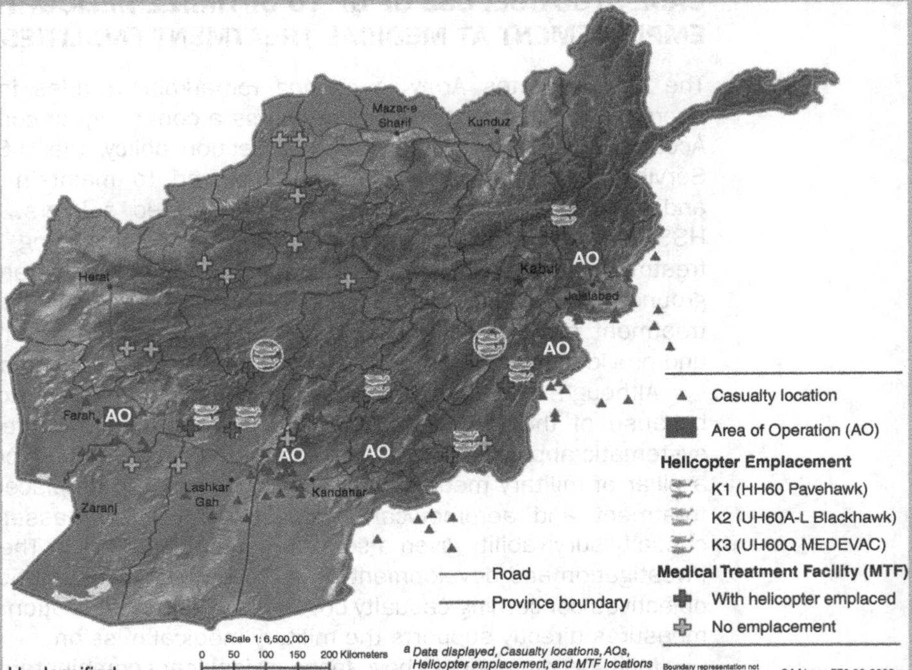

Figure 9.26 Weighted GP Model Solution with MEDEVAC Helicopter Emplacements *Source:* Bastian (2010)

the maximal evacuation site total vulnerability to enemy attack. Figure 9.26 displays the results of the weighted GP model.

This multicriteria decision modeling approach for optimizing aeromedical evacuation asset emplacement has been used by U.S. Army medical planners as a strategic and tactical aeromedical evacuation asset planning tool to help sustain and improve the aeromedical evacuation system.

QUESTIONS AND LEARNING ACTIVITIES

1. *Health, United States* is an annual report on trends in health statistics. It is compiled by the Centers for Disease Control and Prevention (CDC) and available on their website (www.cdc.gov). Select datasets from the report's Chartbook and use Tableau, Spotfire, or JMP software to visually represent them.
2. The Center for Integration of Medicine and Innovative Technology (CIMIT) prize is a national competition open to engineering students

from accredited engineering programs to advance their work in healthcare solutions. Formulate a proposal brief for the CIMIT prize that explores how data visualization can be integrated with other technology.

3. What are the key similarities and differences between multiple-criteria mathematical programming and multiple-criteria selection problems?
4. What are advantages and disadvantages of preemptive goal programming?
5. What are advantages and disadvantages of nonpreemptive goal programming?
6. For weighted goal programming, what are some methods for determining decision-maker weights?
7. Neurological disorders are often treated with medication that should mitigate symptoms such as tremor and bradykinesia. How can the data-mining approach discussed in the case study be extended to support medication effectiveness evaluation?
8. When using DEA to evaluate the efficiency of hospitals in a network, how should the modeler determine the appropriate DEA model?
9. When using DEA, how should the modeler determine the appropriate model inputs and outputs to use?
10. What are the advantages and disadvantages of using DEA?
11. How is DEA different from parametric frontier models? What are some advantages and disadvantages of these parametric models?

REFERENCES

Aigner, W., Miksch, S., Schumann, H., & Tominski, C. (2011). Visualization of time-oriented data. Springer Science & Business Media.

Ancker, J. S., Senathirajah, Y., Kukafka, R., & Starren, J. B. (2006). Design features of graphs in health risk communication: a systematic review. *Journal of the American Medical Informatics Association*, *13*(6), 608–618.

Bastian, N. (2010). A robust, multi-criteria modeling approach for optimizing aeromedical evacuation asset emplacement. *Journal of Defense Modeling and Simulation*, *7*(1), 5–23.

Bastian, N., Fulton, L., Shah, V., & Ekin, T. (2014). Resource allocation decision making in the military health system. *IIE Transactions on Healthcare Systems Engineering*, *4*(2), 80–87.

Bastian, N., Kang, H., Fulton, L., & Griffin, P. (2015). Measuring the effect of pay-for-performance financial incentives on hospital efficiency in the military health system. *IIE Transactions on Healthcare Systems Engineering*. Forthcoming.

Bale, K., Chapman, P., Barraclough, N., Purdy, J., Aydin, N., & Dark, P. (2007). Kaleidomaps: a new technique for the visualization of multivariate time-series data. *Information Visualization, 6*(2), 155–167.

Bate, A., Lindquist, M., Edwards, I. R., Olsson, S., Orre, R., Lansner, A., & de Freitas, R. M. (1998). A Bayesian neural network method for adverse drug reaction signal generation. *European Journal of Clinical Pharmacology, 54*(4), 315–321.

Borgatti, S P., Everett, M. G., & Johnson, J. C. (2013). *Analyzing social networks*. Washington, DC: Sage.

Bui, A., Aberle, D.R., Kangarloo, H. (2007). TimeLine: Visualizing Integrated Patient Records. *IEEE Transactions on Information Technology in Biomedicine, 11*(4): 462–473.

Chaudhry, B., Wang, J., Wu, S., Maglione, M., Mojica, W., Roth, E., … & Shekelle, P. G. (2006). Systematic review: impact of health information technology on quality, efficiency, and costs of medical care. *Annals of internal medicine, 144*(10), 742–752.

Card, S. K., Mackinlay, J. D., & Shneiderman, B. (1999). Readings in information visualization: using vision to think. Morgan Kaufmann.

Charnes, A., Cooper, W., & Rhodes, E. (1978). Measuring the efficiency of decision making units. *European Journal of Operational Research, 2*, 429–444.

Chen, Y. (2012). Data visualization in models for healthcare workflow improvement. Unpublished MS thesis, Penn State University.

Cooper, W., Seiford, L., & Tone, K. (2007). *Data envelopment analysis: A comprehensive text with models, applications, references and DEA-solver software*. New York, NY: Kluwer Academic.

Cousins, S. B., & Kahn, M. G. (1991). The visual display of temporal information. *Artificial intelligence in medicine, 3*(6), 341–357.

Dzemyda, G., Kurasova, O., & Žilinskas, J. (2013). Multidimensional data visualization. Methods and applications series: Springer optimization and its applications, *75*, 122.

Fairfield H. & Bloch M. (2013). "For the Elderly, Diseases That Overlap," The New York Times. Retrieved at: http://www.nytimes.com/

interactive/2013/04/16/science/disease-overlap-inelderly.html?hp&_r=1&

Farrell, M. (1957). The measurement of productive efficiency. *Journal of the Royal Statistical Society, 120*(3), 253–290.

Fayyad, U., Piatetsky-Shapiro, G., & Smyth, P. (1996). From data mining to knowledge discovery in databases. Retrieved from http://www.kdnuggets.com/gpspubs/aimag-kdd-overview-1996-Fayyad.pdf

Few, S., & EDGE, P. (2007). Data visualization: Past, present, and future. IBM Cognos Innovation Center.

Friedman, V. (2008). Data visualization and infographics. Graphics, Monday Inspiration, *14*, 2008.

Friendly, M. (2006). A Brief History of Data Visualization. In: Handbook of Computational Statistics: Data Visualization. Retrieved at: http://www.datavis.ca/papers/hbook.pdf

Fry, B. (2004). Computational Information Design. Master's thesis, Massachusetts Institute of Technology. Retrieved at: http://hdl.handle.net/1721.1/26913

Grinstein, Georges G., et al. "Benchmark development for the evaluation of visualization for data mining." Information visualization in data mining and knowledge discovery (2002): 129–176.

Jackson, M. O. (2008). *Social and economic networks*. Princeton, NJ: Princeton University Press.

Kraschnewski, J. L, Sciamanna, C., Stuckey, H. L., Chuang, C. H., Lehman, E. B., Hwang, K. O., ... Nembhard, H. B. (2013). A silent response to the obesity epidemic: Decline in US physician weight counseling. *Medical Care, 51*(2), 186–192.

Milley, A. (2000). Healthcare and data mining. *Health Management Technology, 21*(8), 44–47.

Muñoz, D., Alonso, W., & Nembhard, H. (2014). A social network analysis-based approach to evaluate workflow and quality in a pediatric intensive care unit. *Proceedings of the 2014 Industrial and Systems Engineering Research Conference*. 1–10.

Newman, M. E. (2010). *Networks: An introduction*. Oxford, England: Oxford University Press.

Rajwan, Y. G., & Kim, G. R. (2010, November). Medical information visualization conceptual model for patient-physician health communication. In Proceedings of the 1st ACM International Health Informatics Symposium (pp. 512-516). ACM.

Roy, B. (1996). *Multicriteria methodology for decision making*. Norwell, MA: Kluwer Academic.

Shearer, C. (2000). The CRISP-DM model: The new blueprint for data mining. *Journal of Data Warehousing, 5,* 13–22.

Tan, P., Steinbach, M. & Kumar, V. (2005). *Introduction to Data Mining*. New York, NY: Pearson.

Tucker, C. S., Behoora, I., Nembhard, H. B., Lewis, M., Sterling, N., Huang, X. (2015). "Machine Learning Classification of Medical Adherence in Patients with Movement Disorders Using Non-Wearable Sensors," *Computers in Biology and Medicine*, 66, 120–134. DOI: 10.1016/j.compbiomed.2015.08.012

Tucker, C. S., Han, Y., Nembhard, H. B., Lee, W.-C., Lewis, M., Sterling, N., Huang, X. (2015). "A Data Mining Methodology for Predicting Early Stage Parkinson's Disease Using Non-Invasive, High Dimensional Gait Sensor Data," *IIE Transactions on Healthcare Systems Engineering*, 5, 4, 238–254.

U.S. Preventive Services Task Force. (2011). USPSTF recommendations. Retrieved from http://www.uspreventiveservicestaskforce.org/uspstopics.htm

Varian, H. (2009). Hal Varian on how the Web challenges managers. *The McKinsey Quarterly, 1.*

Ward, M. O., Grinstein, G., & Keim, D. (2010). *Interactive data visualization: foundations, techniques, and applications*. CRC Press.

Ware, M., Frank, E., Holmes, G., Hall, M., & Witten, I. H. (2001). Interactive machine learning: letting users build classifiers. *International Journal of Human-Computer Studies*, *55*(3), 281–292.

Witten, I., Frank, E., & Hall, M. (2011). *Data mining: Practical machine learning tools and techniques*. Burlington, MA: Morgan Kaufmann.

Yu, P. L. (1985). *Multiple criteria decision making*. New York, NY: Plenum Press.

Zheng, K., Haftel, H. M., Hirschl, R. B., O'Reilly, M., & Hanauer, D. A. (2010). Quantifying the impact of health IT implementations on clinical workflow: a new methodological perspective. *Journal of the American Medical Informatics Association*, *17*(4), 454–461.

Zeleny, M. (1982). *Multiple criteria decision making*. New York, NY: McGraw-Hill.

Zhu, X., & Davidson, I. (2007). *Knowledge discovery and data mining: Challenges and realities* (pp. 31–48). New York, NY: Hershey. ISBN 978-1-59904-252-7.

Chapter 10

Capacity Management

"I violated the Noah rule: Predicting rain doesn't count; building arks does."

—Warren Buffett

Overview

In healthcare settings, there are several resources where effective utilization is important. Examples include scheduling diagnostic equipment for inpatient and outpatient services and scheduling operating rooms for surgeries. Similarly, effective schedules need to be developed for the supporting workforce such as the nursing staff in the ICU. The significant variability and uncertainty that the system faces makes the issue of capacity management a challenge. In this chapter we discuss four capacity management problems: bed management, operating room scheduling, diagnostic scheduling, and nurse scheduling.

10.1 CAPACITY MANAGEMENT CHALLENGES

Capacity decisions can be broken into strategic, tactical, and operational depending on the time horizon of the decision. Strategic decisions essentially define the capacity (e.g., number of operating rooms), and have an impact over multiple years. Tactical decisions are typically concerning how to introduce the strategic decisions into practice, and have an impact from monthly to yearly. Operational decisions define how the capacity will be used on a daily or weekly basis. Our focus in this chapter is around operational issues of capacity management.

There are several challenges in hospital capacity management, which include:

- **High fixed costs:** Many of the resources that a hospital uses including hospital beds, operating rooms, and imaging equipment have very high fixed costs. Further, many of these assets are specific for particular procedures. As such, it is typically not possible to quickly change capacity levels. This puts increased importance in the management of these assets (Green, 2004).
- **Unknown and uncertain demand:** Demand for services can be both unknown and uncertain, both of which lead to a high level of variability. For example, visits to the emergency department (ED) during flu season have both an unknown and uncertain component. It is known that ED visits will increase significantly, though predicting the start and length and severity of that season is a challenging at best (unknown). Further, the actual number of ED visits varies over time (uncertain). Demand variability from either source puts strain on available capacity.
- **Unknown and uncertain procedure times:** The time to complete a procedure such as a surgery has unknown and uncertain components. The uncertainty comes from the fact that each patient is unique as are each team of healthcare professionals. Both lead to natural variability in the time to complete the procedure. The unknown part comes from unanticipated complications that can arise and lead to additional procedures that need to be performed. As with the case of demand, variability in procedure times also significantly impacts capacity utilization.
- **Challenge of healthcare professional staffing:** There is a broad range of hospital professionals for using capacity, each managed in a different way. For example, nurses are typically assigned to shifts, interns serve clinical rotations, and surgical groups can be assigned to blocks of time in an operating room (OR).
- **System complexity and interdependencies:** Units within a hospital are not independent of each other. For example, transferring a patient from the ED to the intensive care unit (ICU) depends on the occupancy of the ICU. If no beds are available, the patient may need to be boarded. Further, resources such as diagnostics are shared across units. Controlling the complex network of activities is therefore challenging.

- **Differences in patient types:** Patients have several different characteristics including type of condition, severity of condition, insurance status, and inpatient versus outpatient requirement. This adds to the complexity of the system.
- **Performance measures:** Hospitals measure themselves on a wide range of measures. These include length of stay, percutaneous coronary intervention received within 90 minutes of hospital arrival, and mortality. Defining the appropriate measures is challenging, and the wrong choice can lead to unintended consequences.

Each of these challenges will be emphasized in the following sections.

10.2 MANAGING NURSING UNITS

One of the most important capacity issues in a hospital is managing inpatient beds in a nursing unit. The number of beds, however, is not something that is easily changed. In the United States, for example, the majority of states have a Certificate of Need (CON) program where hospitals must have requests for new bed acquisitions reviewed and approved by state health planning agencies. The stated purpose of CON programs is to try to limit price inflation that could result from overcapacity. The widely cited Roemer's Law that "in an insured population, a hospital bed built is a filled bed" (Shain & Romer, 1959) provided the initial justification. There is significant debate as to the efficacy of such CON programs, but from an operational perspective it means that changing bed capacity is a time-consuming process. Therefore, additional emphasis is placed on how beds are used in the process. In particular, hospitals typically try to maintain high occupancy rates.

In addition to the ED, there are three types of inpatient units that need to be considered in bed management: *intensive care* units (ICU), *telemetry* (also called "step-down") units, and *medical surgical* (or "med surg") units. Med surg units are often called general care units (GCUs). The relationship among the three units is shown in Figure 10.1. Patients can arrive at these units from a variety of streams including the ED, referrals, operating rooms, and patients with elective procedures.

ICUs typically hold patients with severe or life-threatening conditions. These patients require invasive monitoring and other procedures using specialized equipment. This can include mechanical ventilation if the patient has lost the ability to breathe normally, arterial lines for continuous

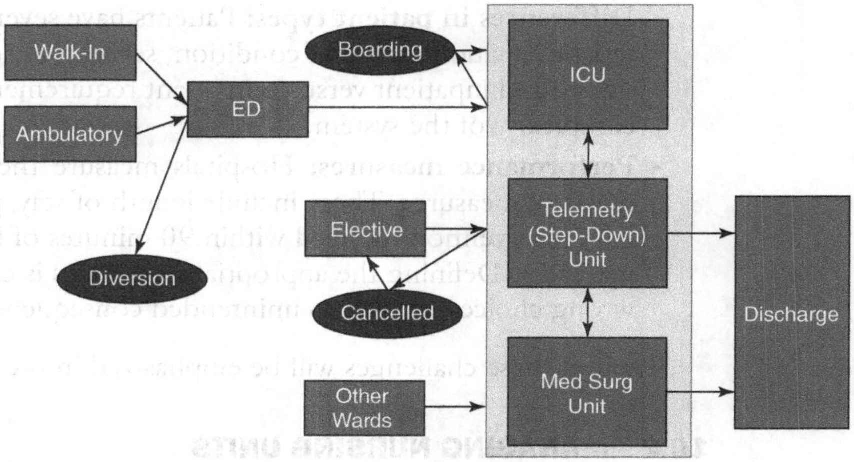

Figure 10.1 Units and Flow in Bed Management

blood pressure monitoring, tube feeding if the patient is unable to eat, and pacemakers if the patient's heart is not beating regularly. There are different types of ICUs including those for premature newborn infants (neonatal ICU or NICU) and those for children (pediatric ICU or PICU). Specialized personnel are required in ICUs.

Patients in telemetry units require continuous electronic monitoring. This includes heart rate and breathing monitoring. Although patients in telemetry units are acutely ill and require continuous monitoring, they are in a less severe condition than those in an ICU. As such, telemetry units often serve as a step down from the ICU. Typical patients may include patients with angina or congestive heart failure or those that have undergone heart valve replacement or coronary artery bypass surgery. An advantage of using remote telemetry is that it allows a nurse to monitor several if not all patients in a unit from a single nursing station.

Med surg units are standard beds in wards that do not require specialized monitoring and cover a broad area of practice from stroke units to burn units to obstetrics units to oncology units.

Some hospitals make use of observation units to hold those patients for whom they are not sure whether they should be admitted as inpatients or not. These units typically have multiple beds in them and have lower nurse to patient ratios than inpatient areas. Patients are typically held in an observation unit for no more than a day.

Patients are typically not discharged from an ICU, but may be from either a telemetry or med surge unit. Although in most cases patient flow is from most severe to least severe units, if their condition deteriorates, a patient may also transfer to a higher critical unit.

Key decisions in inpatient bed management include determining when to transfer a patient from one unit to another and when to discharge a patient. Before discussing these topics, we first present two important issues impacted by bed management, namely, boarding and diversion.

Boarding

When demand is unpredictable, maintaining high bed utilization can lead to regular bed shortages. When patients need a bed and one is not available, this can lead to cancellations of admissions for elective procedures or delaying admissions for emergency patients, as shown in Figure 10.1. It can also lead to delaying transfer between wards. If an admitted patient from the ED cannot be sent to the appropriate unit, a common practice is ***boarding***. This involves leaving the patient on a temporary bed or stretcher in the hallway until an appropriate bed becomes available. Boarding increases the length of stay for a patient. More importantly, it can significantly increase mortality rates.

We illustrate boarding using a simplified queueing model of an ICU (Figure 10.2). In this example patients arrive to the unit and are assigned a bed if one is available. If all of the n beds are occupied, then the patient is boarded. We assume that all patients are of the same type and hence are all from the same ICU length-of-stay (LOS) distribution. Patients arrive to the system according to a Poisson process with rate λ. The length of stay for a patient in the ICU is according to a general distribution with mean ν with variance σ^2. After their time in the ICU, patients are sent to a step-down unit that is not modeled in this example. The example therefore corresponds to an $M/G/n$ queueing system.

There is not an exact form for the waiting time, W, in an $M/G/n$ system. However, a common approach is to use an approximation based on the $M/M/n$ system (where the length of stay is exponentially distributed

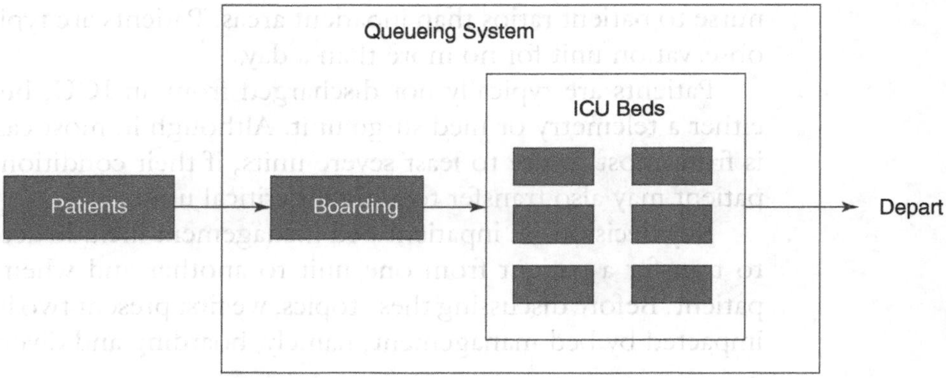

Figure 10.2 ICU Queueing System

with average time $1/\mu$). For this case, the expected number of patients that are boarded, $N_B{}^{M/M/n}$, is defined by:

$$N_B{}^{M/M/n} = \frac{\left(\frac{(n\rho)^n}{n!}\right)\left(\frac{1}{1-\rho}\right)}{\sum_{k=0}^{n-1}\frac{(n\rho)^k}{k!} + \left(\frac{(n\rho)^n}{n!}\right)\left(\frac{1}{1-\rho}\right)} \tag{10-1}$$

where $\rho = \lambda / n\mu$ is the average system utilization (assuming $\rho < 1$). Further, the average time $W^{M/M/n}$ in the system (length of stay plus boarding time) is defined by:

$$W^{M/M/n} = \frac{1}{\mu} + \frac{N_B}{\lambda} \tag{10-2}$$

These can be converted to $M/G/n$ system measures by applying the correction factor $(C^2 + 1)/2$ to the $M/M/n$ formulas, where C^2 is the coefficient of variation of the length of stay (i.e., $C^2 = (\sigma^2 / \nu^2)$). Therefore, the average time in the system (W) and average number in boarding (N_B) is:

$$W = \frac{C^2+1}{2} W^{M/M/n} \quad N_B = \frac{C^2+1}{2} N_B{}^{M/M/n} \tag{10-3}$$

Consider the instance of a 20 beds ICU where the arrival rate is 3.5 patients per day and the length of stay has a mean of 5.3 days with standard deviation of 4.0 days. For this case, the average utilization is 3.5(4.0) / 20 = 0.93. Solving the equations just given we obtain $W = 7.19$ and $N_B = 6.61$. Applying Little's law gives a mean number in the system $L = 3.5(7.19) = 25.16$.

There are several things that can be changed in order to improve the ICU system performance measures. These include increasing the number of beds in the ICU, reducing the length of stay variability (and hence reducing C^2) through better staffing and resource management, and reducing the average length of stay by improving discharge management. Figure 10.3 shows the impact of increasing the number of beds on the total time in the system W keeping all other parameters the same. There must be at least 19 beds or the system utilization would exceed 1. Decreasing the number of beds from 20 to 19 increases W by 87%. Going from 20 to 21 beds decreases W by 15%. Notice, however, that there are diminishing returns to the number of beds (e.g., going from 21 to 22 beds decreases W by only 6%).

Figure 10.4 shows the impact of the coefficient of variation of time in the ICU on the average number of boarded patients. In this figure, the average LOS was 5.3 days and the standard deviation of LOS was allowed to vary. Notice that as the patient arrival rate increases, N_B increases exponentially. This is due to the system utilization approaching 1 as the arrival rate approaches 3.75. Note further that the average number of boarded patients increases in the coefficient of variation. For example, if the arrival

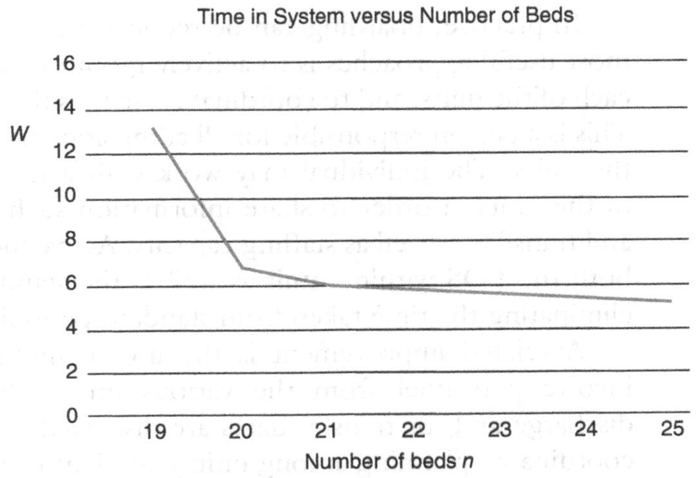

Figure 10.3 Impact of Number of Beds on Average Time in the ICU System

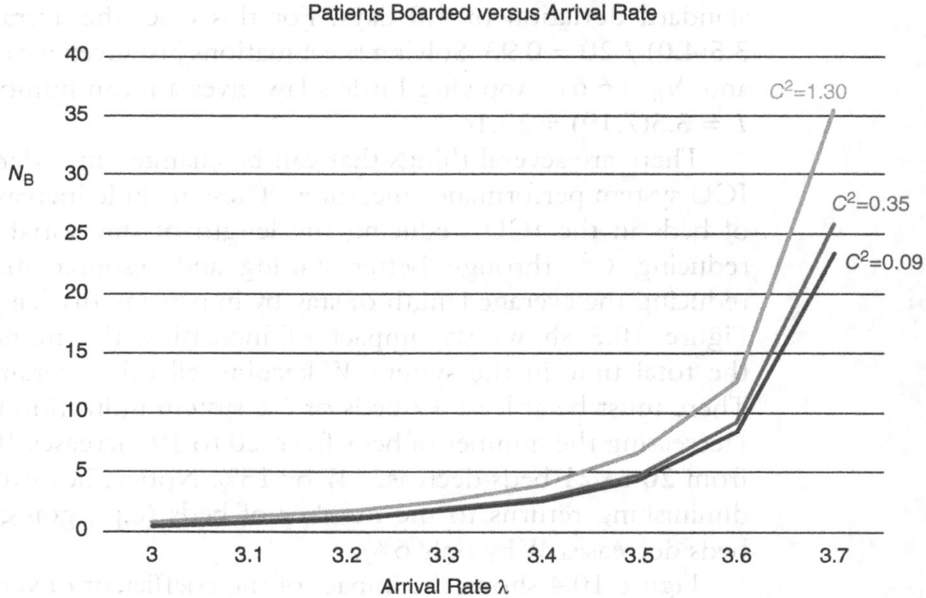

Figure 10.4 Number of Patients Boarded as a Function of Arrival Rate for Various Squared Coefficients of Variation

rate were 3.7, system utilization is 0.98. Decreasing the standard deviation of the time in the ICU can therefore have a significant impact on number of patients boarded.

In practice, boarding can be reduced by several methods. One of the most useful approaches is to actively monitor and manage the patients in each of the units, and to coordinate activities through the use of a *bed czar*. This is a person responsible for all admissions and transfers into and out of the units. The individual may work with a team of individuals for each of the units in order to share information such as anticipated discharges and transfers as well as staffing capacity. Active monitoring helps to reduce both the LOS within a unit as well as the standard deviation of LOS by eliminating the time taken from standard periodic decision making.

A related improvement is the use of multidisciplinary rounds that involve personnel from the various units. During rounds, predicted discharge and/or transfer dates are discussed. This helps establish better coordinated planning among units and identify potential shortages before they occur.

Finally, the policies used for admission, transfer, and discharge can have a big impact on boarding. Several hospitals have found that shifting the peak patient discharge time by a few hours can have a significant impact on boarding. In addition, reserving some capacity in the ICU for certain patient types can also have a positive impact. We examine models for each of these policies in a later section.

Diversion

If there are not enough beds or staff available in the hospital ED to meet the demand in a satisfactory way, the hospital goes on what is called *ambulance diversion* (AD). When this happens, patients transported by ambulance are routed to another hospital not on diversion. Hospitals are typically still required to take walk-in patients to the ED, but the rerouting of ambulances effectively reduces the patient arrival rate to that hospital. Hospitals will then go off of diversion after capacity becomes sufficient again, though typically there is some maximum period of time (e.g., three hours) that a hospital can remain on diversion.

The notification and ambulance rerouting process is typically managed by an area emergency medical service. In some cases, there may be a partial diversion where only the highest-priority patients are transported to the hospital ED, while lower-priority patients are diverted to a hospital close by.

There are several costs when diversion does occur. First, from the patient's perspective, ambulance rerouting can increase the time before they receive care. This can negatively impact patient morbidity and according to some studies patient mortality. Further, they may be rerouted to a hospital that was not their primary choice. Additionally, there is a cost to the hospital on diversion since a large proportion of ED patients are admitted into the hospital, and revenue from inpatient services tends to be the most significant component of overall revenue for most hospitals.

Ambulance diversion is becoming an increasingly important problem. Reducing AD can come from better guideline enforcement and information sharing at the community level and/or an investment in human and facility resources at the hospital level. We illustrate a simple model of AD in the following example (Hagtvedt et al., 2009).

Example 10-1: Diversion in a Two-Bed Unit

Consider the simple case where a unit consists of two beds. If the unit becomes full, then it goes on diversion until it is empty. Patients arrive according to a Poisson process with rate λ, and the length of stay is exponentially distributed with mean $1/\mu$. The birth-death diagram for this example is shown in Figure 10.5. Each node represents the state of system and the transitions are represented by arcs. Nodes 0, 1, and 2 represent the number of beds occupied in the unit and node 1,1 represents the case where the unit is on diversion with one bed empty and one bed remaining; the unit comes off of diversion once that patient leaves.

The balance equations for each of the nodes are given by:

$$\mu P_1 + \mu P_{1,1} = \lambda P_1$$

$$\lambda P_0 = \mu P_1 + \gamma P_1$$

$$\lambda P_1 = 2\mu P_2$$

$$2\mu P_{21} = \mu P_{1,1}$$

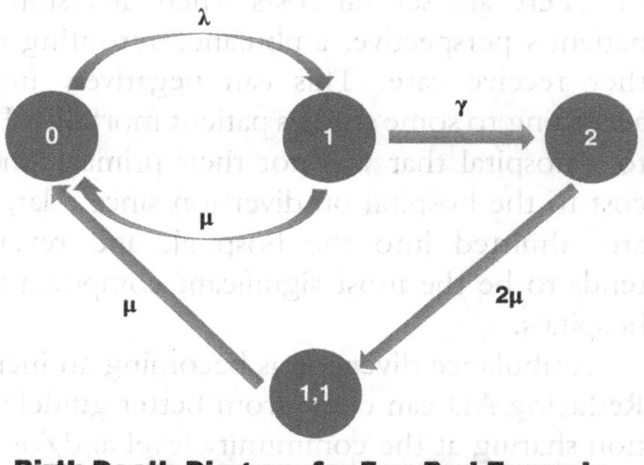

Figure 10.5 Birth-Death Diagram for Two-Bed Example

Solving this system of equations for the probabilities yields:

$$P_0 = \frac{2\mu(\mu+\gamma)}{2\mu(\mu+\gamma+\lambda)+3\gamma\lambda}$$

$$P_1 = \frac{2\mu\lambda}{2\mu(\mu+\gamma+\lambda)+3\gamma\lambda}$$

$$P_2 = \frac{\lambda\gamma}{2\mu(\mu+\gamma+\lambda)+3\gamma\lambda}$$

$$P_3 = \frac{2\lambda\gamma}{2\mu(\mu+\gamma+\lambda)+3\gamma\lambda}$$

If the arrival rate to the unit (λ) cannot be controlled directly, then the only variables that can help to reduce diversion are increasing service rate μ and decreasing the arrival rate γ. The marginal effect of γ can be found by:

$$\frac{\partial P_{1,1}}{\partial \gamma} = \left(\frac{2\lambda}{2\mu\,(\mu+\lambda+\gamma)+3\gamma\lambda}\right)\left[1-\frac{\gamma\,(2\mu+3\lambda)}{2\mu(\mu+\gamma+\lambda)+3\gamma\lambda}\right]$$

Note that this term is clearly positive for any feasible value of the parameter.

Transferring Patients

Changing the capacity of an ICU tends to be a very long-term decision. For this reason, developing effective policies to control the patient flow are of particular importance (Hopp & Lovejoy, 2013). One of the key decisions is when to transfer a patient from the ICU (i.e., a patient step-down). The delay from transfer request to actual transfer can be significant, with an average of around 9 hours and a maximum of over 24 hours (Zhang, 2015). Some hospitals resort to solutions of discharging patients early or rationing admissions, though early discharge may lead to a later readmission. Interestingly, some studies have also shown that keeping a patient for too long in the ICU can also increase mortality rates. Average mortality rates for patients admitted into an ICU range from 10% to 29%.

Transfer policies are most effective for relatively highly utilized ICUs. If the utilization rate is low, then there is no pressure for beds from arriving patients, and so the only driver is transferring a patient out when they are ready. If the utilization is too high, then the policy choice will have little impact since the ICU will always be full.

An example of a transfer policy is where patients with higher severity, but who do not require intensive monitoring, are stepped down. Under given costs related to rejection of ICU and GCU patients, an appropriate threshold can be obtained. That is, when the number of patients reaches a threshold and an ICU transfer request is made for one of the feasible step-down patients. In this way, pressure is removed from the ICU, which can results in higher efficiency. The higher the threshold, the higher the amount of reserved capacity that is held for incoming patients to enter.

The following example models ICU and GCU loading as multisource, multiserver queues, restricted to threshold policies; that is, when the number of patients reaches a threshold, an ICU transfer is conducted (Zhang, 2015). It is important to note that the policy is not globally optimal but only compared to those policies that use a threshold. It is also easily implementable.

Example 10-2: Transfer Policies in an ICU

A nursing unit consists of an ICU and GCU. Both ICU and GCU patient arrivals follow a Poisson process and service times are exponentially distributed. The number of beds in each unit are defined by N_I and N_G, respectively. Queues are allowed to form in front of each unit (which corresponds to boarding). However, when the queue for the ICU exceeds Q_I, then those patients are diverted. Similarly, patients are diverted if the queue exceeds Q_G for the GCU.

There are two stages for patients in the ICU. Newly arriving ICU patients enter stage 1, which represents a critical condition stage and must be treated in the ICU at service rate μ_1 On completion of stage 1 service, patients leave the system with probability h. This corresponds to the case that the patient condition deteriorates and is sent to the OR or the patient dies. With probability $1 - h$, stage 1 patients enter stage 2, which represents the recovery stage. During this stage patients can either be treated in the ICU or GCU. If the patient stays in the ICU, the service rate is μ_2, if the patient is transferred to the GCU, the service rate in the GCU for ICU stage 2 patients is μ_I. It is assumed that $\mu_2 \geq \mu_I$ since the ICU generally provides better care than GCU. Therefore, it takes less time for patients to recover in the ICU than GCU.

The transfer conditions are as follows: If there is a stage 2 patient in the ICU and the total number of patients in the ICU system (including patients admitted and patients in queue) reaches a threshold $M(1 \leq M \leq N_I + Q_I)$ and there is a vacant bed in the GCU, a stage 2 patient will be transferred to the GCU. In this example, we assume that the transfer is instantaneous with no preemption. Due to the memoryless property of the exponential distribution, stage 2 patients transferred to GCU will start a new stay with rate μ_I no matter how long they have stayed in ICU.

1. **Transfer Priority:** We will assume ICU patients always have higher priority than GCU patients, meaning that if there is a bed in the GCU is vacated, an ICU stage 2 patient will be transferred before a GCU patient in the queue is admitted.
2. **Objective function:** Minimize the cost that consists of the penalty for diverting both ICU and GCU patients (patients are not admitted and diverted to other facilities).

The parameters for the example are given in Table 10.1.

Table 10.1 Numerical Parameters When GCU Has Finite Capacity

ICU	$N = 8$	$Q = 3$	$\lambda = 120$	$\mu_1 = 30$	$\mu_2 = 6$
GCU	$N = 11$	$Q = 3$	$\lambda = 60$	$\mu_G = 7$	$\mu_I = 6$

We define a "bumpout" policy as one where a patient in stage 2 is stepped down as long as there is an open bed in the GCU. This is a common practice, and would correspond to $M = 9$ in this example. An alternative policy would be to step down a stage 2 patient whenever a bed becomes available in the ICU.

This corresponds to a threshold of $M = 1$. Note that in the case the ICU rejection probability is reduced by 18.3% by moving from $M = 9$ to $M = 1$. The optimal threshold depends on the cost ratios of rejecting the ICU versus GCU patients and having the ICU versus GCU patients waiting in the queue.

10.3 MANAGING OPERATING ROOMS

Most hospitals have multiple ORs that they control. A common approach for managing the ORs is block scheduling (Hans et al., 2007). In this case, a time segment of an operating room (or block) is reserved for a physician or surgical group (such as cardiology). The "owner" of the block has exclusive rights for how the time is used. If it is known in advance that there will be unused time, then the OR may be made available to other physicians or groups.

Block scheduling is commonly used because it results in groups of similar procedures being done close together. This helps to reduce setup times since similar procedures have similar resource requirements. In addition, there is a variation pooling effect by combining procedures into a block as opposed to assigning procedures individually.[1] Finally, the amount of idle time can be at least indirectly controlled by the proportion of total availability that the OR rooms are blocked. In addition, since a surgical group can batch patients in a block, then if a procedure finishes earlier than expected, the next procedure can be started right after the setup. This helps to reduce the idle time between surgeries. This would be much more difficult operationally if each surgery were individually scheduled without regard to surgical group.

The goal in block scheduling is to develop a method that equitably assign blocks to physicians or surgical groups and achieves high utilization (and/or contribution margin) of the rooms. Note that typical

[1] This is a direct result of the statistical averaging of variation. To see this, suppose X_1 and X_2 are iid random variables that represent time for surgical procedure 1 and 2. In addition, suppose that they are normally distributed with mean μ and variance σ^2. If we combine X_1 and X_2 in a block, the standard deviation of the block time is $\sqrt{2}\sigma$. If X_1 and X_2 are not blocked, then the standard deviation of the sum of the procedures is $2\sigma > \sqrt{2}\sigma$.

surgical groups include general, cardiology, gynecology, orthopedics, and neurology. There are two steps that need to be solved:

- **Stage 1:** Determine the proportion of capacity to be allocated to each surgical group (i.e., blocks). This is typically done based on historical needs, overtime considerations, and margin that the group brings per unit of time (though this last measure is not always explicitly stated).
- **Stage 2:** Assign surgical cases to blocks to achieve high utilization while satisfying the surgical group desires, capacity limits, and other constraints.

Surgeries generally fall into two categories: those that can be planned in advance (elective surgeries) and those that are unplanned (emergency surgeries). Block schedules are clearly only effective for elective surgeries. Note that many hospitals reserve spare capacity in the ORs in order to help deal with complications that may arise in elective surgeries as well as for emergency cases that come up. In addition, some hospitals will block schedule a subset of the ORs and leave some to exclusive handle emergency cases.

Determining Block Times

Both stages 1 and 2 can be solved through the use of optimization. Let's first consider the stage 1 problem, that is, determine the block times from historical data. The following formulation is based on the work of Hosseini and Taafe (in press). There are K surgical groups that used the ORs in a hospital over the past N weeks (where block schedules are used Monday ($i = 1$) to Friday ($i = 5$), and weeks are represented by $j = 1, \ldots, N$). The hours of surgery that group k performed on day i of week j is t^i_{jk}. The maximum available number of OR hours on day i is represented by T_i. Undertime cost u^i_{jk} is incurred if group k was assigned more than t^i_{jk} hours and overtime cost o^i_{jk} is incurred if the group was assigned less. Let h represent the ratio of overtime to undertime costs. We can solve stage 2 by determining the number of hours assigned to group k on day i of the week (x^i_k) in a way that minimizes overtime and undertime costs. The resulting formulation is given by:

$$min \sum_{k=1}^{K} \sum_{j=1}^{N} (u^i_{jk} - ho^i_{jk}) \qquad (10\text{-}4)$$

subject to:

$$\sum_{k=1}^{K} x_k^i \leq T_i \tag{10-5}$$

$$o_{jk}^i \geq t_{jk}^i - x_k^i \ \forall j, k \tag{10-6}$$

$$u_{jk}^i \geq x_k^i - t_{jk}^i \ \forall j, k \tag{10-7}$$

$$x_k^i, o_{jk}^i, u_{jk}^i \geq 0 \ \forall j, k \tag{10-8}$$

This formulation is solved for each day of the week. Constraint (10-5) ensures that available time is not exceeded. Constraints (10-6) and (10-7) ensure that overtime and undertime cannot occur simultaneously.

Example 10-3: Stage 1 Block Scheduling

Consider a hospital that has two surgical groups and two operating rooms that it block schedules. Each operating room is available eight hours a day for five days a week ($h = 1.2$). Over the past three weeks, the following data have been collected for Tuesdays ($i = 2$):

	Week		
Surgical Group	**1**	**2**	**3**
1. Cardiac	9	7	6
2. Orthopedic	9	9	12

The resulting formulation is given by:

$$min \sum_{k=1}^{3} \sum_{j=1}^{4} (u_{jk}^2 - 2o_{jk}^2)$$

subject to:

$$\sum_{k=1}^{3} x_k^2 \leq 16\ \forall j, k$$

$$o_{11}^2 \geq 9 - x_1^2$$

$$o_{21}^2 \geq 7 - x_1^2$$

$$o_{31}^2 \geq 6 - x_1^2$$

$$o_{12}^2 \geq 9 - x_2^2$$

$$o_{22}^2 \geq 9 - x_2^2$$

$$o_{32}^2 \geq 12 + x_2^2$$

$$u_{11}^2 \geq -9 + x_1^2$$

$$u_{21}^2 \geq -7 + x_1^2$$

$$u_{31}^2 \geq -6 + x_1^2$$

$$u_{12}^2 \geq -9 + x_2^2$$

$$u_{22}^2 \geq -9 + x_2^2$$

$$u_{32}^2 \geq -12 + x_2^2$$

The solution to this formulation gives an allocation of 7 hours to the cardiac group and 9 hours to the orthopedic group. Note that the solution did not assign the full 16 hours of available capacity. This is because the cost of overtime is greater than the cost of undertime, and on two of the past three Tuesdays, overtime was required. The remaining 2 hours, then, would be used as spare capacity in the anticipation of overtime from these groups.

Block Scheduling

The stage 2 problem can also be solved through the use of optimization, and several integer programming strategies have been suggested. One approach is to determine the assignment of surgeries to blocks in a way that minimizes the unused OR time (Hans, Nieberg, & van Oosrum, 2007). The inputs for this approach are the maximum total time allocated to the surgical group (T_{max}), the list of available durations of one block (c_l, where $l = 1, \dots, L$), the surgical cases to be performed (S_i), the duration of surgery S_i (d_i, where i = 1, … , S), and the number of times surgery S_i occurs (s_i). Note that T_{max} is determined in stage 1. The decision variables are the number of times that surgery i is assigned to block k (V_{ik}, where $k = 1, \dots, K$) and duration of each block from list L, where a binary variable y_{kl} is used to indicate while duration from the list is chosen for block k. The resulting formulation is:

$$min \sum_{k=1}^{K} \left(\sum_{l=1}^{L} c_l y_{kl} - \sum_{i=1}^{S} V_{ik} d_i \right) \tag{10-9}$$

subject to:

$$\sum_{k=1}^{K} V_{ik} = s_i \ \forall i \tag{10-10}$$

$$\sum_{l=1}^{L} c_l y_{kl} - \sum_{i=1}^{S} V_{ik} d_i \geq 0 \ \forall k \tag{10-11}$$

$$\sum_{k=1}^{K} \sum_{l=1}^{L} c_l y_{kl} \leq T_{max} \tag{10-12}$$

$$\sum_{l=1}^{L} y_{kl} \leq 1 \ \forall k \tag{10-13}$$

$$y_{kl} \in \{0, 1\}, \ V_{ik} \in \{int\} \tag{10-14}$$

Constraint (10-10) ensures each surgery is planned. Constraint (10-11) expressed the nonnegative unused time per block. Constraint (10-12) ensures that the total time isn't exceeded, and constraint (10-13) ensures that only one item from the list is chosen for the block duration. Note that one benefit of this approach is that the surgical groups can influence the result by the choice of setting the block durations.

Block Scheduling in Practice

An example of a simple block schedule is shown in Table 10.2 for four surgical groups in three operating rooms. In this schedule, there are blocks of open times that were not scheduled. As mentioned previously, this can help to buffer against uncertainty that can arise from the procedures (complications, cancellations, etc.). As mentioned previously, there is a benefit of variation pooling that result from blocking, and the larger the block the bigger the benefit. However, large blocks reduce flexibility of the resource. Balancing this trade-off is important for the final schedule.

One of the issues of block scheduling is that updates are frequently required to the schedule. For example, a physician may have a high-priority case that arose after the schedule was developed. Frequent updates to the schedule are therefore required. In order to minimize the frequencies of updates, many hospitals will "freeze" the schedule several days out. Changes are not allowed to the schedule with the freeze time unless the criticality of the surgery meets an urgency threshold. One challenge of this approach is agreement between the physician and the scheduling team of what qualifies as urgent.

Several other issues are important in block scheduling as well. In this section we has assumed that the surgical team includes all of the personnel required including physicians, nurses, and anesthesiologists. In many situations, however, nurses and anesthesiologists are not part of a specific team, and may be thought of as a shared resource. In this case, the scheduling problem becomes more complicated because personnel schedules must also be developed. In addition, resources such as surgical kits must also be managed in the process.

Table 10.2 Block Schedule Example

	Time							
	8–9	**9–10**	**10–11**	**11–12**	**12–1**	**1–2**	**2–3**	**3–4**
OR 1	OBGYN	OBGYN	CARDIO	CARDIO	CARDIO	CARDIO	CARDIO	
OR 2	OPTH		OBGYN	OBGYN	OBGYN		GEN2	GEN2
OR 3	ORTHO	ORTHO	ORTHO	ORTHO		GEN1	GEN1	GEN1

10.4 MANAGING DIAGNOSTIC UNITS

Diagnostic services are part of the ancillary services in a hospital and typically include clinical laboratory/pathology, imaging, and nonimaging departments. The imaging department manages several advanced technologies including magnetic resonance imaging (MRI), x-ray, ultrasound, and computed tomography (CT) scanners. The nonimaging tests include electrocardiograms (ECGs) and electroencephalograms (EEGs).

Most patient diagnosis and treatment determination are determined to a large extent by the results of diagnostic services. Since some of the supporting resources, particularly in imaging, are quite expensive, capacity management is a crucial to effective use of the resources. The focus here then is on imaging units.

Demand for imaging can come from three sources: inpatients, outpatients, and emergencies. Emergency requests typically receive the highest priority and will be immediately scheduled on request. The decision, then, is how to prioritize between an outpatient (which is scheduled) and an inpatient (which is non-scheduled) (Green, Savin, & Wang, 2006). If priority is given to the outpatient, this can lead to a longer length of stay for the corresponding inpatients. Giving priority to an unscheduled inpatient will cause outpatients to wait, which can lead to future lost business due to the inconvenience.

Consider the case where there is a defined set of time slots for a piece of imaging equipment such as an MRI over a day. In order to simplify the analysis, we will assume that they are identical slots (e.g., 30 minutes in length) and that the examination time is independent and identically distributed regardless of patient type. The use of the MRI receives revenue of R_i and R_o for each inpatient and outpatient exam, respectively. In addition, if a service delay leads to a cost per period of W_i and W_o for each patient type. Finally, at the end of the day, if a patient isn't seen, then this leads to a penalty cost of P_i and P_o for inpatients and outpatients. An easily implementable heuristic (critical first) to determine which patient type to prioritize is as follows:

Critical-First Heuristic

- If $C_i = R_i + W_i + P_i > C_o = R_o + W_i + P_i$, then serve inpatients first.
- If $C_i < C_o$, then serve outpatients first.

Green et al. (2006) showed that this heuristic worked extremely well compared to a much more complicated optimal threshold policy under the stated assumptions.

Example 10-4: Priority in MRI Scheduling

Consider the case where $R_i = \$800$, $R_o = \$1,600$, $W_i = \$0$, $W_o = \$80$, $P_i = \$3,000$, $P_o = \$300$. The resulting criticalities are found to be $C_i = 800 + 0 + 3,000 = 3,800$ and $C_o = 1,600 + 800 + 300 = 2,700$. Therefore, inpatients should be given priority over outpatients.

Note that if patient examination times can differ, the heuristic can be easily modified. In this case, let T_j be the expected examination time for inpatient *j* waiting to use the MRI and T_k be the expected examination time for outpatient *k* that is waiting. For each inpatient waiting for service, let $C_{ij} = C_i \,/\, T_j$ and for each outpatient waiting for service, let $C_{oj} = C_o \,/\, T_k$. From all patients waiting, serve the one with the greatest critical ratio.

The use of critical ratios is a simple approach. However, an important issue that can arise with their use is that it is possible to serve later arriving demand before seeing earlier arriving demand. This may add to patient dissatisfaction. Of course, if it is possible to quantify the level of dissatisfaction, then the critical ratio can easily be modified to include it.

10.5 NURSE STAFFING AND SCHEDULING

To this point we have discussed management of physical assets. However, none of the approaches are viable unless the appropriate workforce is available to support it. We therefore give a brief introduction to healthcare workforce staffing and scheduling. In particular, we discuss methods for nurse scheduling.

Most hospitals run three 8-hour shifts, though nurses typically work either 8-hour or 12-hour shifts. In order to accommodate this, we will therefore break the day into four periods: an 8-hour period (P1; 7 A.M.–3 P.M.), followed by a 4-hour period (P2; 3 P.M.–7 P.M.), followed by a 4-hour period (P3; 7 P.M.–11 P.M.), followed by an 8-hour period (P4; 11 P.M.–7 A.M.).

In this way, we can consider five shift types: three 8-hour shifts (P1, P2 + P3, and P4), and two 12-hour shifts (P1 + P2, P3 + P4).

The nurse staffing problem is the determination of the requirements (i.e., number of nurses) needed over each period. This can clearly vary over the day depending on the activities the nursing staff needs to perform. The nurse staffing problem is typically performed by a staffing committee, though some states often have mandatory nurse to patient ratios. Acuity of patients is also important in making staffing decision, and several studies have showed a direct link between staffing levels and patient outcomes, in particular that too low a staffing level will lead to increased morbidity and mortality.

The nurse scheduling problem assigns nurses to specific shifts. Several studies have shown that involvement by the nursing staff in the development of their schedules is important for schedule acceptance (Bailyn et al., 2007). Schedules may be seen as prone to favoritism or simply unfair, which may lead to a high rate of absenteeism and low morale. One way to directly involve nurses is through the use of a self-scheduling system. In this case, nurses use a variation of a signup sheet and can make requests for changes that is typically handled by a nurse manager. Some variations on this approach have used voting or auction mechanisms to help better prioritize preferences.

It is also possible to include preferences (or penalties) in an optimization framework for scheduling. Bard and Purnomo (2005) develop a model to minimize the cost of assignment will meeting needs. The following notation is used:

N = set of nurses to be scheduled

M = penalty cost for outside nurses

D = days in the planning period

S_i = set of schedules considered for nurse i

c_{ij} = the penalty cost for assigning nurse i to schedule j

a_{ijdp} = 1 if schedule j for nurse i contains period p on day d, 0 otherwise

LD_{dp} = lower bound on demand for nurses on day d in period p

UD_{dp} = upper bound on demand for nurses on day d in period p

x_{ij} = 1 if nurse i is assigned to schedule j, 0 otherwise

y_{dp} = number of outside nurses used on day d in period p

s_{dp} = slack variable for day d in period p

The formulation is then given by:

$$min \sum_{i \in N} \sum_{j \in S_i} c_{ij} x_{ij} + M \sum_{d \in D} \sum_{p \in P} y_{dp} \tag{10-15}$$

subject to:

$$-s_{dp} + \sum_{i \in N} \sum_{j \in S_i} a_{ijdp} x_{ij} + y_{pd} = LD_{dp} \; \forall d, p \tag{10-16}$$

$$\sum_{j \in S_i} x_{ij} = 1 \; \forall i \tag{10-17}$$

$$x_{ij} \in \{0, 1\} \; \forall i, j \tag{10-18}$$

$$0 \le s_{dp} \le UD_{dp} - LD_{dp}, \; y_{dp} \ge 0 \; \forall d, p \tag{10-19}$$

Constraint (10-16) the total number of nurses (including added nurses) satisfies the lower bound for demand. Constraint (10-17) ensures that each nurse is assigned to only one schedule. Constraint (10-19) requires that the slack be bounded so that constraint (10-16) will not exceed the upper bound. Note that as long as the upper and lower bounds are integer, the slack and number of outside nurses will also be integer and hence don't need to be declared as integer.

As mentioned earlier, the perception of fairness in a schedule is important. Therefore, the choice of the penalty cost is very important. It is desirable that it be increasing in the number of preference violations and the severity of the preference violations. One approach is to make the penalty cost a function of the severity, where $c_{ij}(v) = 2(v - 1)$. In this case, set $v = 1$ for a simple degree of violation to $v = 4$ for an extreme degree of violation. The model also requires that each nurse submit a base schedule.

Hospitals may have their own specific constraints that they use, and the above model may need to be modified accordingly. Examples may include that no nurse works more than three night shifts in a week or that there must be at least an 8-hour break between shifts. In addition, we have treated all nurses in this formulation as if they are the same. However, nurses have different qualifications (including nondegree, registered nurses, and nurse practitioners). The model may also need to be modified to meet specific qualification requirements.

QUESTIONS AND LEARNING ACTIVITIES

1. Consider an emergency department. What do you believe are the most important measures? Why?
2. How would your answer to the first question differ for an operating room?
3. What are some strategies that a hospital could use to help to reduce boarding?
4. Consider a 10-bed ICU. On average, 3 patients arrive to the ICU each day. The average length of stay in the ICU is 5 days. Assuming that arrivals occur according to a Poisson process and the length of stay is exponentially distributed, what would be the average number of patients that are boarded? What would the average utilization of the ICU be?
5. Suppose now that length of stay for the previous example in normally distributed with an average of 4 days and standard deviation of 2 days. How would your answer change?
6. For problem 4 and 5, determine the average number of patients in the system.
7. For problem 4, what would have a bigger impact on boarding: i) increasing the number of beds in the ICU by 1, or ii) adding a bed czar such that the length of stay would reduce by 5%?
8. For Example 12-4, suppose there is a third surgical group (bariatric). For each Tuesday, they require 2 hours on week 1, 3 hours in week 2, and 4 hours in week 3. All other data stays the same. Redo the formulation and solve.
9. For problem 8, consider the case that ratio of overtime to regular time costs go up to 1.5. How does the solution differ?
10. Discuss some of the problems you foresee with the use of block scheduling.
11. Modify the formulation given in (12-4) to (12-8) to consider specific operating rooms. Discuss implications of this formulation.
12. Modify the formulation in (12-4) to (12-8) to consider revenue. What are the implications of including revenue? Are there any ethical concerns?
13. Consider a simple example where the time it takes to serve a patient on an MRI is 30 minutes (regardless of the patient type). Using the

parameters given in Example 12-5, consider the following arrivals for inpatients: 9 A.M., 9:15 A.M., 11:00 A.M., 11:40 A.M., 12:50 P.M., 2:00 P.M., 4:00 P.M.. Arrivals for outpatients are: 9:30 A.M., 10:00 A.M., 10:45 A.M., 2:15 P.M., 3:00 P.M., and 3:55 P.M.. Determine the order the patients are seen using heuristic. Also, determine the waiting time (if any) for each patient. Determine the profit and comment on the solution.

14. What potential problems do you see with the use of the critical-first heuristic in practice?
15. Suppose the arrival times for inpatients and outpatients were known with certainty (i.e., no variability). Develop a formulation to sequence patients that maximizes profit. Apply the formulation to the data given in problem 13. How does your solution differ from 13?
16. Discuss ways that "fairness" could be addressed in nurse scheduling. How could nurses be involved in the process?
17. Develop a simple optimization model for nurse scheduling using three shifts (8 A.M.–5 P.M., 5 P.M.–12, 12–8 A.M.) where the demand for nurses each day of the week (Sun – Sat) is: 7, 15, 13, 12, 10, 8, and 7. Each nurse works an 8-hour shift, with 5 days on and 2 days off. Your model should satisfy demand using the fewest number of nurses.
18. Modify your formulation in 17 to allow for the use of nurses that work 16-hour shift (3 days on and 4 days off). The cost of a 16-hour shift nurse is 30% higher than a regular nurse.

REFERENCES

Bailyn, L., Collins, R., & Song, Y. (2007). Self-scheduling for hospital nurses: An attempt and its difficulties. *Journal of Nursing Management*, *15*, 72–77.

Bard, J. F., & Purnomo, H. W. (2005). Preference scheduling for nurses using column generation. *European Journal of Operational Research*, *164*(2), 510–534.

Green, L. V. (2004). Capacity planning and management in hospitals. In M. L. Brandeau, F. Sainfort, & W. P. Pierskall (Eds.), *Operations research and healthcare: A handbook of methods and applications*. Boston, MA: Kluwer Academic.

Green, L.V., Savin, S., & Wang, B. (2006). Managing patient service in a diagnostic medical facility. *Operations Research*, *54*(1), 11–25.

Hagtvedt, R., Ferguson, M., Griffin, P., Jones, G. T., & Keskinocak, P. (2009). Cooperative strategies to reduce ambulance diversion. *Proceedings of the 2009 Winter Simulation Conference*, 1861–1874.

Hans, E. W., Nieberg, T., & van Oosrum, J. M. (2007). Optimization in surgery planning. *MET, 15*(1), 20–28.

Hopp, W. J., & Lovejoy, W. S. (2013). *Hospital operations: Principles of high efficiency health care*. New York, NY: Pearson Education.

Hosseini, N., & Taaffe, K. M. (In press). Allocating operating room block time using historical caseload variability. *Health Care Management Science*.

Shain, M., & Roemer, M.I. (1959). Hospital costs related to the supply of beds. *Modern Hospital, 92*(4), 71–73.

Zhang, S. (2015). *Cost-Effectiveness of Health Care Interventions*. Ph.D. Thesis, University Park, PA: Penn State University.

Chapter **11**

Healthcare Logistics

"Would you tell me, please, which way I ought to go from here?"
"That depends a good deal on where you want to get to," said the Cat.
"I don't much care where—" said Alice.
"Then it doesn't matter which way you go," said the Cat.
"—So long as I get SOMEWHERE," Alice added as an explanation.
"Oh, you're sure to do that," said the Cat, "if you only walk long enough."
— Lewis Carroll, *Alice's Adventure in Wonderland*

Overview

In this chapter, we discuss two important areas of healthcare logistics. The first is how to locate healthcare facilities, taking into account population characteristics. The second topic is routing of the home healthcare workforce, a growing area of importance with the aging of the population.

11.1 FACILITY LOCATION

At its heart, facility location considers the trade-off between the cost of opening and operating facilities to the benefit of having access to care closer to potential patients. Although many of the location models developed for traditional facilities such as plants, warehouses, or stores apply, there are some unique aspects for health applications. Examples include targeting specific populations such as those at higher risk and considering the willingness of a patient to travel to a facility a certain distance away, which may depend on car ownership and access to public transportation. We first present an overview of traditional facility location models and then discuss how some of the unique healthcare features can be modeled.

Basic Location Models

In this section, we present some of the foundational formulations of facility location models (Verdat & Lapierre, 2002; Daskin & Dean, 2004). It is in no way meant to be a complete survey. We will apply this material to a health setting in the following section.

Consider the case where a set of n facilities is to be located with respect to a set of m entities such as patients, suppliers, or other providers with which they interact. We first look at the case where $n = 1$. The goal is to determine the location (x^*, y^*) for the facility that best satisfies certain criteria. Two commonly used measures in single facility location is to minimize the sum of the weighted distances (*Minsum*) and to minimize the maximum distance from the facility and any of the entities (*Minimax*).

Let w_i represent the weight of the relationship between facility and entity i in units of distance and d_i represent the distance from the facility to entity i. An example of w_i is the driving cost per unit distance. Distance between two points (x_i, y_i) and (x_j, y_j) can be measured in several ways. Two commonly used metrics are:

- **Euclidean distance:** The straight-line difference between two points:

$$d_{ij} = \sqrt{(x_i - x_j)^2 + (y_i - y_j)^2} \qquad (11\text{-}1)$$

- **Rectilinear (or Manhattan) distance:** The sum of the x and y deviations between the two points, commonly used as an approximation when there are perpendicular lanes or roads such as city blocks or aisles in a ward:

$$d_{ij} = |x_i - x_j| + |y_i - y_j| \qquad (11\text{-}2)$$

Of course, there are circumstances where neither measure is a good approximation. For example, driving between two cities where there is rugged terrain (e.g., between State College and Belleville in Pennsylvania) would lead to very curvy roads. Hence, neither metric would be appropriate.

Recent advances in geospatial mapping software have made it relatively easy to obtain exact distance estimates.

For the Minsum problem, the optimization formulation is:

$$\min \sum_{i-1}^{n} w_i d_i \tag{11-3}$$

where d_i is the distance from entity i to the facility. It is easy to show that the optimal solution from the first order conditions for the Euclidean distance is given by:

$$x^* = \frac{\sum_i w_i x_i / d_i}{\sum_i w_i / d_i}, \quad y^* = \frac{\sum_i w_i y_i / d_i}{\sum_i y_i / d_i} \tag{11-4}$$

Note that since d_i is a function of x^* and y^*, an iterative procedure must be used. This is illustrated in Exercise 11.1.

If the rectilinear distance is used for the Minsum problem, then we have the following formulation:

$$\min \sum_{i=1}^{n} w_i(|x_i - x| + |y_i - y|) \tag{11-5}$$

Note that we can separate this formulation into the x and y components. Therefore, since this is unconstrained then x^* and y^* may be solved for independently. Although we do not prove this property here, it turns out that the median condition holds. Assuming that x is measure horizontally and y is measured vertically, then x^* is located such that no more than half of the total weight is to the left or right of it. Therefore, x^* will come from one of the x values of an entity. Similarly, y^* is located such that no more than half of the total weight is located above or below it, and so its value will also equal to one of the y values an entity. It is interesting that the location of (x^*, y^*) does not depend on distance traveled.

If neither the Euclidean nor rectilinear distances are good approximations to the actual distances, and exact distances can be computed, then another approach can be used. Since it is not possible to compute these distances for all possible locations, then a subset of feasible facility locations must be identified *a priori*. This tends not to be a severe limitation in practice since there typically are only a limited number of m possible locations for a facility. Solving this problem is trivial and involves simply computing the weighted distance for each of the m locations and choosing the one with the smallest value.

The solution to the Minimax problem for the Euclidean distance is equivalent to finding the circle of minimum radius that encloses all of the entities. A simple procedure for solving this problem that involves checking pairs and triplets of entities is given in Exercise 11.2. For the case of rectilinear distance, the solution may not be unique. A procedure (without proof) that will solve this problem is as follows: Place the facility on any point connecting the line segment connecting the points:

$$(x_1, y_1) = 0.5(c_1 - c_3, c_1 + c_3 + c_5)$$
$$(x_2, y_2) = 0.5(c_2 - c_4, c_2 + c_4 - c_5) \tag{11-6}$$

where:

$$c_1 = \min(x_i + y_i)$$
$$c_2 = \max(x_i + y_i)$$
$$c_3 = \min(-x_i + y_i)$$
$$c_4 = \max(-x_i + y_i)$$
$$c_5 = \min(c_2 - c_1, c_4 - c_3)$$

Example 11-1

There are five hospitals in Brooklyn that will be served by a blood bank. Their locations and monthly demand (in terms of deliveries per month) are given in the table below. Distance is measured according to the rectilinear metric.

Hospital	x	y	Demand
H1	5	15	100
H2	12	13	160
H3	3	12	200
H4	9	3	100
H5	6	6	90

If the objective is Minsum, then the optimal location can be found by first sorting in each direction and computing the partial sum of the weekly demands: H3 (200), H5 (290), H1 (390), H4 (490), and H2 (650). The smallest x location that corresponds to a partial sum of at least 650/2=325 is $x^* = 5$. Similarly to determine y^*, the partial sums in sorted y order are: H4 (100), H5 (190), H3 (390), H2 (550), H1 (650), which gives $y^* = 12$. If the objective is Minimax, we first compute c_1 to c_5. From the data we get $c_1 = \min\{5 + 15, 12 + 13, 3 + 12, 9 + 3, 6 + 6\} = 12$. Similarly, $c_2 = 25$, $c_3 = -6$, $c_4 = 10$, and $c_5 = 13$. This gives $(x_1, y_1) = 0.5(12 - (-6), 12 - 6 + 13) = (9, 9.5)$ and $(x_2, y_2) = (7.5, 8)$. Any location on the line segment connecting these points is optimal.

The choice of the proper criteria depends on the application. For many problems in supply chain, which typically involves minimizing inventory and logistics costs from regular truck shipments by locating warehouses and distribution centers, the Minsum criteria are appropriate. For the case where emergency facilities are located such for ambulance dispatching or fire stations, it is desired to not have potential "customers" too far away. In this case, the Minimax criteria are more appropriate.

There are equivalent versions of these problems for multiple facilities location, though we will only discuss the discrete versions. In this case we will assume that there are a set of known feasible locations L of cardinality m from which a subset is chosen to locate the facilities, and hence exact distances will be used. We will also assume that an entity is served by one and only one facility and that a facility has unlimited capacity.

For the Minsum version, the goal is to determine a subset of locations in L that minimizes the sum of the weighted distances. This can be formulated using integer programming as follows:

$$\min \sum_{i-1}^{m} f_i t_i + \sum_{j=1}^{n} w_j \sum_{i=1}^{m} c_{ij} z_{ij} \qquad (11\text{-}7)$$

subject to:

$$\sum_{j=1}^{n} z_{ij} = 1 \ \forall i \qquad (11\text{-}8)$$

$$z_{ij} \leq t_i \ \forall i, j \tag{11-9}$$

$$\sum_{j=1}^{n} w_j z_{ij} \leq W_i \ \forall i \tag{11-10}$$

$$z_{ij} \geq 0, t_i \in \{0, 1\} \tag{11-11}$$

where f_i is the fixed cost of opening a facility in location i, w_j is the weight (or demand) of entity j, and W_i is the capacity of a facility in location i. The decision variables are t_i, which is a binary variable of whether to open a facility or not in location i, and z_{ij}, which equals 1 if entity j is served by facility i. The objective is made up of two components, the fixed cost of opening a facility in location i and the total cost of serving the entities from the assigned locations. In this formulation, there are three key constraints. Constraints (11-8) ensure an entity is served by only one location. Constraints (11-9) restrict assignments to those where a location has an open facility. Constraints (11-10) ensure that the total demand at location i does not exceed its capacity.

Although small to medium-sized problem instances can be solved using standard optimization software, it is important to note that this facility location problem is in the class of NP-hard problems, even without the capacity constraints. For large-problem instances, various solution approaches have been developed. One of the more popular approaches is Lagrangian relaxation and its variants.

For the Minimax version we present a close variant, which is to find the maximum number of entities that can be served from a given number of facilities (p) within a specified maximum distance D. This is called the ***maximal covering location problem*** (MCLP). One way to formulate this as an integer programming is as follows:

$$\max \sum_{i-1}^{p} y_i \tag{11-12}$$

subject to:

$$\sum_{i=1}^{p} z_{ij} \leq 1 \ \forall j \tag{11-13}$$

$$z_{ij} \leq f_{ij} \ \forall i, j \tag{11-14}$$

$$y_i = \sum_{j=1}^{n} z_{ij} \; \forall i \qquad (11\text{-}15)$$

$$z_{ij} \geq 0, t_i \in \{0, 1\} \qquad (11\text{-}16)$$

A capacitated version would be where a facility could serve at most C entities. In this version, the following constraint would be added:

$$y_i \leq C \; \forall i \qquad (11\text{-}17)$$

Example 11-2[1]

Suppose for Example 11-1 the cost of opening a new blood bank facility is $1 million. Once built, a facility will last for roughly 10 years. Further, delivery cost is $1/unit distance traveled/trip, and rectilinear distance is used. There are three possible locations for blood banks—(4,11), (3,6), (12,5)—and a blood bank can make 450 deliveries per month at most. We wish to determine which locations to choose that minimizes the total cost. In order to define fixed and variable costs in the same terms, we need to compute the monthly payment for the facility over the 10-year life. If an annual discount rate of 8% is used, then the monthly payment per facility is PMT(8%/12,12*10,1000000) = $12,133. The distances between the potential blood bank facility locations and the hospital locations are computed and given in the following table:

	H1	H2	H3	H4	H5
Location 1	5	10	2	13	7
Location 2	11	16	6	9	3
Location 3	17	8	16	5	7

[1]Although this instance is small enough to easily be solved by hand, we formulate it as an integer program to illustrate the material presented in this section.

If all of the hospital demand must be satisfied by a single blood bank, then the formulation is given by:

$$\min 12133(t_1 + t_2 + t_3) + 100(5z_{11} + 11z_{21} + 17z_{31})$$
$$+160(10z_{12} + 16z_{22} + 8z_{32})$$
$$+200(2z_{13} + 6z_{23} + 16z_{33}) + 100(5z_{14} + 11z_{24} + 17z_{34})$$
$$+90(5z_{15} + 11z_{25} + 17z_{35})$$

subject to:

$$z_{11} + z_{21} + z_{31} = 1$$
$$z_{12} + z_{22} + z_{32} = 1$$
$$z_{13} + z_{23} + z_{33} = 1$$
$$z_{14} + z_{24} + z_{34} = 1$$
$$z_{15} + z_{25} + z_{35} = 1$$
$$z_{11} \le t_1, z_{12} \le t_1, z_{13} \le t_1, z_{14} \le t_1, z_{15} \le t_1$$
$$z_{21} \le t_2, z_{22} \le t_2, z_{23} \le t_2, z_{24} \le t_2, z_{25} \le t_2$$
$$z_{31} \le t_3, z_{32} \le t_3, z_{33} \le t_3, z_{34} \le t_3, z_{35} \le t_3$$
$$100z_{11} + 160z_{12} + 200z_{13} + 100z_{14} + 90z_{15} \le 450$$
$$100z_{21} + 160z_{22} + 200z_{23} + 100z_{24} + 90z_{25} \le 450$$
$$100z_{31} + 160z_{32} + 200z_{33} + 100z_{34} + 90z_{35} \le 450$$

The solution to this formulation is to open a blood bank in locations 1 (4, 11) and 3 (12, 5). Hospitals 1 and 3 will be served by the blood bank in location 1 and the remaining hospitals will be served by the blood bank in location 3. The total monthly cost for this solution is $27,576.

Note that if we do not restrict all of the demand from a hospital to be served from a single blood bank, we will need to modify the formulation. In this case let x_{ij} be the amount of demand from hospital j served from blood bank i. Note that it is an integer variable.

This gives the following formulation:

$$\min 12133(t_1 + t_2 + t_3) + 5x_{11} + 11zx_{21} + 17x_{31} + 10x_{12} + 16x_{22}$$
$$+8x_{32} + 2x_{13} + 6x_{23} + 16x_{33} + 5x_{14} + 11x_{24} + 5x_{15} + 11x_{25} + 17x$$

subject to:

$$x_{11} + x_{12} + x_{13} + x_{14} + x_{15} \leq 450t_1$$
$$x_{21} + x_{22} + x_{23} + x_{24} + x_{25} \leq 450t_2$$
$$x_{31} + x_{32} + x_{33} + x_{34} + x_{35} \leq 450t_3$$
$$x_{11} + x_{21} + x_{31} \geq 100$$
$$x_{12} + x_{22} + x_{32} \geq 160$$
$$x_{13} + x_{23} + x_{33} \geq 200$$
$$x_{14} + x_{24} + x_{34} \geq 100$$
$$x_{15} + x_{25} + x_{35} \geq 90$$

In this formulation, the objective remains the total monthly cost made up of the facility cost and the delivery cost. The first set of constraints serves two purposes. First, they ensure that the total monthly deliveries served from a blood bank do not exceed the capacity of 450. Second, they ensure that if any demand is served from a location, a blood bank must be located there. Note that the left-hand side is restricted to zero until t_i equals 1. The remaining constraints ensure all of the demand is met. The solution to this formulation ends up being the same.

Location of Healthcare Facilities

The location of healthcare facilities presents some interesting challenges when compared to traditional facility location. In the traditional case, travel typically occurs from the facility to the corresponding customers. In healthcare delivery, however, the customers (e.g., patients) typically travel to the facility for care at no cost to the provider. In some cases, including home healthcare or emergency care, the provider travels to the patient and does incur a cost.

Demand estimation for healthcare delivery is also a complicated problem. A key reason is that it depends not only on need, but also the likelihood of patient travel, and patient coverage. Patients may eliminate or postpone visits if they are far away from providers such as in rural areas or if they don't have a car or easy access to public transportation, which can occur in rural or urban areas. In addition, if they do not have health insurance, the potential out-of-pocket costs may prevent them from seeking care.

Alternative sources of care can exist for a patient. For example, if a patient is feeling nauseous, she may contact her primary care physician or she might decide to go to a hospital emergency department. The patient does not necessarily pick the most appropriate source of care.

Finally, in many applications, there is a concern of equity of care. By equity, we mean that healthcare services are provided in a way that does not promote disparities in care. For example, if providers were located only in highly populated areas, then patients in a rural community would be at a disadvantage.

We first discuss some approaches that can be used for modeling these from an optimization framework and then demonstrate a full model that determines the best locations for community health centers in order to improve access for at-risk populations.

Willingness-to-Travel Constraints

A patient's willingness to travel to a healthcare provider is based primarily on the distance from the patient's home to the provider location. That is, conditioned that the patient has a healthcare need, the likelihood that the patient travels to the provider decreases monotonically by distance. A simpler approach is to make the decision binary. As long as the distance from patient i to provider j (d_{ij}) is less than or equal to a specified value D, then the patient is willing to travel (i.e., $Z_{ij} = 1$). This can be modeled by the addition of the following patient constraint:

$$d_{ij} \leq DZ_{ij} \ \forall i, j \tag{11-18}$$

In constraints (11-18), Z_{ij} can only be set to 1 (i.e., patient is willing to travel to j) if $d_{ij} \leq D$; otherwise, the constraint would be violated.

One problem with this approach is that a large number of binary constraints are required. It can therefore be useful to preassign each patient to a specific geographic region. These could be done using predefined regions such as cities, counties, or census tracts with the centroid of each region ($\mu_j = \{x_j, y_j\}$) defined. An alternative approach is to define the geographic regions though the use of distance-based clustering such as *k*-means.[2] In this case, *k* different regions are defined. The method assigns patients *i* into one of the clusters C_j based on the objective of minimizing the within cluster sum of the squares:

$$\min_C \sum_{j=1}^{k} \sum_{x \in C_j} (x - \mu_j)^2 \qquad (11\text{-}19)$$

where x is the location vector of the patients in cluster C_j, C is set of clusters, and μ_j is the average of the location of the patients in the cluster. Each geographic region can be defined by its mean location μ_j. In this case, we can generalize the binary decision by using the parameter P_l for distance level l. For a given set of distance levels, P_l is the probability that a patient with need travels at least distance l for care. Consider 4 levels of 0 miles, 15 miles, 30 miles and 50 miles with $P_0 = 1.0$, $P_{15} = 0.8$, $P_{30} = 0.4$, $P_{50} = 0.0$. In this case, the probability that a patient would travel at least 30 miles to receive care is 0.4. From a population perspective, the P_l values represent the proportion of patients with need that will travel at least distance l for care. If we limit the potential locations of the healthcare facilities to a known set of potential locations, then we can use the constraints:

$$\sum_j y_{ij} I_{ijl} \le P_l n_i \; \forall i, l \qquad (11\text{-}20)$$

where I_{ijl} equal 1 if the distance between region (cluster) i and potential facility location j is less than or equal to distance level l, n_i is the number of people in region (cluster)i with health need, and y_{ij} is the number of people from region i served by location j. Note that I_{ijl} is not a decision variable, but can be preprocessed since the potential locations are known in advance.

---

[2] Note that *k*-means requires the triangle inequality to hold, and so Euclidean distance is the typical metric that is used. Note that several software packages exist for doing *k*-means clustering, and so we do not discuss a specific algorithm here.

Choice Constraints

When multiple options for a source of care exist for a patient, then the selection (or choice) must be explicitly modeled. Economists typically assume that when individuals are making choices that they are using a process that maximizes their individual utility. One approach to this is to develop a probability-based model for a set of discrete choices. Let P_{ij} be the probability that individual i chooses alternative j from a given set of alternatives. Each individual has certain attributes such as insurance status, gender, and education level. Further, alternatives may have attributes that an individual considers. Distance may be one of the attributes, but there can be several others such as reputation, expected waiting time, and past experience. The expected fraction of individuals choosing alternative j is simply:

$$E[P_j] = \frac{\sum_i p_{ij}}{N} \tag{11-21}$$

where N is the total number of individuals making a "choice." The key, then, is estimating the probability values. One approach for doing this is to assume that the utility that individual i chooses alternative j (U_{ij}) is of the form:

$$U_{ij} = \beta_j' x_i + \varepsilon_{ij} \tag{11-22}$$

where x_i is a vector of observed variables for individual i, and ε_{ij} captures the unobserved factors that influence an individual's choice. In this case, an individual chooses alternative j if the utility for that alternative is greater than the utility for any other alternative. Further, we assume the selection depends on the attributes of the individual but there are no attributes of the alternatives. In this case, the multinomial logit procedure in a statistical software package may be used, and the choice probabilities are given by:

$$P_{ij} = \frac{exp(\beta_j x_i)}{\sum_k exp(\beta_k x_i)} \tag{11-23}$$

It is necessary to normalize for one of the alternatives. This is typically done by setting $\beta_1 = 0$.

If it is the case that the choice depends on the actual attributes of the alternatives, then the utility can be defined by:

$$U_{ij} = \beta z_{ij} + \varepsilon_{ij} \tag{11-24}$$

where z_{ij} is a vector of observed variables relating to alternative j for individual i. The conditional logit procedure in a statistical software package may be used, and the choice probabilities are given by:

$$P_{ij} = \frac{exp(\beta z_{ij})}{\sum_k exp(\beta z_{ik})} \tag{11-25}$$

Note that the material presented in this section assumes that there is no correlation in the unobserved factors over the set of alternatives. If this assumption does not hold, there are a number of alternatives that allow for modeling the correlation such as nested-logit or C-logit models.

Equity Modeling

In the area of healthcare delivery logistics one topic that often arises is the issue of "fairness" in the allocation of resources, which we will refer to as equity. From a social perspective, equity with regard to resources helps to eliminate certain disparities between groups that may be due to factors such as geographic or sociodemographic differences.

One approach for dealing with equity is to use goal programming. The idea is to set a target for the maximum difference of services provided between two groups. Since the populations will typically differ between these groups, we can scale by the population. Let x_{ij} represent the quantity of service i delivered to group j, n_j be the population of group j, and d_{ij} be the demand for service i from group j. Further, let G_i be the maximum scaled difference goal between any two groups for service i, C_i be the total capacity for service i, β_{ij} be the cost of delivering a unit of service i to group j and γ the penalty per unit of unmet demand. We can formulate this model as follows:

$$\min_{x_{ij}} \sum_{j=1}^{n-1} \sum_{k=j+1}^{n} \alpha_{jk}(s_{jk} + t_{jk}) + \sum_{i=1}^{m}\sum_{j=1}^{n} \beta_{ij}x_{ij} - \gamma \sum_{i=1}^{m}\sum_{j=1}^{n}(d_{ij} - x_{ij}) \tag{11-26}$$

subject to:

$$\frac{x_{ij}}{n_j} - \frac{x_{ik}}{n_k} - s_{jk} + s'_{jk} \leq G_i \; \forall i, k > j \tag{11-27}$$

$$-\left(\frac{x_{ij}}{n_j} - \frac{x_{ik}}{n_k}\right) - t_{jk} + t'_{jk} \leq G_i \; \forall i, k > j \tag{11-28}$$

$$\sum_{j=1}^{n} x_{ij} \leq C_i \; \forall i \tag{11-29}$$

$$x_{ij} \leq d_{ij} \; \forall i, j \tag{11-30}$$

$$x_{ij}, s_{jk}, s'_{jk}, t_{jk}, t'_{jk} \geq 0 \tag{11-31}$$

In this formulation, constraints (11-27) and (11-28) define the target goals. If the goal G_i cannot be strictly met, then the corresponding slack variables s_{jk} or t_{jk} are strictly positive and penalized in the objective by weight α_{jk}. Constraints (11-29) ensure that capacity of services is not exceeded and constraints (11-30) ensure that supply of services does not exceed demand for those services.

Providing comprehensive healthcare services to all the members in a community is important for the achievement of health equity and for increasing a community members' quality of life. However, there are many disparities that exist in healthcare services that affect not only individuals but also the entire community. Two important measures of disparity are not having a primary source of care, or lack of access, and a persistent lack of insurance coverage. The Federally Qualified Health Center (FQHC) initiative is one program designed to improve access of primary care, particularly for needy populations. These centers provide primary and preventive healthcare, outreach, dental care, some mental health and substance abuse treatments, and prenatal care, especially for people living in rural and urban medically underserved communities. Over 90% of FQHC patients live with incomes below 200% of the federal poverty limits, and over 40% of FQHC patients are uninsured. The following case (Griffin, Lee, Scherrer, & Swann, 2014) illustrates how several of the concepts discussed in this section can be modeled.

CASE STUDY 11: LOCATION OF FEDERALLY QUALIFIED HEALTHCARE CENTERS

Health status disparities can be better addressed if the population is prioritized according to current access and coverage status. Table 11.1 shows six populations grouped according to their current access status (served and underserved) and coverage status (private, public, and persistent lack of insurance (uninsured)). We define an individual as being underserved if they live in a health professional shortage area (HPSA) as defined by the Health Resources and Services Administration. An individual is defined as having a persistent lack of insurance if they have been uninsured for at least one year. From this point on we simply use the term uninsured. The primary components of public insurance are Medicaid and Medicare.

Table 11.1 Population Groups by Access and Coverage

Access \ Coverage	No Insurance	Public Insurance	Private Insurance
Underserved	⊖	⊜	⊛
Served	④	⑤	⑥

We introduce a multiobjective model to decide the optimal FQHC locations considering population groups with different priorities. Demand is estimated based on current access and coverage status. Details of demand estimate are given in Griffin et al. (2014). The following are the indices, parameters, and decision variables used in the model.

Model Indices

i : FQHC location

z : Population location

j : Service type (general, OBGyn, dental, mental)

k : Capacity (small, medium, large)

l : Distance level (0, 10 miles, 20 miles, 30 miles)

g_1: Insurance group (private, public, uninsured)

g_2: Access (access, no access)

Model Parameters

FL : Fixed cost per location

FS_k : Fixed cost per capacity level

VS_j : Variable cost per service

RB_{g_1}: Reimbursement rate

CAP_{jk}: Number of patients of service type j that can be served at level k

w_j: Weight by service type j

P_l: Maximum percentage of z's population that can be served at distance level l

$n_{zjg_1g_2}$: Demand for service j in county z of insurance and access group

$m_{izjg_1g_2}$: Maximum demand of county z can be served CHC located county i

($=P_l n_{zjg_1g_2}$, if the distance between i and z corresponds to level l, 0 otherwise)

l_{izl} : Binary parameter equal to 1 if the distance level between i and z is greater than l, 0 otherwise

Model Decision Variables

y_{izj} : Number of encounters from county z served by FQHC in county i for service j

s_{ijk} : Binary variable equal to 1 if county i has FQHC with service j at capacity k

c_i : Number of FQHC centers in location i

$y_{izjg_1g_2}$: Number of encounters by insurance group g_1 and access group g_2 in county z served by FQHC in county i for service j

We categorize demand by insurance and access group, which makes it possible to use multiple objectives based on the groups. We set the first priority to maximize insurance coverage (11-13), which is the sum of encounters of the uninsured population ($g_1 = 3$). The second priority is to maximize access (11-14), which is from the underserved population ($g_2 = 2$). Finally, we maximize the utilization of FQHCs by providing the most weighted services (11-15). These weights were based on a logistic regression model performed on nationally representative data from the National Health Nutrition and Examination Survey (NHANES), where self-reported general health (poor, fair, good, excellent) was used as the dependent variable. The weights used are shown in Table 11.2, and the details of the regression are provided in Griffin, Scherrer, & Swann (2008).

Table 11.2 Adjusted Weights for the Four Service Types

	Primary (w_1)	OBGyN (w_2)	Dental (w_3)	Mental (w_4)	Total
Weights	0.88	1.20	0.07	0.05	2.20
Normalized Weights	0.40	0.55	0.03	0.02	1.00

Model Formulation

Objectives:

$$\text{(Max Coverage)} : \max \sum_{i,z,j,g_2,g_1=3} w_j y_{izjg_1g_2} \qquad (11\text{-}13)$$

(Max Access):

$$\max \sum_{i,z,j,g_1,g_2=2} w_j y_{izjg_1g_2} \qquad (11\text{-}14)$$

(Max Utilization):

$$\max \sum_{i,z,j,g_1,g_2} w_j y_{izjg_1g_2} \qquad (11\text{-}15)$$

subject to:

$$y_{izjg_1g_2} = y_{izj} \times \frac{n_{zjg_1g_2}}{\sum_{g_1g_2} n_{zjg_1g_2}} \ \forall i,z,j,g_1,g_2 \qquad (11\text{-}16)$$

$$\sum_i \mathrm{FL}\, c_i + \sum_{ijk} FS_k\, s_{ijk} + \sum_{izjg_1g_2} \mathrm{VS}_j\, RB_{g1}\, y_{izjg_1g_2} \le B \qquad (11\text{-}17)$$

$$\sum_z y_{izj} \le \sum_k CAP_{jk}\, s_{ijk} \ \ \forall i,j \qquad (11\text{-}18)$$

$$\sum_k s_{ijk} \le c_i \ \forall i,j \qquad (11\text{-}19)$$

$$\sum_i l_{izl} y_{izj} \le P_l \sum_{g_1,g_2} n_{zjg_1g_2} \ \ \forall l,z,j \qquad (11\text{-}20)$$

$$y_{izj} \le \sum_{g_1,g_2} m_{izjg_1g_2} \ \forall i,z,j \qquad (11\text{-}21)$$

To define decision variable $y_{izjg_1g_2}$, we assume that the proportion of each group in FQHC encounters will follow the same rate of estimated demand at the population location (11-16). This variable is defined by the ratio of each group in the estimated demand ($n_{zjg_1g_2}$) at the location to the total number

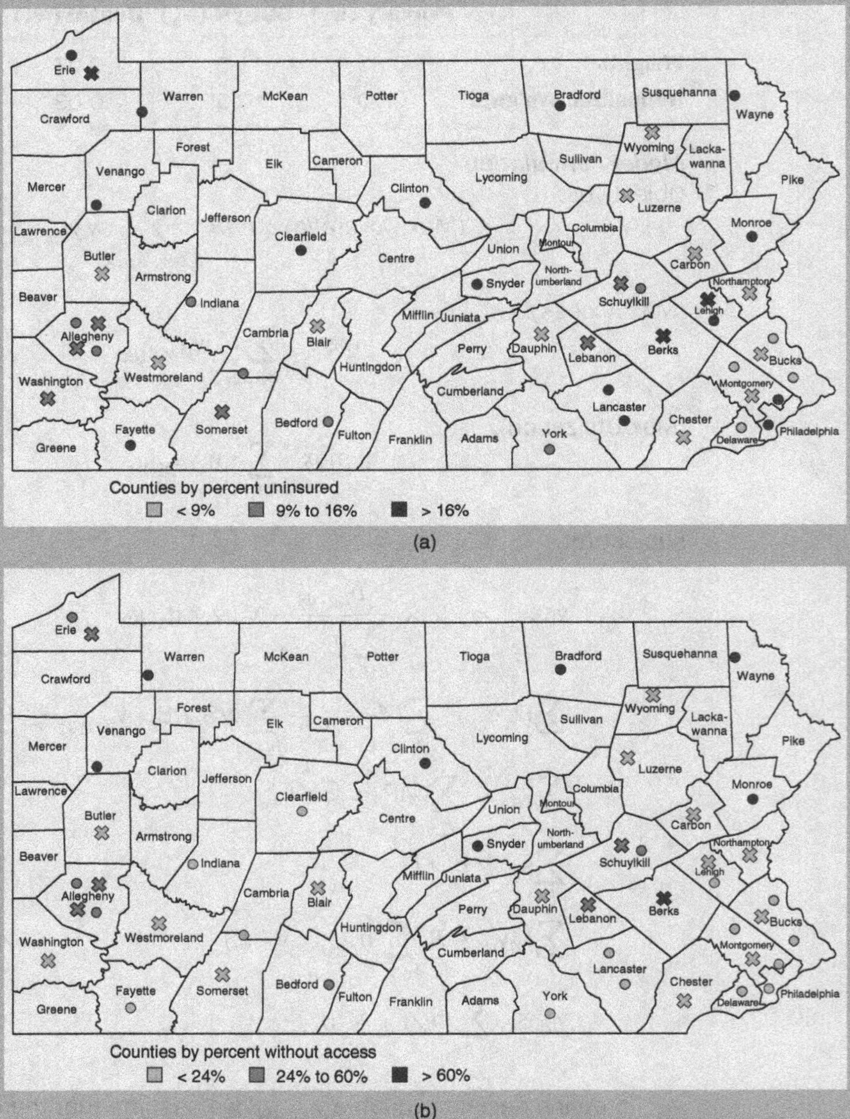

Figure 11.1 FQHC Optimal Locations Comparing (a) Access and (b) Coverage Status

of encounters (y_{izj}). Constraint (11-17) restricts the total spending on fixed costs for locating FQHCs, fixed costs for capacity, and variable costs for services to be no greater than a given budget B. Constraint (11-18) ensures that for each service type (General, OBGyn, Dental, and Mental) the number of patients served does not exceed the capacity for that service. Constraint (11-19) ensures that services can only be provided at a location if there is capacity established at the location for that service. Constraint (11-20) ensures that service is provided to a patient only if they are eligible for that service and live close enough to the facility to access it. It is implied in this constraint that the likelihood a patient visits an FQHC decreases linearly in distance from the facility. Finally, constraint (11-21) ensures that the number of patients served in a county for each service type does not exceed the number of persons that require that service.

We apply the model for data for the state of Pennsylvania and use a preemptive optimization approach based on the order given for the objective functions. The optimal locations given for the FQHCs for Pennsylvania from this example are shown in Figure 11.1. In this example, the budget is \$300M, and the resulting served population is 1,456,641 persons. The "•" locations are from the multi-objective solution and the "x" solutions are for the case where only the single objective of utilization (Eq. 11-15) were used. In part (a) of this figure, access status is shown by county. The darker the shading, the poorer is the access for an average person in the county. Similarly, part (b) of this figure shows coverage; darker-shaded counties. As can be seen from the Figure, the optimal solution using the multiobjective tends to place FQHCs in darker-shaded regions and clearly outperforms the single objective solution, and shows the value of a multicriteria framework.

11.2 HOME HEALTHCARE ROUTING AND SCHEDULING

Home healthcare (HHC) is provided in the patient's home by healthcare professionals such as visiting nurses, and typically includes some combination of professional healthcare services such as medical or psychological assessment, disease education and management, and physical or occupational therapy (Chi, 2015). These services may also assist with daily tasks such as meal preparation, medication reminders, laundry, light housekeeping, errands, shopping, transportation, and companionship. HHC is often an integral component of the post-hospitalization recovery process, particularly during the initial weeks after discharge when the patient still requires some level of regular physical assistance. Compared to other forms of care, it has many advantages, including fewer expenses, better morale, faster recovery, a higher level of personal independence, and more support of family and friends.

HHC has seen a growing share of the healthcare market. In the United States, 2012 expenditures were US$77.8 billion, accounting for 2.8% of total health spending. Further, roughly 12 million individuals received care from over 33,000 HHC providers, who drove over 5 billion miles and made 428 million visits. As a comparison metric, the United Parcel Service (UPS) drove just over 2 billion miles globally over that same time period.

Fundamental to effective HHC delivery is the determination of effective routes. On any particular day, several patient homes need to be visited by a set of HHC workers, each with a capacity defined by available time. A care visit to a home patient is characterized by the expected duration of the visit (determined by an assessment of the patient's needs), a time window (earliest and latest visit start time by the patient), and required qualifications of the care worker. A visit is only allowed if the qualification level of the worker providing the care is at least as high as required. The patients to be treated are dispersed geographically and may require multiple care visits on the same day.

The travel time between patients depends on the route of the HHC worker and travel conditions, which may vary based on time of day (e.g., rush hour traffic). We will make the assumption that the visit duration is not a function of the route. In this case, there are two decisions that need to be made: (1) which HHC staff should be used on a particular day (HHC rostering), and (2) how the staff should be assigned to the home patients (routing/allocation). We discuss the case the staff is given and the routing/allocation problem needs to be solved (Chi, 2015).

HHC Routing/Allocation

Let N denote the set of care visit nodes, and visit i has a duration time d_i, a prestarting time window $[A_i, B_i]$, and a skill requirement set $S_i = \{s_i^1, s_i^2, s_i^3, \ldots\ldots\}$, the elements of which are binary. For instance, visit i requires skill 2, 4, then $s_i^2 = 1, s_i^4 = 1$ and the rest of skill elements equal to zero. Let M denote the set of nurses and H denote the set of nurse nodes, and nurse k has two nurse nodes d_k(departure) and r_k (return), a departure time window from her or his home $[A_{d_k}, B_{d_k}]$ and an return time window back home $[A_{r_k}, B_{r_k}]$. Similarly, nurse k also has a skill set $S_k =$

$\{s_k^1, s_k^2, s_k^3, \ldots\ldots\}$, and if visit i is made by nurse k, the following constraint must be satisfied: $S_i \leq S_k$.

Let B denote the set of lunch break node, and nurse k has a dummy lunch break node b_k with a staring time window $[A_{b_k}, B_{b_k}]$. The travel times between lunch nodes and visit nodes are zero. The duration of the lunch break is set without loss of generality to one hour.

Two types of variables are used in the mathematical formulation:

1. Binary flow variables: x_{ij}^k

$$x_{ij}^k = \begin{cases} 1, \textit{if HHC worker k visits node j immediately after node i}; \\ 0, \textit{otherwise}. \end{cases}$$

   where $i, j = d_k, r_k, b_k, 1, \ldots, n$, $k = 1, \ldots, m$ and $i \neq j$.

2. Time variables: T_i

   T_i is the start time of the event at i (leaving home, arriving at home, service, break), $i = d_k, r_k, b_k, 1, \ldots, n$, $k = 1, \ldots, m$

The HHC routing and allocation problem can be defined on a complete directed graph noted $G = (V, A)$ with: $V = \mathrm{N \cup H \cup B}$ the set of nodes, and A the set of arcs. The travel time and cost on the arc (i, j) are t_{ij} and c_{ij}. The goal is to find the optimal routes of care workers such as that each visit is made by exactly one care worker and the constraints are satisfied. The objective is to minimize the total cost of the routes followed by care worker and to assign the time window using the variance associated with the route.

As mentioned previously, there are a fixed number of nurses and the tasks are known. The objective is to determine the best routes with the least total working time. Further, a time window should be assigned to each patient. Time is measured in units of minutes, so everything must happen within an 8-hour shift [0, 1,440 min]. The mathematical programming model for this case is given by:

$$\min \sum_{k \in M} (T_{r_k} - T_{d_k}) \tag{11-22}$$

subject to:

$$\sum_{j \in N} x_{d_k j}^k = 1, k \in M \tag{11-23}$$

$$\sum_{i \in N} x^k_{i\, r_k} = 1, k \in M \tag{11-24}$$

$$\sum_{i \in N} x^k_{i\, d_k} = 0, k \in M \tag{11-25}$$

$$\sum_{j \in N} x^k_{r_k\, j} = 0, k \in M \tag{11-26}$$

Constraints (11-23) and (11-24) ensure that each care worker is used exactly once and constraints (11-25) and (11-26) ensure that the first and last stops are their homes.

$$\sum_{i \in V} \sum_{k \in M} x^k_{ij} = 1, j \in N, k \in M \text{ and } i \neq j \tag{11-27}$$

Constraint (11-27) ensures that each visit should be made exactly once.

$$\sum_{i \in V} x^k_{ip} - \sum_{j \in V} x^k_{pj} = 0, p \in N, k \in M \text{ and } i \neq p, p \neq j \tag{11-28}$$

Constraints (11-28) ensures that once a nurse visits a patient, then she or he must also depart from the same patient.

$$\sum_{i \in N} x^k_{i\, b_k} = 1, k \in M \tag{11-29}$$

$$\sum_{j \in N} x^k_{b_k j} = 1, k \in M \tag{11-30}$$

Constraints (11-29) and (11-30) ensure each HHC worker takes a lunch break.

$$A_i \leq T_i \leq B_i, i \in N \cup H \cup B \tag{11-31}$$

$$T_{b_k} \geq T_{d_k} + 120, k \in M \tag{11-32}$$

$$T_{b_k} \leq T_{r_k} - 120, k \in M \tag{11-33}$$

Constraints (11-31) are time window constraints, and constraints (11-32) and (11-33) are for the break start times.

$$T_i + d_i + t_{ij} - T_j \leq \left(1 - x^k_{ij}\right) Q, i, j \in V, k \in M, i \neq j \tag{11-34}$$

Constraints (11-34) ensure the compatibility schedule requirements and serve as linking constraints, where Q is a large number.

$$\left(\sum_{i \in v} x_{ij}^{k}\right) S_j \leq S_k, j \in N, k \in M \qquad (11\text{-}35)$$

Constraints (11-35) are for the skill qualification level, where $S_i = \{s_i^1, s_i^2, s_i^3, \ldots\ldots\}$ and $S_k = \{s_k^1, s_k^2, s_k^3, \ldots\ldots\}$.

From the model, we can determine the routes and start times at each node. Figure 11.2 shows the results in the form of Gantt charts of applying the model to the case of 4 nurses and 20 patients. Note that the 20 patients are drawn from a population of 30.

HHC Heuristics

It is important to note that the HHC routing formulation will not work well for very large instances. In this case, it may be necessary to use heuristics. One very popular routing heuristic is the Clarke-Wright Savings Algorithm (CWSA). The key driver of the heuristic is to determine the savings from the merger of two routes into a single route, assuming the combination is feasible based on capacity. Let c_i be the cost of route $(0,\ldots,i,0)$ and c_j be the cost of route $(0,j,\ldots,0)$, where 0 is the start/end position and cost is based on distance or travel time. Further, let d_{ij} be the cost of traveling from node i to node j. If the two routes can feasibly be merged into a single route $(0,\ldots,i,j,\ldots,0)$, the savings (s_{ij}) is determined by the difference in cost of the original two routes minus the cost of the combined route:

$$s_{ij} = (c_i + d_{i0} + c_j + d_{0j}) - (c_i + d_{ij} + c_j) = d_{i0} + d_{j0} - d_{ij} \qquad (11\text{-}36)$$

The basic algorithm works at follows:

Clarke-Wright Savings Algorithm (CWSA)

Step 1. Initialization Step

Compute s_{ij} for all $i \neq j$.

Create n initial routes $(0,i,0)$ for all i.

For a savings list by ordering the s_{ij} values in a nonincreasing fashion.

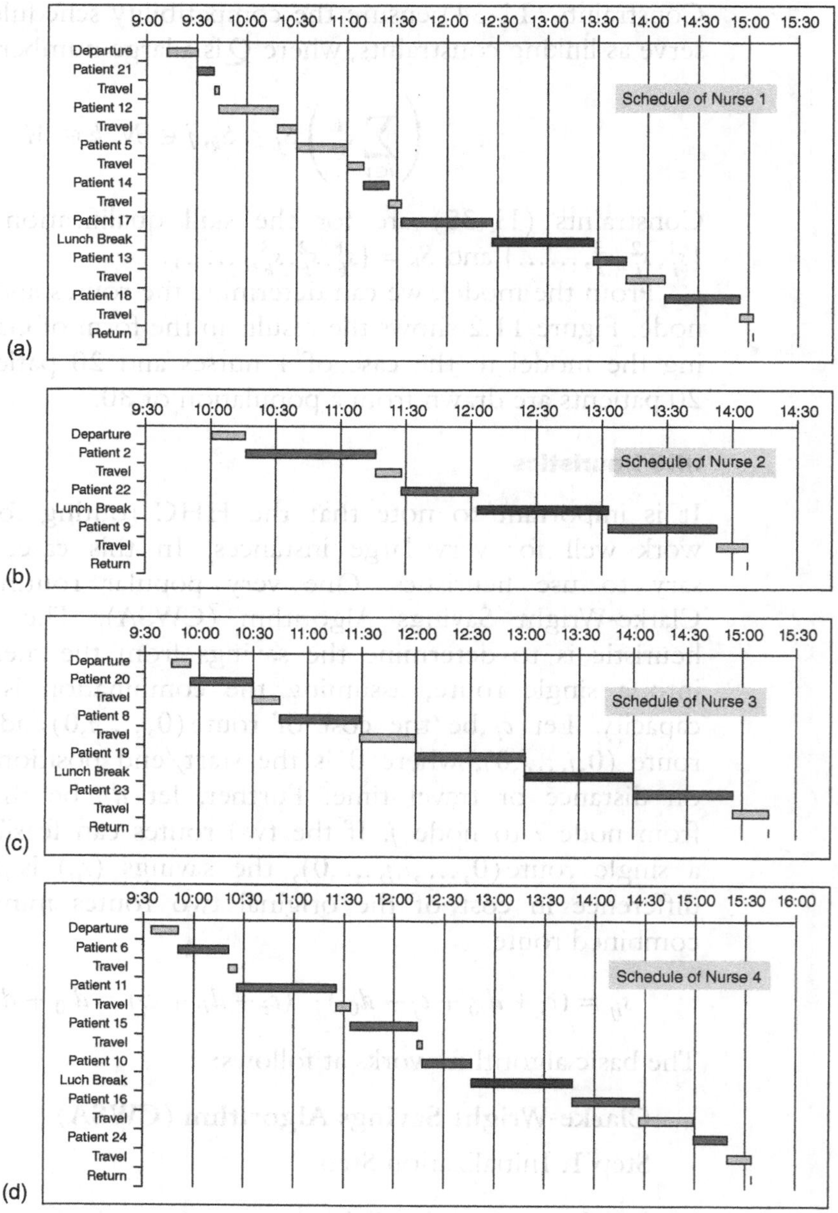

Figure 11.2 Solution of a 4-Nurse 20-Patient Example

Step 2. Route Merge Step

Starting from the top of the savings list, execute the following:

For the given s_{ij}, determine whether the route staring with $(0,j)$ can be merged with the route ending with $(i,0)$.

If the merger is feasible, combine the two routes by deleting $(0,j)$ and $(i,0)$ and introducing (i,j).

Delete s_{ij} from the list.

In order to modify this for time windows (Liu & Shen, 1999), let's define D_i to be the departure time for a nurse vising patient i. Further, let LT_i represent the latest departure time when patient i can be feasibly visited on the route. If $s(i)$ represents the immediate successor of patient i, then we can define LT_i by:

$$LT_i = \{B_i, LT_{s(i)} - t_{i,s(i)}\} \qquad (11\text{-}37)$$

where B_i is the latest time patient i can be served and $t_{i,s(i)}$ is the travel time from patient i to $s(i)$. We then need to add the two following conditions to CWSA to determine the feasibility of an insertion:

$$D_i + t_{i,k} \leq b_k \qquad (11\text{-}38)$$

$$max\{a_k, D_i + t_{i,k}\} + t_{k,j} \leq LT_j \qquad (11\text{-}39)$$

The first condition ensures that the nurse arrives at patient k before its specified latest starting time. The second condition requires that the new arrival time to patient j does not exceed its latest departure time, thereby ensuring time feasibility for the successors of j. We illustrate this heuristic in the practice problems.

QUESTIONS AND LEARNING ACTIVITIES

1. What are key factors to go into any facility location problem?
2. What are specific factors unique to healthcare facility location problems?

3. An asthma clinic is considering locating in Northwest Georgia to serve Dade, Walker, Chattooga, Catoosa, and Floyd counties. The prevalence of asthma in each county is 6.1%, 5,3%, 7.2%, 3.9%, and 4.8%, respectively.
   a. What are the factors that they should consider?
   b. Estimate the centroid of each county, and determine the population of each county.
   c. Using the information in b, determine minsum solution assuming Euclidean distance.
4. Solve problem 3 using the rectilinear minsum formulation.
5. Solve problem 3 using the Euclidean minimax formulation.
6. For the data given in problem 3, do you believe the rectilinear minsum or Euclidean minimax is a better formulation? Why? What are some of the main problems with either formulation?
7. For exercise 11.2, assume there is a sixth hospital (H6) that is 4 units from location 1, 6 units from location 2, and 5 units from location 3. Further, suppose that both location 1 and location 3 can't be chosen concurrently. Reformulate the problem and solve.
8. Consider 10 patients with location: (1, 4), (15, 8), (9, 9), (18, 3), (9, 15), (2, 8), (8, 19), (4, 14), (14, 12), and (15, 6). There are 4 potential clinic locations: (4, 4), (9, 9), (12, 16), and (13, 4). No patient is willing to travel over 7 units (assume travel is according to rectilinear distance). What is the minimum number of clinics that can be opened to serve the population? Where should they be located?
9. For problem 8, suppose patients 1-5 are uninsured and patients 6-10 are privately insured. If you are to locate a single facility, where should it be located to maximize profit if revenue is twice as high for an insured patient compared to an uninsured one?
10. Repeat problem 9 accounting for equity between the two types of patients. Discuss your solution.
11. Given the patient locations in problem 8. Each patient has a visit time of 1 hour each day. Travel time is 10 minutes per unit distance. Assume that there is a home health care facility located at (8,8). A nurse work no more than 8 hours in a day, including travel time back to the facility. Use the CWSA to find the routes.

12. For problem 11, assume patients 1, 2, and 3 require a special skilled nurse. Resolve using this additional requirement.
13. Suppose for problem 11, the latest time patient 2 can be seen is 11am, that patient 4 can be seen is 1pm and that patient 6 can be seen is 1pm. All other patients must be seen by 3pm. Nurses can leave the facility no earlier than 8am. Use the modified heuristics of Liu and Shen.

REFERENCES

Chi, C. (2015). Mathematical Programming Approaches to Home Healthcare Nurse Routing Problem and Truckload Transportation Procurement via Combinatorial Auctions. Ph.D. Thesis, University Park, PA: Pennsylvania State University.

Daskin, M. S., & Dean, L. K. (2004). Location of health care facilities. In M. L. Brandeau, F. Sainfort, & W. P. Pierskall (Eds.), *Operations research and health care: A handbook of methods and applications*, 43–76. Boston, MA: Kluwer Academic Publishers.

Griffin, P. M., Scherrer, C. R., & Swann, J. L. (2008). Optimization of community health center locations and service offerings with statistical need estimation. *IIE Transactions, 40,* 880–892.

Griffin, P. M., Lee, H., Scherrer, C. R., & Swann, J. L. (2014). Balancing investments in federally qualified health centers and Medicaid for improved access and coverage in Pennsylvania. *Health Care Management Science, 17,* 348–364.

Liu, F.-H., & Shen, S.-Y. (1999). A method for vehicle routing problem with multiple vehicle types and time windows. *Proceedings of the National Science Council, ROC(A), 23*(4), 526–536.

Verdat, V., & Lapierre, S. D. (2002). Location of preventive health care facilities. *Annals of Operations Research, 110,* 123–132.

Chapter **12**

Health Supply Chains

"As scarce as truth is, the supply has always been in excess of the demand."

—Josh Billings

Overview

Coordinating the ordering and distribution of materials and supplies across the various entities that make up the healthcare delivery system is an important and challenging process. These entities include the various delivery sites such as hospital networks, regional hospitals, clinics and the patient home as well as the supply and distribution sites. The collection of these entities is called the supply chain, and Figure 12.1 illustrates the relationships.

In this chapter, we discuss some of the most important supply chain functions (Chopra & Meindl, 2013; Ravindran & Warsing, 2013). These include forecasting, inventory control, and distribution planning. We also discuss various strategies that can help to coordinate activities.

12.1 FORECASTING DEMAND

Forecasting is the attempt to predict future events in the presence of uncertainty. Healthcare agencies use analytical forecasting tools to predict patient outcomes, patient mix, and patient demand for services. Effective forecasting can help a healthcare organization in planning and allocating capacity such as number and type of beds, surgical suites, and equipment. Developing effective forecasts, however, is a challenging process. There are several reasons for this. First, the basic assumption in forecasting is that past data is useful to predict future events. However, environmental factors that influence forecasts change over time. An example would be

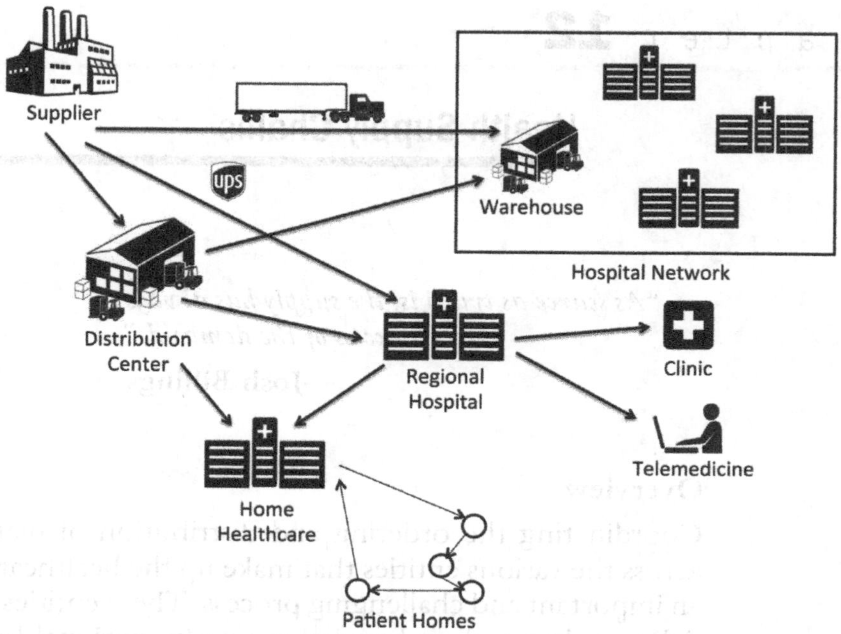

Figure 12.1 Healthcare Supply Chain

the change in sociodemographics of geographic regions over time. If these future conditions are significantly different, then historical data will have much less value. Second, in order to forecast, there must be some pattern to the factor of interest. This may be a trend or seasonal effect. If the change in the factor is simply due to random effects, then no forecasting tool can be accurate. Finally, the factor we wish to forecast is often a function of several other factors that may be difficult to measure or include in the model.

Forecasting methods in general fall into two categories: quantitative and qualitative techniques. Common quantitative techniques include time series models such as exponential smoothing and causal models such as regression. Qualitative techniques rely heavily on the use of expert opinion. The appropriate model depends on what is being forecasted and the time horizon of the forecast. Qualitative methods tend to be used for long-term strategic forecasting, while quantitative techniques are more useful for shorter-term forecasts.

Effective causal models tend to be very specific to the context. For example, an important measure for emergency departments (EDs) is the start and duration of the flu season, since demand for ED services goes up significantly over this period. Google Flu Trends (GFT) used the premise that the use of search terms on flu-related topics was a useful factor for predicting actual illness. That is, there was a causal relation between search and illness.[1]

Time series models may be more generally applied, and so we present a simple and commonly used technique here. We also present performance measures that can be useful in modifying the forecasting procedure.

Time Series Models

In time series models, we use historical data to predict the future. We will use demand in period t (D_t) as our factor of interest, though of course this method can be applied to other factors. The basic time series relation we will use is:

$$D_t = a + bt + S_t + \varepsilon_t \tag{12-1}$$

where a is the base demand, b is the trend, S_t is the seasonal effect, and ε_t is the random effect. Note that this relation assumes the trend is linear and that the data is not auto correlated. There are more sophisticated models than what is presented here that can account for nonlinearity and autocorrelation.

We assume that a season occurs over P periods. Common examples are 7 days in a week, 12 months in a year, or 52 weeks in a year. This depends on what measure of demand is of interest (i.e., hourly, daily, weekly, monthly). For example, daily patient volumes in an ED tend to be higher over weekends than during the week. We could in this case define a season as a week ($P = 7$ days).

In order to use this relation, we need to be able to estimate a, b, and S_t, and update them over time. A very popular approach to achieve this is the Holt Winter method. In this method we assume that demand is distributed with mean a and standard deviation σ. If we ignore trend and seasonal effects, then we have the simplified relation:

$$D_t = a + \varepsilon_t \tag{12-2}$$

We can update a by averaging the most current measure of demand with historical demand. We will let α ($0 < \alpha < 1$) be the weight that we put on

[1] GFT was not very successful at predicting flu prevalence. Their estimate in 2012–2013 was roughly double the prevalence figures provided by the Centers for Disease Control and Prevention (CDC) over that period.

the current measure. In time period t, then, the updating procedure for a is (we use the "hat" to represent an estimate of the actual value):

$$\hat{a}_t = \alpha D_t + (1 - \alpha)\hat{a}_{t-1} \tag{12-3}$$

The estimate of future demand is then:

$$\hat{D}_{t+1} = \hat{a}_t \tag{12-4}$$

The larger the value of α, the more weight we put on current data. This idea can be easily extended to include trend and seasonality. We modify our initial time series relation to:

$$D_t = (a + bt)S_t + \varepsilon_t \tag{12-5}$$

In this case, we need to update a, b, and S_t. We use α, β, and γ to weight current observations of base demand, trend, and seasonality. The updating procedure is then:

$$\hat{a}_t = \alpha\left(D_t/\hat{S}_{t-P}\right) + (1 - \alpha)(\hat{a}_{t-1} + \hat{b}_{t-1}) \tag{12-6}$$

$$\hat{b}_t = \beta(\hat{a}_t - \hat{a}_{t-1}) + (1 - \beta)\hat{b}_{t-1} \tag{12-7}$$

$$\hat{S}_t = \gamma\left(D_t/\hat{a}_t\right) + (1 - \gamma)\hat{S}_{t-P} \tag{12-8}$$

We need initial estimates for a and b. These values can be obtained for the n historical data points from:

$$\hat{a}_0 = \frac{6}{n(n+1)}\sum_{t=-n+1}^{0} tx_t + \frac{2(2n-1)}{n(n+1)}\sum_{t=-n+1}^{0} x_t \tag{12-9}$$

$$\hat{b}_0 = \frac{12}{n(n^2-1)}\sum_{t=-n+1}^{0} tx_t + \frac{6}{n(n+1)}\sum_{t=-n+1}^{0} x_t \tag{12-10}$$

where x_t is the deseasonalized demand. These values are obtained by solving for the first order conditions of the sum of the squared error, namely:

$$\min Z = \sum_{t=-n+1}^{0}\left(x_t - (\hat{a}_0 + t\hat{b}_0)\right)^2 \tag{12-11}$$

We illustrate this approach with an example.

Example 12-1

Quarterly demand for the past three years (year 1 to year 3) has been collected for new and replacement mattresses for a hospital network. We wish to forecast demand for year 4. The historical demand is given in the following table:

	Q1	Q2	Q3	Q4
Year 1	93	108	122	100
Year 2	106	110	130	111
Year 3	113	130	141	115

The first step is the develop estimates of the seasonal factors. This is done by comparing the quarterly demand to annual averages as shown in the following table.

Year	Quarter	Period	Demand	Avg 1	Avg 2	Centered	Estimate of S_t
(1)	(2)	(3)	(4)	(5)	(6)	(7) = ((5) + (6))/2	(8) = (4)/(7)
1	1	−11	93				
1	2	−10	108				
1	3	−9	122	105.75	109	107.375	1.14
1	4	−8	100	109	109.5	109.25	0.92
2	1	−7	106	109.5	111.5	110.25	0.96
2	2	−6	110	111.5	114.25	112.875	0.97
2	3	−5	130	114.25	116	115.125	1.13
2	4	−4	111	116	121	118.5	0.94
3	1	−3	113	121	123.75	122.375	0.92
3	2	−2	130	123.75	124.75	124.75	1.05
3	3	−1	141				
3	4	0	115				

Columns (5) and (6) are annual averages. Two averages are computed since *P* is even and hence is not centered in a year. The first value in column (5) is obtained by averaging the demand in periods −11 to −8 in column (4). The first value in column (6) is obtained by averaging the demand in periods −10 to −7 in column (4). We can estimate a centered average by average the numbers in columns (5) and (6). The seasonal effect can be estimated by comparing the actual demand to the centered demand. A value above 1 means that demand was above the annual average for that quarter and below 1 means it was below the annual average. We can refine the estimate by averaging the estimates from the same quarter in column (8) and normalizing over all of the estimates. This is shown in the following table:

Quarter	Average S_t	Normalized S_t
1	0.941	0.939
2	1.010	1.008
3	1.133	1.130
4	0.926	0.924
Total	4.010	4.000

We can next determine the initial estimates of *a* and *b*. We first deseasonalize the data by dividing the *demand* by its corresponding seasonal factor. Using the formulas for initial estimates then gives $a_0 = 127.83$ and $b_0 = 2.36$. We can now forecast for year 4. For the first two quarters:

$$D_1 = (127.83 + 2.36(1))\,0.939 = 122.25$$

$$D_2 = (127.83 + 2.36(2))\,1.008 = 133.61$$

Similarly, $D_3 = 152.45$ and $D_4 = 126.84$. To illustrate how to update *a*, *b*, and *S*, suppose actual demand in the first quarter of year 4 is 125 and that we use $\alpha = 0.25$, $\beta = 0.08$, and $\gamma = 0.30$. Then we get:

$$\hat{a}_1 = 0.25\left({}^{125}/_{0.939}\right) + (1 - 0.25)(127.83 + 2.36) = 130.92$$

$$\hat{b}_t = 0.08(130.92 - 127.83) + (1 - 0.08)2.36 = 2.42$$

$$\hat{S}_t = 0.30\left({}^{125}/_{130.92}\right) + (1 - 0.30)0.939 = 0.944$$

These new values could be used to reforecast for the rest of the year.

Forecast Error

There are several useful measures for forecast accuracy. Of particular interest is an estimate of the bias and variance of the forecast compared to the true values. A commonly used measure that gives information about the bias is the mean absolute deviation (MAD):

$$MAD = \frac{1}{n}\sum_{i=1}^{n} |\hat{D}_t - D_t| \tag{12-12}$$

and for the variance is the mean squared error (MSE):

$$MSE = \frac{1}{n}\sum_{i=1}^{n} (\hat{D}_t - D_t)^2 \tag{12-13}$$

In addition, a tracking signal T_t can be easily constructed to help monitor the bias. The tracking signal is a cumulative sum:

$$T_t = T_{t-1} + (\hat{D}_t - D_t) \tag{12-14}$$

If at any point $|T_t| > 0.4$, then this gives evidence that there is excessive bias in the forecast and the model should be reconsidered.

12.2 INVENTORY CONTROL

One of the key decisions in supply chain management is around inventory. Specifically, it is important to determine how much inventory to hold at a location, and when to replenish. Three types of inventory enter into the decision: cycle stock, safety stock, and pipeline stock. Cycle stock is inventory that is used to satisfy average demand. In most situations, demand is not known with certainty and disruptions can occur in supply. Safety stock is therefore held to buffer against this variability. Finally, inventory that is being delivered (e.g., by truck) is pipeline stock.

In this section we present three basic models for managing inventory. We first look at models for managing inventory when demand is known (i.e., deterministic). We then show a commonly used approach to manage inventory when only the demand distribution is known (i.e., stochastic). Finally, we present a special case for perishable items called the newsvendor model.

Deterministic Models

Consider the case where a location holds inventory to satisfy demand. The location can order items from a supplier. There is a lead time L between when the order is placed and when the items arrive. The fundamental trade-off in inventory is between the order cost A (\$/order) and the holding cost h (\$/time period). Holding cost may be defined as $h = vr$ where v is the item value (\$/item) and r is the time value of money (\$/\$value of item/time period). Ordering more frequently means that there is less average inventory, and hence a lower total holding cost, but that the total ordering cost increases. Our goal is to determine the order quantity that minimizes the total cost.

We will assume that the forecast provides demand per period (D_i) over the next n periods. Orders may be placed in any period (Q_i) and the lead time is zero (i.e., instantaneous replenishment). We also assume that the value of an item, the order cost, and the time value of money does not change over the horizon. Holding cost is applied to inventory carried from one period into the next, and an order arrives at the beginning of the period. Finally, we assume that all of the demand must be met and that the supplier has sufficient stock to meet all of our orders. Let x_i be a binary variable equal to 1 if an order is placed in period i. The optimization model to minimize total cost is given by:

$$\min TC = \sum_i Ax_i + \sum_i vrI_i \tag{12-15}$$

subject to:

$$I_i = I_{i-1} + Q_i - D_{i-1} \ \forall i \tag{12-16}$$

$$Q_i \leq Mx_i \tag{12-17}$$

$$I_i \geq 0 \ \forall i \tag{12-18}$$

$$I_0 = I \tag{12-19}$$

The first term in the objective function is the order cost and the second term is the holding cost. Note that the purchase price is not included as an additional term since regardless of the ordering solution, we will always order the same total quantity. The first constraint is the inventory constraint. The second constraint is the linking constraint, which ensures that

Q_i is positive only if an order is placed (i.e., $x_i = 1$), where M is a large number. The third constraint ensures that demand is met in each period. The last constraint sets the initial inventory level.

Example 12-2

Demand for an item over the next five periods is forecasted to be 10, 20, 15, 16, and 18. There are currently 20 items on hand. The order cost is $100 per order, and each item costs $50. The time value of money per period is 0.1. The largest an order could be would be the sum of the total demand minus initial inventory, and so M can be set to 59. The resulting formulation is:

$$\min 100\sum_i x_i + 50(0.1)\sum_i I_i$$

subject to:

$$I_1 = I_0 + Q_1 - 10$$
$$I_2 = I_1 + Q_2 - 20$$
$$I_3 = I_2 + Q_3 - 15$$
$$I_4 = I_3 + Q_4 - 16$$
$$I_5 = I_4 + Q_5 - 18$$
$$Q_1 \leq 59x_1$$
$$Q_2 \leq 59x_2$$
$$Q_3 \leq 59x_3$$
$$Q_4 \leq 59x_4$$
$$Q_5 \leq 59x_5$$
$$I_0 = 20$$

Solution of this formulation gives an order for 25 items in period 2 and for 34 items in period 4. The total cost is $415.

A very special case of deterministic inventory is when demand is constant. This simplifies the problem greatly. If we assume that r is given over the horizon rather than per period (i.e., \$/\$ value of item/horizon length), demand D is total demand over the horizon, an order may be placed at any time and not just the beginning of period, and that lead time is zero, then we have an unconstrained optimization problem. In this case, total cost equals total order cost over the horizon plus inventory cost over the horizon. The order quantity will always be the same since demand is constant and hence the average inventory in this case will be Q/2. Therefore, we get the following unconstrained problem:

$$\min TC = A\frac{D}{Q} + vr\frac{Q}{2} \tag{12-20}$$

First-order conditions yield:

$$Q = \sqrt{\frac{2AD}{vr}} \tag{12-21}$$

This formula is called the economic order quantity (EOQ). It is very commonly (though often inappropriately) applied.

Stochastic Models

In the previous section, we assumed that demand was known by period. This assumes that our forecast is perfect, which is rarely the case in practice. In this section, we present two simple models that can be used if the distribution of demand is known. The first model, called the continuous review model, assumes that inventory levels can be continuously monitored. If it is not possible to continuously review inventory, then a periodic review model may be used instead.

The basic idea behind the **continuous review** model is to establish a trigger point T for ordering based on the demand distribution and desired service level. We will assume that lead time for delivery is constant. The variable that will be tracked in called inventory position, which is equal to the on-hand inventory plus the inventory that is on order. The system orders Q items when the inventory position falls below T. Since the mean demand is assumed to not change over the horizon, then the EOQ formula may be used for Q. The key then is to determine T.

Let D be the average annual demand for the item and σ the standard deviation of annual demand from forecast error. Forecast error is assumed to be normally distributed, with a mean of zero (unbiased) and standard deviation of σ_L over the lead time L. The mean demand over the lead time L (in years) and the standard deviation of demand over L are given by:

$$d_L = LD$$

$$\sigma_L = \sqrt{L}\sigma$$

We will define the service level *SER* for this model as one minus the probability that our inventory position falls below zero (i.e., stock out of inventory). Let s be the safety factory, and set the trigger point T to the following:

$$T = d_L + s\sigma_L$$

In this expression, the second term is the safety stock quantity, since it is defined by the variability. Using this value for T implies that SER = P{Demand over lead time $\leq d_L + s\sigma_L$}. Converting this to standard normal form gives SER = $\Phi(s)$, where $\Phi(\cdot)$ is the cumulative density function for the standard normal distribution. Therefore, for a given service level, $s = \Phi^{-1}(\text{SER})$.

Example 12-3

Annual demand is forecasted to be 10,000 units with a standard deviation of 2,000 units. Lead time for a replenishment is 1 week. Order cost is \$200, item value is \$40, and the time value of money is 0.2 \$/\$ value of item/year. A service level of 0.95 is desired. The EOQ for this example is:

$$Q = 2(200)(10000).2(40) = 707$$

The trigger point for the continuous review system is given by:

$$d_L = \frac{1}{52}10000 = 192.3$$

$$\sigma_L = \sqrt{\frac{1}{52}}2000 = 277.4$$

$$s = \Phi^{-1}(0.95) = 1.645$$

$$T = 192.3 + 1.645(277.4) = 648.6$$

Therefore, when the inventory position hits 648, we order 707 units.

If inventory cannot be reviewed continuously, but instead every P periods (P in the same units as L), then it is relatively straightforward to modify the continuous review model to obtain a **periodic review** model. In this model, we will check inventory every P periods and then order up to level U. Therefore, an order is placed every review period that is equal to U minus the inventory position. In this case, we need to protect inventory over $L + P$. We therefore set U by:

$$U = d_{L+P} + s\sigma_{L+P} \qquad (12\text{-}25)$$

where s is determined by the service level as previous (i.e., $s = \Phi^{-1}(\text{SER})$).

Newsvendor Model

In this section, we consider a special case for perishable items. It is motivated by a newsvendor example as follows. A vendor buys a number papers Q at a cost of c per paper to be sold during the day at a price of p per paper. This is the only opportunity for purchase. Based on historical data, she knows the distribution of demand (f_D), but not the actual demand for that day. During the day, she sells the papers. If at the end of the day there are leftover papers, she gets a small salvage value s per paper since old news has little value. If she runs out during the day, then she loses the potential sales. The trade-off in this case is between having too few and too many papers.

In order to determine the optimal value of Q we can consider what happens at the margin. Suppose she orders $Q' - 1$ papers. If we increase this by one paper to Q', then the marginal cost is c. The marginal revenue of this increase depends on where there is demand for the additional paper or not. If there is (i.e., $D \geq Q$), then she receives p. If not (i.e., $D < Q$), then she receives sells the paper for salvage s. The optimality condition is when the marginal cost equals the marginal revenue. If we define $F_D(Q)$ as the cumulative demand probability (i.e., $F_D(Q) = \text{P}\{D < Q\}$), then setting marginal cost equal to marginal revenue gives:

$$c = p(1 - F_D(Q)) + s(F_D(Q)) \qquad (12\text{-}26)$$

Note that $(1 - F_D(Q))$ is the probability that the item is sold and $F_D(Q)$ is the probability it isn't. Rearranging terms gives:

$$F_D(Q) = \frac{p - c}{p - s} \qquad (12\text{-}27)$$

The optimal order quantity is equal to the value of Q that makes this relation hold. Several items that are stocked in healthcare settings are perishable including blood, vaccines, and drugs.

Example 12-4

Platelets are supplied to a hospital every three days. The cost per unit is $c = \$100$. At the end of the three-day period, the hospital sells unused platelet units back to the supplier at $s = \$40$ per unit (the reduced amount is due to the average shelf life of platelets being only five days). The average revenue received by the hospital is $p = \$200$ per unit. Demand for units of blood over a three-day period is uniformly distributed between 120 and 190 units. To determine the optimal order quantity, we set:

$$F_D(Q) = \frac{200 - 100}{200 - 40} = 0.625$$

The cumulative density function for the uniform distribution on interval [120, 190] is:

$$F_D(Q) = \frac{Q - 120}{190 - 120}$$

Setting these equal gives $Q = 164$.

12.3 HEALTHCARE DISTRIBUTION

The models presented in the previous section are useful for a single location. For the health supply chain, however, these decisions must be made over a network. In addition, the transportation decisions must also be determined. Figure 12.2 shows a network example of a distribution center (DC) supplying a set of hospitals. In this case, there are several decisions that need to be made:

- How will each hospital control its inventory (when to order and how much)?
- Will inventory replenishment be decided by the DC or the hospitals?

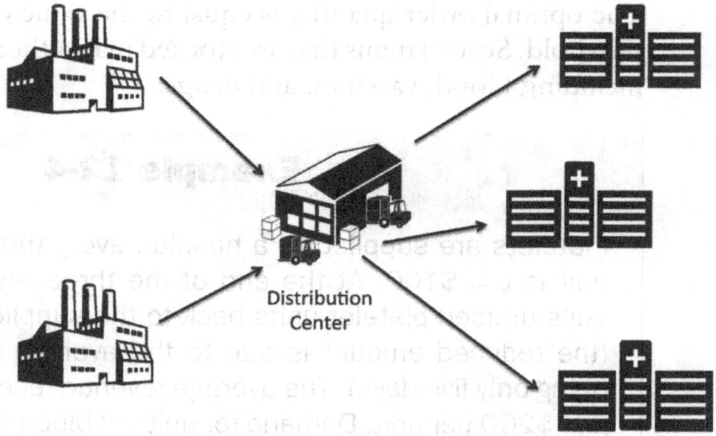

Figure 12.2 Healthcare Distribution Network Example

- What mode of transportation will the DC use to supply the hospitals (full truckload, less than truckload, contract carrier)?
- What will the timing of the deliveries be?
- Will special processes such as cold storage be required?

We discuss several methods that can be used to help address these issues.

Distribution Requirements Planning

In the previous section, we discussed how ordering decisions can be made for a location given demand forecasts. Let's consider now how the DC as shown in Figure 12.2 would respond with deliveries to orders from this set of hospitals and place its own orders. Part of the answer to this question depends on whether information is shared between the hospitals and the DC. If the DC has access to the ordering system of the hospital, then it can plan its own orders accordingly. This is illustrated by the following example.

Example 12-5

Three hospitals face (forecasted) demand for cases of surgical masks over the next six weeks. The DC can provide same day delivery. For each hospital, $A = \$400$, $v = \$300$, and $r = \$0.01/\$$

value of item/week, and suppose each hospital has 80 cases of inventory on hand. The demand for each hospital is given in the following table, along with the solution from the deterministic optimization model presented earlier:

		Week 1	Week 2	Week 3	Week 4	Week 5	Week 6
Hospital 1	Demand	50	90	120	70	140	110
	Orders	0	60	190	0	250	0
Hospital 2	Demand	110	40	20	150	120	40
	Orders	90	0	0	150	160	0
Hospital 3	Demand	90	30	98	190	70	80
	Orders	40	0	98	260	0	80

The demand for the DC can be determined by summing the orders for each hospital. Suppose that the DC has 250 cases on hand and the lead time from their supplier is one week. For the DC, $A = \$700$, $v = \$300$, and $r = \$0.02/\$$ value of item/week. The same optimization model can be used. The solutions shown in the table below give when the orders should be received at the DC. Since there is a one-week lead time, the orders need to be placed one week earlier.

	Week 1	Week 2	Week 3	Week 4	Week 5	Week 6
Demand	130	90	288	410	410	80
Orders Received	0	0	258	410	490	0
Orders Placed	0	258	410	490	0	0

If the DC didn't have visibility into the hospital orders, it is not able to plan in advance. Instead, it would need to react to individual hospital orders. Using forecasted average total demand from the hospitals, it employs a continuous or periodic review system to determine when to place orders. From a system perspective, however, this would require additional safety stock at the DC, which would increase the total inventory cost for the system.

Transportation Planning

In addition to determining when to place orders, the DC needs to determine how to deliver orders the hospitals. This is made up of two decisions: (1) which mode of transportation to use, and (2) when to make deliveries.

For transportation mode, the decision typically only involves comparing the cost across the set of alternatives. Standard alternatives are full truckload (FTL), less than truckload (LTL), and contract carrier such as UPS or DHL. There are two potential complications to this. This first occurs if the healthcare network manages its own fleet of trucks. In this case, the routing decision also impacts the best mode to be used. The second occurs if a rail or ocean carrier is used as a mode since they have significantly longer delivery lead times compared to trucks. Longer lead times lead to an increase in safety stock and pipeline stock requirements, so inventory must be included in the analysis. However, neither of these complications tends to occur in healthcare supply chains in practice.

Example 12-6

Consider the data given in Example 12-5. In addition, suppose that the distances of the DC to the hospitals are 80 miles, 130 miles, and 70 miles, respectively. There are three choices of delivery the DC can choose:

- Contract carrier: DHL charges \$5.80 per case. All hospitals are in the same delivery zone, so the price is the same for each.
- LTL: The rate is \$2.45 per mile traveled plus a handling fee of \$2.50 per case.
- FTL: The rate is \$1.40 per mile traveled plus a handling fee of \$400. The handling fee is paid regardless of the number of containers. The capacity of a truck is 200 cases.

If we consider week 4, then the total cost (TC) for each mode is:

- Contract carrier: $TC = 19.80(150 + 260) = \$2,378.00$
- LTL: $TC = 2.50(150 + 260) + 2.45(80 + 130) = \$1,539.50$
- FTL: $TC = 400 + 1.40(80) + 2(400 + 1.40(130)) = \$1,676.00$

In this case, using LTL is the cheapest. Note that we could use mixed modes. For example, we could use FTL for hospital 1 and then use FTL for the first 200 units of hospital 2 and LTL for the remaining 60 units. This would give $TC = 400 + 1.40(80) + 400 + 1.40(130) + 2.50(60) + 2.45(130) = \$1,562.00$, which would still be slightly more expensive than LTL.

The other key decision to be made by the DC is whether to combine orders into a single shipment. Although this would not have any impact on contract carrier or LTL since these costs are purely based on per item movement, it could impact FTL costs. The trade-off in this case is between reduced transportation costs from the combined shipment and the increase in inventory holding costs from early delivery of part of the shipment.

Example 12-7

Consider the data given in Examples 12-5 and 12-6. The DC could combine the orders for Hospital 3 for weeks 3 and 4. The total demand would be 358 cases and so two full truckloads would be required. Note that 260 cases of this total would arrive one week early, and so there would be additional carrying cost for the hospital. Let's compare the two alternatives:

- Combined shipment: $TC = 2(400 + 1.4(70)) + 260(300)(0.01) = \$1,776.00$
- Separate shipments (use LTL for week 3 and FTL and LTL for week 4): $TC = 2.50(98) + 2.45(70) + 400 + 1.4(70) + 2.50(60) + 2.45(70) = \$1,236.00$

In this case, the additional cost from holding inventory was too high to justify the combined shipments, and so it would be more economical to keep them separate.

An additional issue that can impact the transportation planning is how the product pricing is done. Free on board (FOB) pricing generally occurs in two forms. The customer can either take ownership at the supplier (FOB origin) or at their location (FOB destination). In the first case, the customer still needs to pay for the transportation and also must consider this pipeline stock in their inventory calculations. In the second case, transportation costs are typically included in the price, and hence the supplier owns the pipeline stock. Suppliers may also offer price breaks for larger order quantities due to economies of scale. In this case the inventory ordering models must be modified to account for this. Since the product price can vary, purchase price must also be included in the objective function.

Vendor Management Inventory

In Example 12-5, the hospitals drove the decision making for the DC (or supplier). An alternative approach is to let the DC decide when and how much to stock each hospital. This type of control is called *vendor-managed inventory* (VMI). The potential benefit to the DC for using VMI is that it can coordinate activities rather than responding to each hospital request. The potential benefit to the hospitals is that they no longer have to support that inventory control function.

The key to successful VMI implementation is the contractual agreement between DC and hospitals. For example, if the supplier is paid for each item delivered to a hospital, it is in the supplier's interest to delivery those items as early as possible since they are not paying for the holding costs at the hospital. However, the hospital would like their inventory to be as low as possible while still maintaining their service level. In order to incentivize the supplier to not overstock the hospital, a simple contract could be established, made of two parts. First, a minimum inventory level is established at the hospital to ensure service. Second, the hospital will not reimburse the supplier until after the inventory item is used. As long as these two items are met, the supplier can stock according to whatever policy works best for them. In this way, the supplier can still reap some of the benefits of coordinated replenishments across their customer base but will refrain from overstocking.

Cold Chains for Healthcare Products

For many healthcare applications, temperature control is an extremely important aspect of distribution. Controlled environments may be maintained in both the storage and transportation of goods. This includes refrigerated warehouse storage and trucks or containers with cooling systems. Special packaging may also be required. Common examples of products that require temperature control include vaccines, drugs, blood and blood products, film products, and so on. Many products require temperature along their entire distribution path. Such a temperature-controlled supply chain is referred to as a "cold chain."

Global demand of cold chains for healthcare products has grown significantly. In 2013, cold chain services in healthcare were US$7.3 billion. It is projected to grow to US$11.4 billion by 2018.

Effective cold chains need to be continuously monitored and controlled. radio-frequency identification (RFID) and other tracking devices may be used to monitor temperature at the item level. Documentation at each step in the process is essential; however, if a continuous cold chain is required for the product.

12.4 COORDINATING ACTIVITIES IN THE SUPPLY CHAIN

With so many activities that take place in a supply chain, coordination of those activities is extremely important. In this section, we discuss three coordination strategies. Pooling is a technique that can help to reduce the impact of variability through different types of aggregation. Information sharing helps to reduce the impact of lead times so that entities may engage in proactive strategies rather than reactive ones. Finally, contracting can be used to help share risk between parties, which can lead to overall better performance.

Pooling

The two key reasons for inventory in a supply chain are from high variability and long lead times. To illustrate, consider the following simple case. A warehouse faces constant demand of 4 items per day. The warehouse places a replenishment order every three days, and stockouts are not allowed. If replenishments are instantaneous (lead time is zero), then the warehouse would order 12 items every three days and average inventory would be 6 items. Suppose now that variability is introduced into the system in the following way (though nothing else changes). Average daily demand is 4 items, but half the time it is zero items and half the time it is 8 times, though the warehouse doesn't know which it will be for any particular period. In this case, the warehouse will order on average 12 items every 3 days. However, it is possible that demand could be as high as 24 units if all three days are peak. To ensure there is no stockout, then the warehouse will have to order up to 24 items each three days. Therefore, on average, there will be 12 units of safety stock and 6 items of cycle stock, for a total average of 18 items. Therefore, even though demand on average stayed the same, the variability led to a tripling of average inventory. Finally, suppose that replenishment lead time is actually 1 week. This adds inventory in two ways. First, the on-order inventory would be pipeline stock for the warehouse. In addition, since variability is occurring over the lead time, then this variability would also need to be protected for.

Pooling is a strategy to try to reduce the impact of lead time and variability on inventory levels. To common types of pooling are location pooling and item pooling. Location pooling is used in situations when there are long lead times from a supplier or set of suppliers. Local stocking of inventory is used to effectively reduce the lead time to the demand points (e.g., hospitals and clinics), which can lead to an overall reduction of inventory.

Example 12-8

A supplier provides cases of bandages to three hospitals in the Atlanta area. The lead time from the supplier to each hospital is 12 days. Each hospital has daily use of 10 cases with a standard deviation of 5. Inventory is managed using a continuous review policy with a service level of 95%. The average safety stock at each hospital using this policy would be equal to $ss = \sigma_L \Phi^{-1}(0.95) = (12)^{1/2}(5)(1.645) = 28.5$. Total safety stock across all three hospitals would then be 85.5. Consider now the use of lead time pooling by locating a local warehouse in the Atlanta area. Suppose lead time to the local warehouse from the supplier is 11 days and from the warehouse to the hospital is 1 day. The safety stock at each hospital would now be $ss = \sigma_L \Phi^{-1}(0.95) = (1)^{1/2}(5)(1.645) = 8.2$. The warehouse would face average daily demand of $3(10) = 15$ with a standard deviation of $(3(5^2))^{1/2} = 8.66$. If the warehouse also used a continuous review system with a 95% service level,[2]

[2]Note that the even though the hospitals and warehouse all use a service level of 95%, the overall system service level would not be 95%. This is because the warehouse may not always have sufficient on-hand inventory to satisfy the hospital orders (i.e., may periodically stock out).

then average safety stock in the warehouse would be ss = $(11)^{1/2}(8.66)(1.645) = 47.2$. Total safety stock across all three hospitals and the warehouse would be 73.2, which is over a 16% reduction is safety stock.

Lead time pooling can also lead to a reduction in logistics costs since orders are aggregated from the supplier to the local warehouse. The method is beneficial, however, only if the additional costs from the warehouse operations are less than the savings from inventory and logistics.

Item pooling is accomplished by looking for commonality across items and taking advantage in the reduction in total variation by aggregating these items. For example, a hospital may keep suture kits from various suppliers due to different preferences of the staff. If the hospital were to standardize on one kit type, less inventory would be required for the same level of service.

Example 12-9

A hospital has three suture kits from different suppliers. Each kit performs the same function, and the replenishment lead time for each is 10 days. Daily demand for each is normally distributed with a mean of 10, 20, and 13, respectively, and a standard deviation of 6, 10, and 7, respectively. If a continuous review system is used with a 95% service level for each kit, then total safety stock for the three kits would be ss = $(10)^{1/2}(6 + 10 + 7)(1.645) = 119.6$. If instead the hospital standardized to a single kit, the average daily demand would have a mean of 43 and standard deviation of $(6^2 + 10^2 + 7^2)^{1/2} = 13.6$. The safety stock for an equivalent continuous review system would be ss = $(10)^{1/2}(13.6)(1.645) = 70.7$, which is a 69% reduction in safety stock.

Information Sharing

Information sharing (or *transparency*) in a supply chain can bring many benefits. For example, if a supplier has visibility into the stocking levels of a hospital, it can better predict when restocks will need to take place, which helps with planning and coordinating activities. Some key examples of information sharing in a supply chain include:

- Sharing demand forecasts
- Sharing inventory information
- Sharing item-tracking data

In the absence of information sharing, entities in a supply chain are forced to react to events such as an order rather than planning for them.

One well-studied example of a negative consequence of reactive supply chains is what has come to be known as the "bullwhip effect." Simply stated, the bullwhip effect is that variability in inventory (and ordering) increases as one looks farther upstream in a supply chain. This is illustrated in Figure 12.3. The bullwhip effect is exacerbated by longer lead times. It can also be exacerbated by batching and incentives for larger order quantities. For example, a promotion on an item will cause a customer to purchase a larger than normal items in an order. If the demand the customer face remains the same, then they are essentially forward buying (i.e., buying in advance of the demand). This leads to "lumpy" demand for their supplier, which continues to get passed upstream.

We can quantify the bullwhip effect for a simple example. Consider the case of a supply chain consisting of a DC and a hospital. When the hospital places an order, it receives the order after a fixed lead time L. The hospital uses a moving average to forecast over p periods and manages its inventory using a continuous review system. Although we do not derive the results here, it can be shown that if $p = 5$ and $L = 1$, the ratio of the variance of the order quantity to the variance of the demand it faces is at least 1.4. This means that the variance of the orders placed by the retailer is 40% higher than the variance of the actual demand. This ratio increases as L increases.

The bullwhip effect leads to increased levels of safety stock, reduced service levels, and increased transportation costs (arising from expediting orders). The phenomenon has been observed in practice even when customer demand is smooth. Reducing lead times can help to reduce the impact of the bullwhip effect as can information sharing. Two common approaches are to provide visibility for end user demand and collaborative planning, forecasting, and replenishment (CPFR). CPFR is a process of cooperatively managing inventory by joint visibility and replenishment of

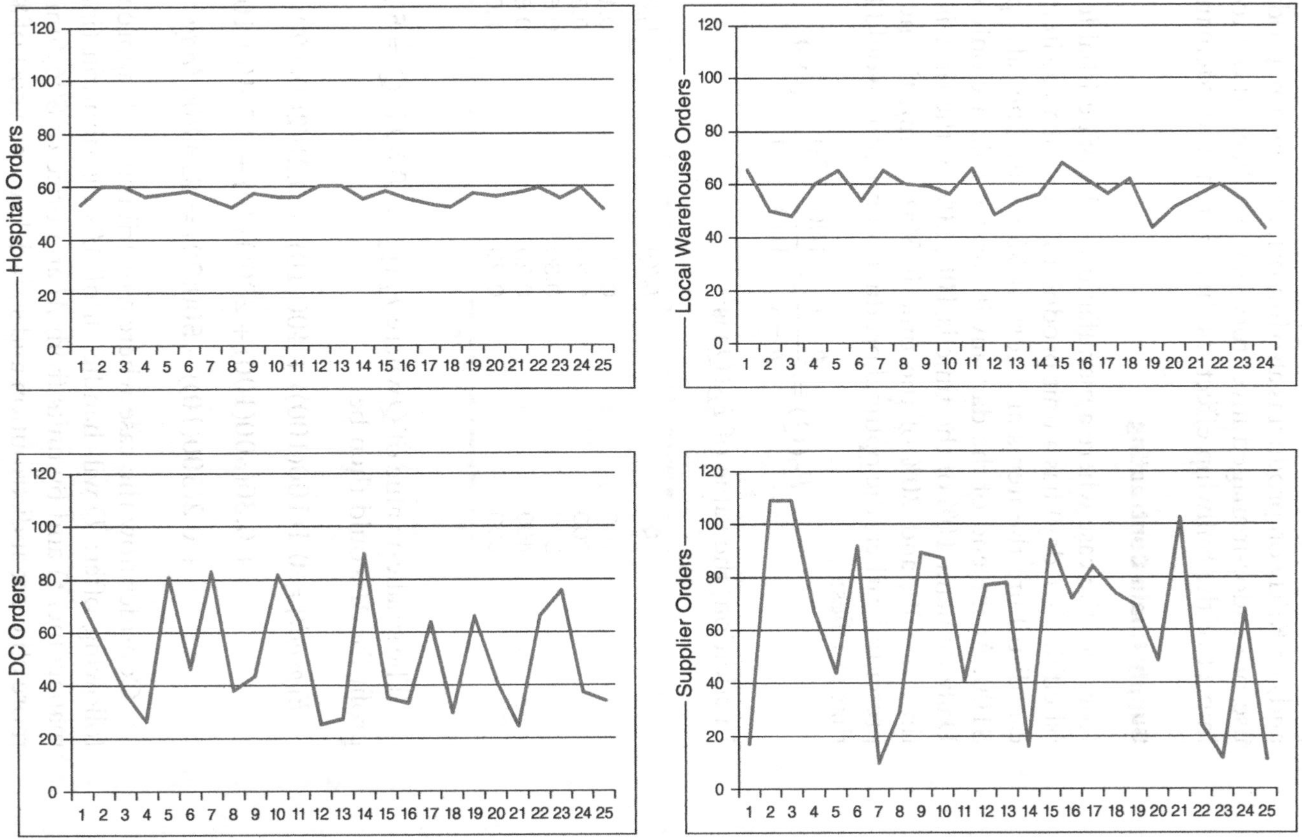

Figure 12.3 Illustration of the Bullwhip Effect

products through the entire supply chain. The extensive use of bar-coding and/or RFID on product is also important to ensure product visibility. The use of vendor-managed inventory has also been show to help reduce the impact of the bullwhip effect, as there is reduced "demand distortion."

Supply Chain Contracting

Consider the case where a manufacturer (M) of perishable medical supplies (shelf life is 1 day) has a unit production cost of \$15/item. In addition, M can distribute the items at a cost of \$5/unit. The sales price for a unit is \$100. At the end of the day, any items not sold are only worth \$10/unit. Daily demand 10% of the time is 100, 20% of the time is 200, 30% of the time is 300, and 20% of the time is 500. Using the newsvendor model presented earlier, the optimal production quantity would be the value of Q that satisfies:

$$F_D(Q) = \frac{p-c}{p-s} = \frac{100-20}{100-10} = 0.89$$

In addition, the value of $F_D(Q)$ is:

Q	$f_D(Q)$	$F_D(Q)$
100	0.10	0.10
200	0.20	0.30
300	0.30	0.60
400	0.20	0.80
500	0.20	1.00

The smallest value of Q where $F_D(Q) \geq 0.89$ is $Q = 500$. The expected profit for M would then be:

$$\begin{aligned} E[profit] &= 0.1(100(100)) + 400(10)) + 0.20(200(100) + 300(10)) \\ &+ 0.30(300(100) + 200(100)) + 0.2(400(100) + 100(10)) \\ &+ 0.2(500(10)) - 500(20) = \$23{,}800/day \end{aligned}$$

Consider now the case where a distributor (D) approaches M with the following offer: D will handle all of M's distribution (including demand management) and by units from M at a price of \$96/unit (note that D is more efficient at distribution and so it delivers at a cost of \$1/unit). Should M accept this offer?

If M does accept the offer, the D must determine the optimal order quantity. From D's perspective, the newsvendor formulation would yield:

$$F_D(Q) = \frac{p - c}{p - s} = \frac{100 - 97}{100 - 10} = 0.03$$

The optimal order quantity would then be 100. In this case, the expected profit for D would be $100(100 - 97) = \$300$ and for M would be $100(96 - 15) = \$8,100$. The total profit between them would be \$8,400. M would be much worse off in this case. The reason is that since D has a cost of \$96/unit rather than the \$15/unit of M, then there is a much greater cost for unsold items, that is, is facing greater downside risk. The question is whether a contract can be established that would make both parties better off.

One possible way to accomplish this is for M to lower the selling price to D. In return, D will share a portion of their revenue with M. Suppose M reduces the wholesale price to \$25. In this case, the newsvendor formulation for D would be:

$$F_D(Q) = \frac{p - c}{p - s} = \frac{100 - 27}{100 - 10} = 0.81$$

which gives an optimal order quantity of 500. The expected profit for D in the case is equal to \$20,300, and for M would be \$5,000. If D were to give M \$19,500 (or 96% of its revenue), the D's profit would be \$800 (up from \$300) and M's profit would be \$24,500 (up from \$23,800). Both parties are then made better off. The key is that this revenue sharing arrangement allowed both parties to share the risk.

This is an example of supply chain contracting. In many cases they can improve the performance of the supply chain. Revenue sharing is perhaps the most common approach. Another common approach is the buy-back contract where the supplier agrees to "buy back" any unsold/unused items from the buyer. This encourages the buyer to order larger quantities.

QUESTIONS AND LEARNING ACTIVITIES

1. In what ways is forecasting health demand different than forecasting demand for a product? In what way is it the same?
2. How is forecasting for chronic conditions different that forecasting for infectious disease?

3. Apply basic exponential smoothing to the data given in Example 12-1 (use a smoothing factor of 0.7) using only the first two years of data, and forecast the third year. Determine the MAD and MSE by comparing the forecast to the actual third year. Comment on the results.
4. For example 12-1, suppose the fourth year of demand is 120, 138, 145, and 115. Use the Holt Winter method to forecast demand for year 5.
5. Redo example 12-1 using use $\alpha = 0.20$, $\beta = 0.10$, and $\gamma = 0.35$.
6. Demand over the next 10 periods is: 100, 98, 123, 118, 173, 189, 143, 129, 119, and 142. Holding cost is $2/unit/period. Order cost is $200/order.

   a. Using EOQ, determine the order quantity. Determine the actual cost (holding plus order) over the time horizon.
   b. Use the optimization formulation and compare your answer to a.
   c. How would your answer to b change if the most inventory that can be held is 350 units?

7. Modify the optimization For problem 6.b if the leadtime for a replenishment is 2 periods.
8. Consider a continuous review system with a desired service level of 95%. Replenishment leadtime is 1 week, and annual demand is 20,000 units (with a standard deviation of 2,000 units). Item cost is $68, and holding cost rate is 18%. Order cost is $1000. Determine the reorder point. What is the average safety stock in case?
9. Using the data given in problem 7, sketch a tradeoff curve of service level (from 80% to 99.9%, using at least 10 values) versus safety stock value.
10. Based on your solution to 9, what do you believe a reasonable service level should be? Why?
11. A hospital orders vaccines from a supplier once a month. They pay $15 per unit. Demand is normally distributed with mean of 100 and standard deviation of 15 per month. The price that the hospital charges for the vaccine is $30. There is no salvage at the end of the month. How many vaccine units should they order?
12. For problem 10, what factors should the hospital use to define the price (i.e., value) of the vaccine for the newsvendor model?

13. In what ways would problem 10 violate the newsvendor assumptions in practice?
14. Consider example 12-5. Demand in hospitals 1 and 2 double from given, and demand for hospital 3 remains the same.

    a. Resolve for the optimal order quantities.
    b. Determine the resulting demand for the DCs.
    c. Determine the optimal order placement.

15. Apply the data in Example 12-6 to your answer to problem 14 to determine the best delivery modes. What is the total distribution and inventory cost to your solution?
16. For the supply chain contracting example in the Chapter, why might it be better to use revenue sharing rather than cost sharing?
17. Develop a scheme for buy-back contracting (*hint*: think of using the salvage value as the coordinating mechanism). Illustrate the approach using the same data as in the revenue sharing contract.
18. Compare and contrast the buy-back coordinating scheme found in problem 17 to the revenue sharing scheme.

REFERENCES

Chopra, S., & Meindl, P. (2013). *Supply chain management: Strategy, planning, and operation* (5th ed.). Upper Saddle River, NJ: Prentice Hall.

Ravindran, A. R., & Warsing, D. P. Jr. (2013). *Supply chain engineering: Models and applications*. Boca Raton, FL: CRC Press.

Chapter **13**

Infection Control

"I sing and play the guitar, and I'm a walking, talking bacterial infection."

—Kurt Cobain

Overview

In the United States, 100,000 people die annually from hospital acquired infections (HAIs),[1] including urinary tract infections, pulmonary infections, and surgical site infections, and 10% of patients acquire them. HAIs significantly increase patient length of stay. The resulting annual cost is estimated to range from US$5 billion to US$11 billion. In Europe, the infection rates are somewhat lower, ranging from 6% to 10%. Infection control practices are of extreme importance in combating HAIs. In this chapter, we first provide some historical perspective on infection control and then describe how they are classified. We then discuss the importance of administrative controls such as checklists before presenting some mathematical models for quantifying the benefits of infection control interventions.

13.1 HISTORICAL PERSPECTIVE

Hungarian physician Ignaz Philipp Semmelweis is considered by many to be the pioneer of infection control. In 1847, Semmelweis observed that the incidence of puerperal fever (also known as childbed fever) was reduced from 18% to 2% when healthcare workers washed their hands

[1] Note that these were originally called nosocomial infections and are also commonly called hospital-associated infections.

with a chlorinated lime solution.[2] This was perhaps the first study that used a control trial (albeit not randomized); he compared one ward with hand washing to another ward without hand washing. He published his findings, but they received little attention because he could not "scientifically" explain the effect. It wasn't until Pasteur's "germ theory of disease" was published many years later that Semmelweis's work received attention. Semmelweis was so upset by the medical community's lack of acceptance of his recommendations that he wrote letters to prominent obstetricians and accused them of being irresponsible murderers. Many, including his wife, felt Semmelweis was losing his mind. He was committed to a mental institution, where he died shortly thereafter, after being beaten by guards.

British nurse Florence Nightingale showed the importance of hygiene, including clean water and the washing of bed linens, in healthcare facilities during the Crimean War (1854–1856) and in India under the British occupation. She developed several statistical graphics such as the polar diagram shown in Figure 13.1 to support her findings. In this figure, the red wedges show deaths from wounds, and the blue wedges correspond to deaths from preventable causes. The chart on the right shows deaths from 1854 to 1855, and the chart on the left from 1855 to 1856 after implementation of her recommended interventions. Nightingale was an important proponent of systematic data collection, and her visual graphics were also extremely effective in swaying public opinion.

It wasn't until the 1950s and 1960s that hospitals started to develop programs that focused on reducing infections acquired in the hospital. This was in large part a response to the staphylococcal epidemics of the 1950s. At this time, Dr. Charles Rammelkamp demonstrated that the main mode of transmission of *Staphylococcus aureus* was by direct contact (as opposed to by air), and that hand washing by healthcare workers between patient contacts reduced that likelihood of transmission. In the 1960s, epidemiologists from the Centers for Disease Control and Prevention (CDC) also started analyzing public health surveillance data to identify HAIs. The field of infection control was established shortly thereafter. In 1972, the Association of Practitioners in Infection control was established, and in 1980 they developed a certification program.

In the 1970s, the CDC led the Study on the Effectiveness of Nosocomial Infection Control (SENIC) to determine the impact of CDC-recommended practice on HAIs. The results showed a significant

---

[2] Specifically, Semmelweis believed that the medical interns were transmitting the disease by treating women after handling cadavers as part of their studies.

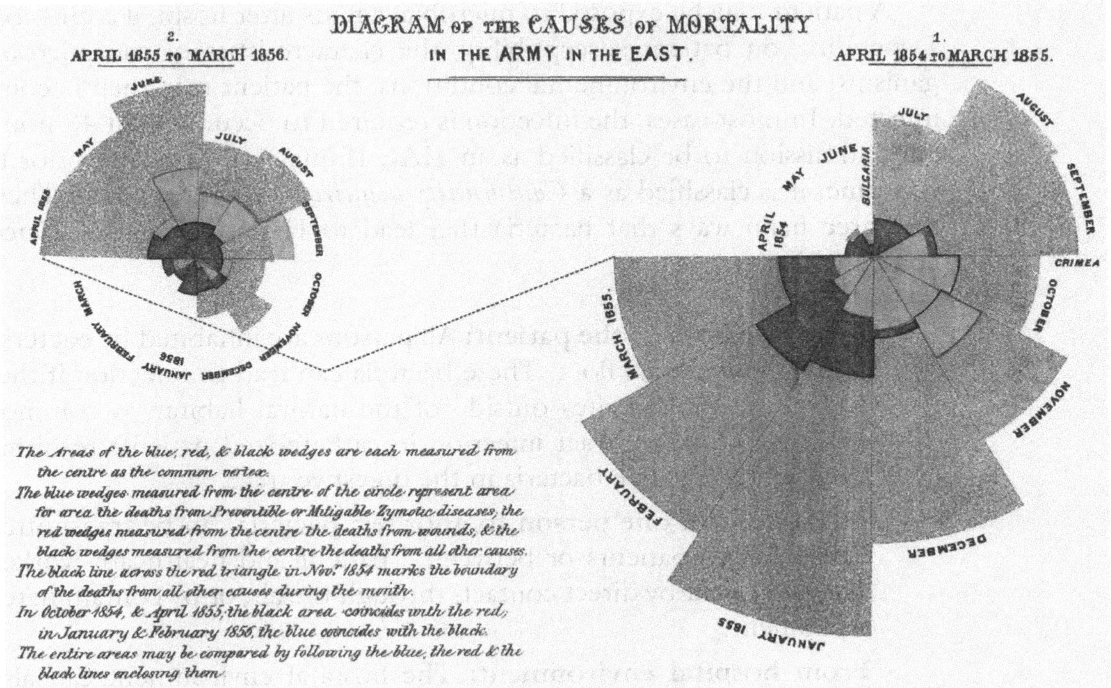

Figure 13.1 Coxcomb Diagram Showing Causes of Death Over Time during the Crimean War

impact, and as a result a Joint Commission on Accreditation of Hospitals mandated that infection control programs must be implemented in hospitals in order to receive accreditation. More recently, several states have mandatory reporting requirements of HAIs, and the Centers for Medicare & Medicaid Services (CMS) requires reporting of several HAIs such as central line–associated bloodstream infection (CLABSI).

13.2 INFECTION CONTROL CLASSIFICATION

The World Health Organization (WHO) uses the following definition for an HAI:

> An infection acquired in hospital by a patient who was admitted for a reason other than that infection. An infection occurring in a patient in a hospital or other healthcare facility in whom the infection was not present or incubating at the time of admission. This includes infections acquired in the hospital but appearing after discharge, and also occupational infections among staff of the facility.

A patient may be exposed to microbial agents after hospital admission. Depending on patient susceptibility, the characteristics of the microorganisms, and the environmental conditions, the patient may then become infected. In most cases, the infection is required to occur at least 48 hours after admission to be classified as an HAI. If infection occurred prior to this time, it is classified as a *Community-acquired Infection (CAI)*. There are three main ways that bacteria that lead to HAIs can spread (Ducel et al., 2002):

- **From the flora of the patient:** All persons are inhabited by bacteria, called the bacterial flora. These bacteria can lead to infection if they are transmitted to sites outside of the natural habitat. A common example is urinary tract infection in catheterized patients resulting from gram-negative bacteria in the digestive tract.
- **From flora of one person to another:** Bacteria can be transmitted between two patients or between a patient and healthcare worker. This can occur by direct contact, through the air, or by contaminated equipment.
- **From hospital environment:** The hospital environment can also support microorganisms. Examples include damp environments, bed linens, and food. Patients coming into contact with these environmental factors can become infected.

Although there are many different classifications of types of HAIs, WHO defines five main categories:

- **Urinary tract infections:** This is the most common HAI and is usually a result of catheters (catheter-associated urinary tract infection; CAUTI). It has lower morbidity that other HAIs, but can lead to death.
- **Surgical site infections:** This is also a fairly common HAI and is typically acquired during the surgery. This can come from hospital environment or from flora on the skin or operative site.
- **Respiratory infection:** This most commonly occurs in patients on ventilators (ventilator-associated pneumonia [VAP]). It can be a result of contaminated equipment or from flora of the digestive system infecting the lungs.

- **Vascular catheter infection:** This is a bloodstream infection that occurs in patients with central vascular catheter and typically results from infected equipment (CLABSI). The resulting morbidity is quite high.
- **Septicemia:** Sepsis can occur from any of the above infections. It is associated with high morbidity and mortality.

In response to HAIs, several infection control programs have been developed. The CDC promotes evidence-based guidelines. For the case of CAUTI, their recommendations are:

- Insert catheters only for appropriate indications.
- Leave catheters in place only as long as needed.
- Ensure that only properly trained persons insert and maintain catheters.
- Insert catheters using aseptic technique and sterile equipment (acute care setting).
- Follow aseptic insertion and maintain a closed drainage system.
- Maintain unobstructed urine flow.
- Comply with CDC hand hygiene recommendations and Standard Precautions.

By far the most important intervention has been hand washing. However, for several reasons including skin irritation, a perceived lack of time, and insufficient emphasis from hospital administration, compliance can be an issue. Another important intervention is the use of checklists, which we briefly discuss in the next section.

13.3 CHECKLISTS FOR INFECTION CONTROL

One of the most useful tools that have emerged in the past several years to help reduce the incidence of HAIs is the use of a simple checklist. In 2003, Dr. Peter Pronovost and colleagues initiated a cohort study in several intensive care units (ICUs) in the state of Michigan (Pronovost et al., 2006), called the Keystone Initiative. They specifically examined the impact that a set of simple interventions including a checklist would have on catheter-related bloodstream infection rates. Prior to the study, the median rate of catheter-related bloodstream infection ranged from 1.8 to 5.2 per 1,000 catheter-days in ICUs. The average cost of care for an infected patient was approximately US$45,000.

The five items in the checklist that doctors should follow are shown in Table 13.1. The results of the study over an 18-month period showed that the use of the interventions led to a decrease of incidence-rate ratios from 0.62 to 0.34. This led to an estimated savings of over US$100 million and an estimate of 1,500 lives saved over the time of the study. The study received significant press, including a widely read article by Dr. Atul Gawande (2007) in *The New Yorker* magazine.

In practice, however, the idea of physician checklists has been met with some resistance in their broader adaptation. One of the stated reasons is that physicians and nurses do not like being monitored or are simply offended at the notion that they needed a checklist. In Gawande's *New Yorker* article, Pronovost is quoted as saying:

> The fundamental problem with the quality of American medicine is that we've failed to view delivery of healthcare as a science. The tasks of medical science fall into three buckets. One is understanding disease biology. One is finding effective therapies. And one is insuring those therapies are delivered effectively. That third bucket has been almost totally ignored by research funders, government, and academia. It's viewed as the art of medicine. That's a mistake, a huge mistake. And from a taxpayer's perspective it's outrageous. (Gawande, 2007)

More recently, checklists have gained wider acceptance in hospitals, and are being used across a broad range of hospital applications including surgical site infections and CLABSIs. We look at the case for sepsis in the next section, and also introduce the notion of infection control bundles.

Table 13.1 Checklist for Reducing Catheter-Related Bloodstream Infection Rates

1. Wash their hands with soap.
2. Clean the patient's skin with chlorhexidine antiseptic.
3. Put sterile drapes over the entire patient.
4. Wear a sterile mask, hat, gown, and gloves.
5. Put a sterile dressing over the catheter site.

13.4 THE CASE OF SEPSIS

One example of the challenges faced with infection control is the case of sepsis. Sepsis is defined as an infection plus the systemic manifestations thereof that trigger a cascade of inflammatory responses in the body. Sepsis is increasing in incidence and is the primary cause of death from infection despite medical advances, such as vaccines, antibiotics, and acute care. Mortality rates with severe sepsis (sepsis plus acute organ dysfunction) and septic shock (severe sepsis plus refractory hypotension) in U.S. ICUs range from roughly 15% to 50%. In the United States, sepsis is the third-leading cause of death, causing more than 250,000 deaths each year (more than breast cancer, prostate cancer, and HIV combined) and representing the leading cause of death among ICU patients.

Septic patients take up approximately 25% of ICU bed capacity, making up over a million hospitalizations annually in the United States. Early recognition, treatment, and management of sepsis can significantly improve outcomes. For example, survival rates decrease by 7.6% for each hour of delay in antimicrobial administration at the onset of septic shock.

The efficient and effective transfer of sepsis patients into and out of the ICU is a key component to reducing patient harms. The slow transfer of patients into the ICU has been shown to lead to increased morbidity and mortality. Each hour of delay into the ICU increases ICU mortality by 8%, and patients with certain high-risk vital signs delayed by 18 to 24 hours were found to have a 52% mortality rate in the ICU, significantly higher than their nondelayed counterparts.

Patient transfers into the ICU often lead to medical errors due to information loss and other forms of communication gaps such as a lack of appreciation of significant patient problems. In addition, the transport of ICU patients out of the ICU for diagnostics can lead to patient harms and causes stress to ICU nurses due to the perceived vulnerability of the patients. There can also be significant variability regarding how to admit patients into the ICU, even within the same unit. Further differences in clinical training between residents, nurse practitioners, and physician assistants in critical care settings lead to differences in how information is sought (e.g., patient-based information seeking versus source-based information seeking) and information technology (IT) tools used.

Unexpected events during ICU transfers are common, occurring 67% of the time. These include equipment errors (39%), patient/staff management issues (61%), and serious adverse outcomes (31%), including major physiological derangement (15%) and death (2%). Transfer of patients to ICUs with high volume have better outcomes compared to those with low volume, however, ICUs with strained capacity have been shown to have worse patient outcomes pointing to the importance of effective decision making in bed management systems.

Communication lapses are also common during patient handoff and over shift changes due in large part to increased memory load at those transitions. These lapses include medication errors, omission of pending tests, and lack of responsibility handoff. Fatigue of nurses and interns common in night shifts can also lead to harm. Team performance frameworks have been developed to improve communication; however, standardized and reliable measurement tools are still needed.

Several programs have been developed to reduce sepsis morbidity and mortality. The Surviving Sepsis Campaign, which established international guidelines for severe sepsis management, is a well-known approach. The guidelines include building awareness, improving diagnosis, facilitating data collection, and using evidence-based healthcare bundles to help simplify complex patient care. Implementation of these guidelines has reduced mortality by as much as 50%. Although resuscitation bundles exist for sepsis patients, compliance varies. Understanding the sociotechnical systems of the ICU will help elucidate the barriers to effective implementation and identify additional interventions.

Applying a daily goals check sheet to improve physician-to-physician communication in the ICU has also shown a positive impact. There has been a recent call for checklists to be developed for intrahospital transports, including ICU transfers. The use of specialized intrahospital transport teams has shown benefits (albeit preliminary) compared to traditional transport teams. Despite the different processes in place to improve outcomes, there are still barriers to the management of these patients. An example of a checklist developed by the U.S. Department of Health and Human Services is shown in Figure 13.2.

Built environment can play an important role in patient harms. For example, higher noise levels are associated with increased stress among staff in the ICU. Additionally, the ICU sound environment such as ambient alarms significantly impacts sleep and delirium in ICU patients, and

TOP TEN EVIDENCE BASED INTERVENTIONS				
PROCESS CHANGE	IN PLACE	NOT DONE	WILL ADOPT	NOTES (RESPONSIBLE AND BY WHEN?)
Adopt a Sepsis Screening tool/system in the ED and/or in one inpatient dapartment.	☐	☐	☐	
Screen every adult patient during triage in the ED and/or once a shift in one identified inpatient department.	☐	☐	☐	
Develop an "Alert" mechanism to provide for prompt escalation and action from care providers with defined roles and responsibilities.	☐	☐	☐	
Develop standard order set or protocol linking blood cultures and lactate lab draws (blood culture = lactate level).	☐	☐	☐	
Develop a process to have lactate results within 45min. Make a lactate of > 4mmol/L a CRITICAL result for prompt notification.	☐	☐	☐	
Place broad-spectrum antibiotics in the ED medication delivery system to allow for antibiotic administration within 1 hour (collaborate with Pharmacy and infectious Disease Specialist for appropriate selection).	☐	☐	☐	
"Protocolize" fluid administration for sepsis patients to achieve goal of 30mL/kg crystalloid for rapid resuscitation.	☐	☐	☐	
Develop an order-set or protocol for 3-hour resuscitation bundle and the 6-hour septic shock bundle that uses an "opt-out" process instead of an "opt-in" for all bundle elements with the explicit end goals of therapy.	☐	☐	☐	
Ensure resources available for prompt performance of necessary imaging studies to confirm potential source of infection and intervene within 12 hours.	☐	☐	☐	
Utilize a "TIME ZERO" method that also displays visual cues for the healthcare team for timing of interventions for the sepsis bundle (identification time).	☐	☐	☐	

Figure 13.2 Severe Sepsis/Septic Shock Top 10 Checklist

the use of single patient rooms versus multiple patient rooms significantly improves patient and family satisfaction. Little work, however, has been done on the impact of built environment in directly reducing harm in patient transfers and handoffs.

The use of smart technology in the ICU has been successful at reducing harms. Examples include smart pump technology to reduce medication infusion errors and automated patient workflow management. A key concern, however, is the number of false alarms, as well as alarm fatigue. The development of algorithms and systems to reduce false alarms will improve patient safety. In addition, even with supportive technology, a high level of communication is required during handoffs, and as with many implementations, supportive technology actually increases length of ICU stay.

13.5 MATHEMATICAL MODELING OF HOSPITAL INFECTION CONTROL

In modeling infections in hospital settings, three frameworks are typically considered. The first is a model of indirect patient-to-healthcare worker (HCW) transmission. In this case, colonized patients are admitted into the hospital and interact with HCWs. The second is a model with direct patient-to-patient transmission between randomly mixing individuals. This model typically includes interaction between the hospital and the community. The third model is a meta-model that captures individuals who migrated between different institutions and the community. In this section, we present examples of the first two types. The first case will be to model the impact of hand washing on ***methicillin-resistant Staphylococcus aureus*** (MRSA) in ICUs. The second case will be to model the effectiveness of controls on the spread of tuberculosis in clinics. We first introduce some modeling basics.

Compartment Models

Epidemic models are used to describe the transmission of disease in a population. The SIR model is the simplest of these and classifies an individual as belonging to one of three groups in a closed system: $S(t)$—the number of individuals who are not infected at time t (susceptible); $I(t)$—the number of individuals infected at time t; and $R(t)$—the number of individuals who have been infected and removed from the system. If we assume that the flow goes from susceptible to infected to removed, and that the number of individuals in the population (N) is fixed, and that each individual has the same probability of infection (with rate = β) and probability of recovery/death (with rate = γ), then the transitions between groups can be defined the following system of differential equations:

$$\frac{dS}{dt} = \frac{-\beta S(t) I(t)}{N} \tag{13-1}$$

$$\frac{dI}{dt} = \frac{\beta S(t) I(t)}{N} - \gamma I(t) \tag{13-2}$$

$$\frac{dR}{dt} = \gamma I(t) \tag{13-3}$$

If we define $s(t) = S(t)/N$, $i(t) = i(t)/N$, and $r(t) = R(t)/N$, then we can apply Euler's method to obtain approximate solutions. Euler's method works as follows: Given a point (t_o, y_o) and a slope at that point

dy/dt (at $t = t_o$) $= f(t_o, y_o)$, then the tangent line at the point (t_o, y_o) is given by:

$$y = y_o + f(t_o, y_o)(t - t_o) \qquad (13\text{-}4)$$

We can then approximate the value of y_1, by substituting t_1 for t in (13-4). Applying Euler's method to (13-1) to (13-3) yields:

$$s_n = s_{n-1} - \beta s_{n-1} i_{n-1} \Delta t \qquad (13\text{-}5)$$

$$i_n = i_{n-1} + (\beta s_{n-1} i_{n-1} - \gamma i_{n-1}) \Delta t \qquad (13\text{-}6)$$

$$r_n = r_{n-1} + \gamma i_{n-1} \Delta t \qquad (13\text{-}7)$$

Example 13-1: Infection in a Ward

Consider a ward with 100 patients. At time 0, there are 98 susceptible patients and 2 infected patients with flu. In addition, it takes on average 45 days to recover from the flu ($\gamma = 1/5 = 0.2$) and the infection rate is $\beta = 1$. In this case, after one day, we can solve the system of equations and get s(1) = 0.9604, *i*(1) = 0.0356, and *r*(1) = 0.004, which corresponds to 96.04 susceptible, 3.56 infected, and 0.4 recovered patients. If we plot this over time, we get the figure below. Note that the peak of infection is at day 8.

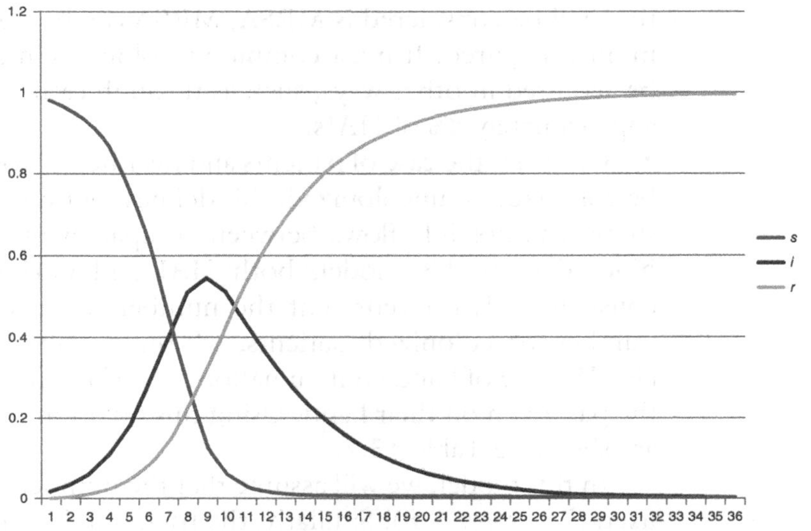

The SIR model is called a "compartment" model, where each patient class is a compartment, and we model the flows between compartments. We will extend this model to consider hand washing and the dynamics of HAIs in the next section. We first introduce the concept of the reproduction number.

The reproduction number R_0 is a measure of the number of infections that one case generates on average over the course of its infectious period. If $R_0 < 1$, then the infection will die out in the long run, but if $R_0 > 1$, then it can spread through the population (assuming the population is homogeneous). The greater the reproduction number, the harder it can be to control the epidemic (depending on treatments). Some example figures include $R_0 \subseteq [1.5, 2.5]$ for Ebola, $R_0 \subseteq [4, 7]$ for mumps, and $R_0 \subseteq [12, 18]$ for measles. From the SIR compartment model, R_0 is determined by:

$$R_0 = \frac{\beta N}{\gamma} \tag{13-8}$$

Note that in Example 13-1, the value of the reproduction number is $R_0 = 1(100)/0.2 = 500$. In this example, almost all patients in the ward were infected by day 13.

Compartment Model Applied to Hand Washing

The compartment model can be modified to consider the transmission dynamics of HAIs and the impact of hand washing. The particular HAI that will be considered is MRSA. MRSA can be hospital acquired or community acquired. It most commonly colonizes in the nostrils but can also be acquired in other ways, such as through catheters. MRSA accounts for approximately 8% of HAIs.

Consider the case of patients and healthcare workers, each of which can be colonized or uncolonized; this defines four compartments. Figure 13.3 shows the possible flows between compartments (Cooper et al., 1999). Note that in this model, both HAI and CAI versions of MRSA are considered. Let x represent the number of uncolonized patients, y the number of colonized patients, x' the number of healthcare workers (HCW) free of hand contamination, and y' the number of HCWs carrying the pathogen on their hand. Using this notation, the events and their rates are shown in Table 13.2.

In this model, we will assume that the number of patients in the ward is fixed (i.e., $x + y = n$). Similarly, the total number of carriers is $x' + y' = n'$.

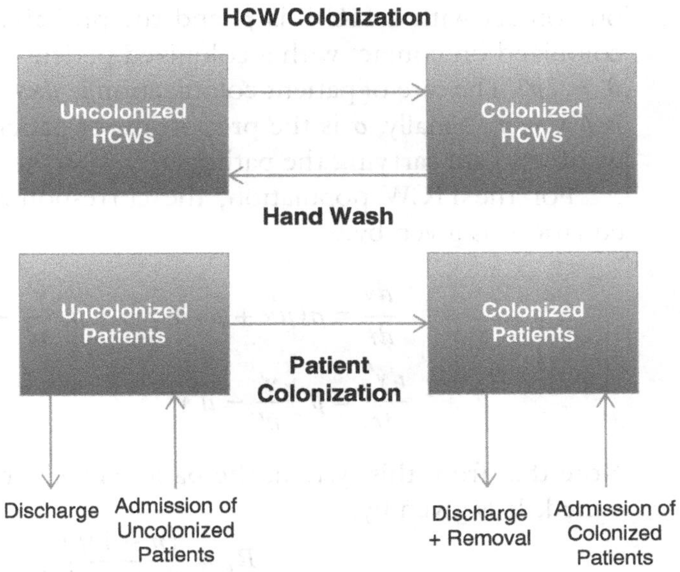

Figure 13.3 Flow Diagram of Compartment Model

Table 13.2 Events and Rates

Event	Rate of Event
Patient removal (when no colonization detected)	$\mu(x+y)$
Detection of colonized patient and removal	γy
HCW hand wash	$\mu'(x'+y')$
Removal of contamination from hands of HCW	$\mu' y'$
Carer-patient contact	$c(x+y)$
Carer-patient transmission	$\beta x y'/n'$
Patient-carer transmission	$\beta y x'/n'$
Admission of uncolonized patient	$(1-\sigma)(\mu x+\mu y+\gamma y)$
Admission of colonized patient	$\sigma(\mu x+\mu y+\gamma y)$

The removal rate (death, discharge, or transfer to another ward) for patients known to be carrying MRSA is μ. The detection rate of patients colonized with the pathogen is γ. This then represents an additional removal rate for colonized patients. The hand-washing rate (which is assumed to clear the HCW of colonization) is μ'. The mean number of contacts required by a patient each day is c. The probability that a patient becomes colonized on contact with a HCW is p, and the probability that a HCW becomes colonized on contact with a colonized patient is p'. We define $\beta = cp$ and $\beta' = cp'$. The rate of patient colonization is $\beta xy'/n'$ and carer colonization is $\beta' x' y/n'$. Finally, σ is the proportion of patients newly admitted to the ward who are carrying the pathogen.

For the HCW population, the corresponding system of differential equations is given by:

$$\frac{dy}{dt} = \sigma(\mu x + \mu y + \gamma y) + \beta x \frac{y'}{n'} - (\mu + \gamma)y \tag{13-9}$$

$$\frac{dy'}{dt} = \beta' y \frac{x'}{n'} - \mu' y' \tag{13-10}$$

Note that from this system, the basic reproduction number can be determined. It is given by:

$$R_o = \frac{(n-1)\beta\beta'}{(\mu + \gamma)n'\mu'} \tag{13-11}$$

Using the initial parameters given in Table 13.3 (Cooper et al., 1999), the effect of hand washing can be determined assuming that the HCW is homogeneous. Applying Euler's method to (13-9) and (13-10), we can determine that once the hand-washing rate exceeds roughly 20%, a dramatic decrease in ward level prevalence ensues. In this model, once the rate hits roughly 40%, the prevalence reaches close to a minimum and doesn't change thereafter. This is in large part due to the assumption that hand washing was 100% effective in removing the pathogen.

Tuberculosis Infection Control

Tuberculosis (TB) is caused by *Mycobacterium tuberculosis* (MTB), and usually attacks the lungs, being spread from one person to another through the air.[3] Approximately one-third of the world's population is infected with TB bacteria. It is one of the greatest killers worldwide from a single infectious agent. In 2010, 1.4 million died from TB. The largest number of new TB cases occurred in Asia, accounting for 60% of new cases globally. Sub-Saharan Africa, however, carried the greatest disease rate, with over 270 cases per 100,000 (Yan, 2014).

---

[3] Material in the section is based on the work of Yan (2014).

The Centers for Disease Control and Prevention (CDC) published the *Guidelines for Preventing the Transmission of* Mycobacterium tuberculosis *in Healthcare Facilities.* According to these guidelines, all healthcare settings should have an infection-control program designed with an emphasis on prompt detection, airborne precautions, and treatment of persons with suspected or confirmed TB disease. Such policies are of particular importance in clinical settings with medium to high risk, where the TB control policies should be developed and reviewed periodically, and evaluated for effectiveness to minimize the risk of TB transmission.

The wide implementation of the measures has shown success as evidenced by a decrease in the number of TB outbreaks in healthcare settings and a continuous reduction in healthcare-related transmission of MTB to patients and healthcare workers (HCW) (CDC, 2015). Following the changes in epidemiology, the CDC updated the TB infection-control guidelines for healthcare settings (CDC, 2015(CDC, 2005). In the report, emphasis is placed on actions to eliminate the threat to HCWs, who are susceptible to patients with unsuspected and undiagnosed infectious TB disease.

The CDC guidelines recommend a three-level hierarchy of control measures based on a risk assessment classification of healthcare facilities. Those three levels are administrative, environmental, and respiratory-protection control. The primary administrative measures include conducting a TB risk assessment of the setting, implementing effective work practices for the management of patients with suspected or confirmed TB disease, training and educating HCWs regarding TB, and screening and evaluating HCWs who are at risk for TB disease or infection. Environmental controls prevent the spread and reduce the concentration of infectious droplet nuclei in the air. Primary measures control the source of infection by using local exhaust ventilation and general ventilation. Secondary environmental controls manage the airflow to prevent contamination in areas adjacent to the source. Examples include an airborne infection isolation (AII) room, high-efficiency particulate air (HEPA) filtration, or ultraviolet germicidal irradiation (UVGI). Finally, respiratory control is accomplished through the use of respiratory protective equipment in situations where there is a high likelihood of exposure to MTB.

Analysis of MTB infection risks under different control strategies within healthcare settings is needed, including environmental control devices and respiration equipment utilized, screening program designed, as well as isolation strategy implemented, and so on. To achieve such goal, a mathematical model for infection risk estimation is required. With such a model, a personalized and cost-effective intervention strategy can be established based on the specific conditions of the healthcare setting.

Riley, Murphy, and Riley (1978) established a simple Poisson model for indoor airborne infection that provides a reasonable explanation for observed experimental data. The model includes the probability, R, that a susceptible individual breathes in one or more quanta of airborne infection and thus becomes infected:

$$R = 1 - e^{-D} = 1 - e^{\left(-I \cdot q \cdot p \cdot \frac{t}{Q}\right)} \tag{13-12}$$

A quantum is defined as the number of infectious airborne particles required to initiate infection, which may include one or more airborne particles. In Equation (13-12), I is the number of patients in the infectious stage, or infectors; q is quanta of airborne infection produced per infector; p is pulmonary ventilation rate; Q is room ventilation rate with germ-free air. Thus, $I \cdot q$ is the total quanta produced per unit time; and $I \cdot q/Q$ is the equilibrium concentration of quanta in the air at steady state. Therefore, the cumulative number of quanta (D) inhaled during the time of stay would be the concentration $I \cdot q/Q$ times the volume of air breathed in $p \cdot t$.

The total number of infection cases, C, appearing in the next generation is equal to the sum of the individual probabilities of infection over all susceptible patients, S. Hence,

$$C = \sum_{i=1}^{S} R_i \tag{13-13}$$

The model assumes:

- A steady-state concentration in the room, that is, a constant release rate of infectious quanta and constant ventilation rate, and the build-up period and biologic delay period was neglected.
- Complete air mixing, that is, quanta of infection are evenly dispersed throughout the air of a confined space.

- Equal susceptibility to a quantum of infection.
- No difference between individuals.

Respiration protection is the most commonly used environmental control measures in clinical settings. These include mechanical ventilation (MV), HEPA filtration, UVGI, surgical masks, and particulate respirator. The efficacy of these measures can be either expressed as an equivalent ventilation rate that alter the value of *Q*, or the quanta releasing rate *q*, or the fraction of the cumulative dose that is actually breathed in. In this section, we explain how these measures are incorporated into the model.

The efficacy of UVGI can be expressed as an equivalent ventilation rate in the lower room by:

$$\frac{1}{\Delta K_{Lw}} = \frac{1}{\dot{V}/V}\left(\frac{V_L}{V}\right) + \frac{1}{\Delta K_{Uw}}\left(\frac{V}{V_U}\right) \tag{13-14}$$

where $\dot{V}$ is volume of air coming from the upper room per unit of time, *VL* is the volume of the nonirradiated lower room, *VU* is the volume of the irradiated upper room, *V* is the total volume of the room. ΔK_{Lw} and ΔK_{Uw} are the air exchange rates for lower and upper room respectively. As the UVGI is installed in the upper room, ΔK_{Uw} can be expressed as:

$$\Delta K_{Uw} = 3600 \cdot Z \cdot IR \tag{13-15}$$

where *Z* is the susceptibility of UV and *IR* is the irradiance of UV. The equivalent increase in ventilation rate due to upper room UVGI, therefore, is defined by:

$$\Delta Q = \Delta K_{Lw} \cdot V_L \tag{13-16}$$

which alters the denominator of the risk equation (13-1). The MV system increases the ventilation rate in unit volume, *A*. HEPA supplementation further increases the effective ventilation by the factor *F*. These two systems alter the denominator of the risk equation (13-12) by:

$$Q' = A \cdot V \cdot F \tag{13-17}$$

Surgical masks and particulate respirators have different efficiencies, which differ by patient group. In addition, leakage depends on the face seal. For individuals infected with TB, masks and respirators are used for source and emitter control. They change the rate at which MTB quanta are released. This can be expressed as:

$$q' = q \cdot (1 - E_e^*) \tag{13-18}$$

$$E_e^* = E_e \cdot (1 - L) \tag{13-19}$$

where E_e^* is the effective "efficiency" of emitter control, which as a function of efficiency of the source control (E_e) and leakage rate (L) as expressed in Equation (13-19).

For susceptible individuals, the mask and respirator work in a similar way. They alter the effective pulmonary ventilation rate and therefore, adjust the effective number of MTB quanta being inhaled. The adjusted pulmonary ventilation rate can be expressed as:

$$p' = p \cdot (1 - E_s^*) \tag{13-20}$$

$$E_s^* = E_s \cdot (1 - L) \tag{13-21}$$

where E_s^* is the effective "Efficiency" of protection, which is as a function of efficiency of protection (E_s) and leakage rate (L) as presented in equation (13-21).

Note that when the same kind of protective equipment is used for individuals that infected with TB and those without, the effective efficiencies of the two groups expressed in Equations (13-19) and (13-21) have the same value. Based on the above discussion for the modifications of the risk model developed by Riley et al., the following model is proposed for the estimation of an individual's cumulative infection risk:

$$R = 1 - \exp(-D) = 1 - \exp\left(-\frac{I \cdot q \cdot p \cdot t \cdot \left(1 - E_e^*\right) \cdot (1 - E_s^*)}{A \cdot V \cdot F + \Delta K_{Lw} \cdot (w \cdot l \cdot h_{UV})}\right) \tag{13-22}$$

To this point, we have assumed complete air mixing in the room. We can use a sequential box model to divide the space into different segmentations to estimate the infection risk within each segment. The model assumes complete air mixing within each box, while movement of air with infectious particles could occur between adjacent boxes.

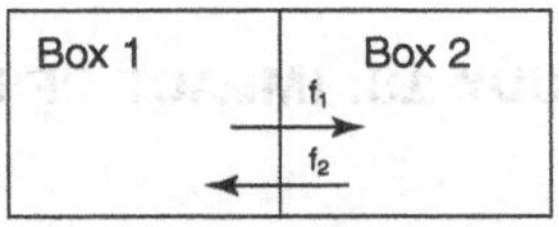

Figure 13.4 Sequential Box Model for the Two-Segmentation Case

First, consider the case where the space is divided equally into two boxes shown as Figure 13.4. In this case, infectious patients will always sit in box 1, while the susceptible patients will sit randomly in the two boxes. In each of the two boxes, the air is completely mixed. The fraction of air transfers between the two boxes per hour is f_i (m^3/hour). The concentration in each box, C_i, can be calculated as:

$$\left[f_1 + A, -\frac{f_2 \cdot V_2}{V_1} \right] \begin{bmatrix} C_1 \\ C_2 \end{bmatrix} = \begin{bmatrix} \frac{S_1}{V_1} \\ \frac{S_2}{V_2} \end{bmatrix} \tag{13-12}$$

where A is the air exchange rate per hour, and S_i is the strength of the infectious patient in quanta/hour.

This can be generalized to four boxes (Figure 13.5). Note that air transfer only happens between the adjacent two boxes. The calculation of concentration matrix, C_i, is expressed by:

$$\begin{bmatrix} 2f_1 + A & \frac{-f_2 \cdot V_2}{V_1} & \frac{-f_3 \cdot V_3}{V_1} & 0 \\ \frac{-f_1 \cdot V_1}{V_1} & 2f_2 + A & 0 & \frac{-f_4 \cdot V_4}{V_2} \\ \frac{-f_1 \cdot V_1}{V_3} & 0 & 2f_3 + A & \frac{-f_4 \cdot V_4}{V_3} \\ 0 & \frac{-f_2 \cdot V_2}{V_4} & \frac{-f_3 \cdot V_3}{V_4} & 2f_4 + A \end{bmatrix} \begin{bmatrix} C_1 \\ C_2 \\ C_3 \\ C_4 \end{bmatrix} = \begin{bmatrix} \frac{S_1}{V_1} \\ \frac{S_2}{V_2} \\ \frac{S_3}{V_3} \\ \frac{S_4}{V_4} \end{bmatrix} \tag{13-13}$$

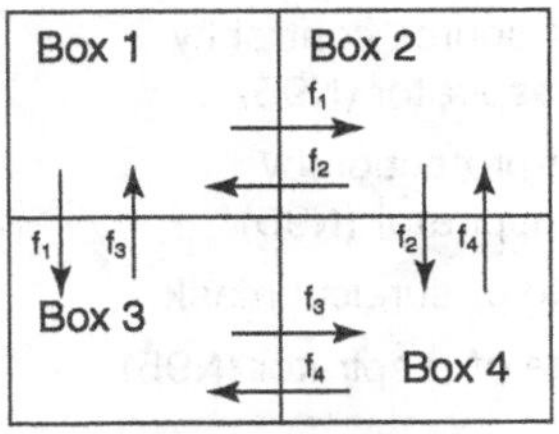

Figure 13.5 Sequential Box Model for the Four-Segmentation Case

CASE STUDY 13: IMPACT OF INTERVENTIONS ON TB INFECTION IN A CLINIC

Consider a community clinic that has on average 30 TB patient visits per year, with a general waiting room sized 6 × 6 × 3 m^3 (Yan, 2014). The variables for the infection risk model are provided in Table 13.3. The average number of susceptible patients who are not infected with TB before their visit is 30 per day. The actual number of patients, with and without TB disease, visiting each day follows a Poisson process. The length of stay for a visitor stays follows a gamma distribution, which arises naturally in processes for which the waiting times between Poisson distributed events are relevant. The variables and their values for the hypothesized waiting room are summarized in Table 13.4.

Table 13.3 Summary of Variables in the Infection Risk Model

Input	Index	Unit	Value
Breathing rate	p	m^3/hr	0.34
MTB quanta released per patient	q	qph	2, 13 or 108 with equal possibility
UV susceptibility	Z	$\mu W^{-1}s^{-1}cm^2$	1.2 to 3
UV irradiance	IR	$\mu W/cm^2$	partial operation:12; fully operation: 42
Height of UV installation	h_{uv}	m	2
Air exchange rate in lower room due to UVGI	K_{Lw}	hr^{-1}	Eq. (3.3)
Height of UV installation	h_{UV}	m	2
Air exchange rate by mechanical ventilation	A	hr^{-1}	6
HEPA increased rate of ACH	F	/	3
Efficiency of source control by wearing surgical mask	E_e^m	%	58%
Efficiency of protection by wearing surgical mask	E_s^m	%	58%
Efficiency of source control by wearing respirator (N95)	E_e^r	%	95%
Efficiency of protection by wearing respirator (N95)	E_e^r	%	95%
Leakage rate of surgical mask	L^m	%	20%
Leakage rate of respirator (N95)	L^r	%	10%

Table 13.4 Summary Variables for the Example Clinical Waiting Room

Parameter	Index	Unit	Value
# of TB patient visits per year	I_0	People/day	Poisson (30/365)
Room width	w	m	6
Room length	l	m	6
Room height	h	m	3
# of susceptible patients	S	People/day	Poisson (20)
Estimated patients exposure time	t	min	Gamma (0.71, 16.94)

Thirty HCWs were assigned to the waiting room, each of whom works 8 hours per day, 219 days per year. Initially, none of the HCWs are infected or have active TB. Once an HCW is detected to be positive for TB through a screening program, she would be removed from her position and replaced by a new HCW.

As mentioned previously, infected HCWs are not infectious to others until active TB disease is developed. Studies revealed that after infection, 10% of individuals would develop TB disease over a lifetime, 50% of these individuals develop TB in one year following infection, and 80% within two years following infection. Therefore, the daily risk of developing TB disease in the first year following infection is approximately 1.4×10^{-4}, and in the second year following infection is approximately 8.8×10^{-5}.

Using this information, we may conduct a dynamic simulation on the occupational TB infection. Due to the delay of detection of active TB disease, the diseased HCWs will become the secondary source of infection and transmit MTB to other HCWs and patients. The time of delay depends on the screening program the hospital designed. The simulation process for each hospital day is as follows:

- For each day, the number of susceptible patients, number of TB patients, waiting time for each patient, and whether or not each HCW is on duty that day is generated based on the appropriate distribution. Both TB patients and the diseased HCWs are potential sources of infection for exposed individuals. Source strength for each of them is approximately 2 qph for chemotherapy-treated TB, 13 qph for active TB, and 108 qph for highly infectious TB.
- The cumulative daily risk (R) for each patient and HCW is then calculated. A uniformly distributed random value (U_R) between 0 and 1 is then generated. If $U_R \leq R$, the individual is infected. HCWs can only be infected once during the two years of simulation.

- For infected HCWs, a uniform distributed random value (U_{TB}) between 0 and 1 is generated. During the first year after infection, if $U_{TB} \leq 1.4 \times 10^{-4}$, the HCW will become infected with TB; during the second year, if $U_{TB} \leq 8.8 \times 10^{-5}$, the HCW will then develop TB.
- The HCWs who have active TB disease will remain on job until the disease is detected by the TB screening program. According to the CDC, the frequency of screening is annual for medium risk settings, and 8 to 10 weeks for TB transmission ongoing settings.
- The above steps are repeated each day over a period of two years, and a total of 1,000 iterations were conducted.

We examine the impact all possible combination of the intervention devices and equipment, without adopting any screening program or isolation strategy. The interventions considered and results are presented in Table 13.5 and Figure 13.6. Note that the annual cost is calculated by the sum of installation (discounted over 10 years), maintenance cost per year, and mask and respirator cost per year.

Table 13.5 Different Intervention Options and Results

Scenario	Intervention Options						Equivalent ACH	Annual Infection Cases	
	MV (ACH)	HEPA	UV	Surgical Mask	Surgical Mask	N95		Patients	Health-care Workers
P1	6				√		6	1.3 ± 0.6	7.4 ± 2.1
P2	3	√			√		9	0.9 ± 0.4	5.1 ± 1.4
P3			√		√		10	0.8 ± 0.4	4.5 ± 1.3
P4	6			√	√		6	0.4 ± 0.2	4.1 ± 1.2
P5	3		√		√		11	0.7 ± 0.3	4.1 ± 1.2
P6			√	√	√		10	0.2 ± 0.1	2.5 ± 0.7
P7	3		√	√	√		11	0.2 ± 0.1	2.3 ± 0.7
P8	3	√		√	√		9	0.2 ± 0.1	2.8 ± 0.8
P9	6					√	6	1.3 ± 0.6	2.2 ± 0.6
P10	3	√				√	9	0.9 ± 0.4	1.5 ± 0.4
P11			√			√	10	0.8 ± 0.3	1.3 ± 0.4
P12	6			√		√	6	0.4 ± 0.2	1.2 ± 0.3
P13	3		√			√	11	0.7 ± 0.3	1.2 ± 0.3
P14			√	√		√	10	0.2 ± 0.1	0.7 ± 0.2
P15	3		√	√		√	11	0.2 ± 0.1	0.6 ± 0.2
P16	3	√		√		√	9	0.2 ± 0.1	0.8 ± 0.2

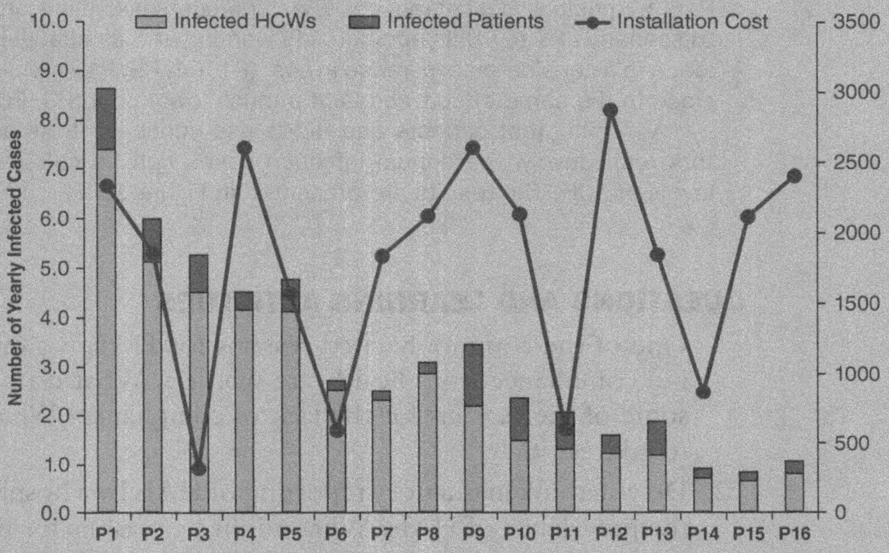

Figure 13.6 Number of Infection Cases and Cost of Different Intervention Options

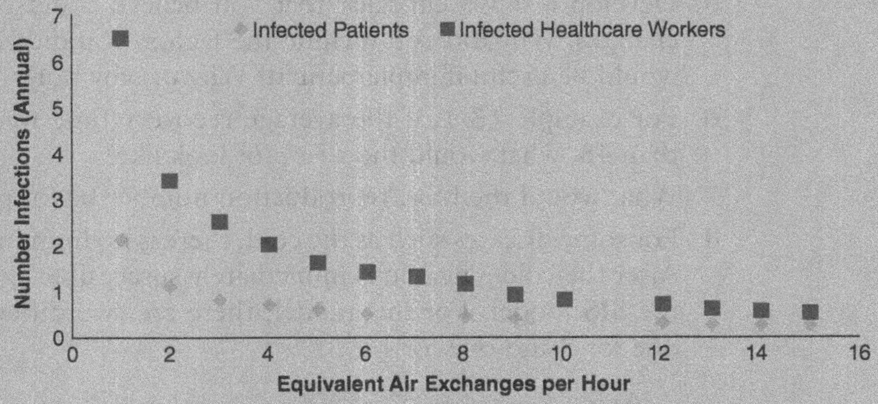

Figure 13.7 Annual Infections as a Function of Equivalent Air Exchange

Requiring all HCWs to wear a particulate respirator, which is scenario P9 to P16, surprisingly did not reduce HCWs' infection risk significantly as compared to scenarios P1 to P8. In addition, MV working at 6 air changes per hour (ACH) leads to a considerable increase in cost (P1, P4, P9, P12), while UVGI achieves close to the same effectiveness at a much lower cost (P3, P6, P11, P14).

Assuming that patients and HCWs use surgical masks and N95 respirators respectively, the annual infection cases can be calculated at different levels of ACH. The results are presented in Figure 13.7.

QUESTIONS AND LEARNING ACTIVITIES

1. One of the commonly cited reasons for HAIs is a lack of handwashing compliance from healthcare workers. What do you think may be some of the reasons for the lack of compliance? What could be done to address it?
2. Do you think mandatory reporting of HAIs by a hospital helps or hurts the prevalence of HAIs? What might be reasons for both?
3. Consider technological interventions that could help with handwashing?
4. What is the role of the EHR in reducing HAIs?
5. Develop a sepsis checklist that you believe could be useful for shift changes. Why did you include the factors you did? Do you believe it would be useful if implemented? Why or why not?
6. For example 13-1, if the average recovery time were 19 days rather than 45, what would the *s-i-r* plot look like?
7. What would the basic reproduction number be for problem 6?
8. For some diseases such as the cold, there is no immunity that develops. After the cold a person is immediately susceptible again. This is called the SIS model. For this model, there are two differential equations, one for S and one for I.

   a. Write out the model.
   b. Use Euler's method to obtain the approximate solutions.

9. Explain how you would parameterize the handwashing compartment model presented. Also, apply Euler's method to (13-9) and (13-10) to obtain the approximate solution.

10. Consider the model by Riley et al. (13-12) and (13-13). There is a room that contains 1 infected individual and 99 uninfected, and there is uniform mixing of the air in the room. The breathing rate of each person is 0.35 cubic meters per hour, an infected patient releases 3 quanta per hour, and the room ventilation rate is 15 cubic meters per hour. On average, how many people would be infected after 4 hours if no one left the room?
11. Consider the case where each person wears a mask that is 93% effective. How would your answer to 10 change?
12. Suppose in Table 13.3, the air exchange rate were 10, and all other data remained the same. Determine the cumulative risk for a healthcare worker on an 8-hour shift that doesn't wear a mask.
13. Repeat problem 12 for the case that a healthcare work wears a mask.
14. What are some of the technical challenges of using the interventions discussed in the TB example?
15. We discussed the notion of genome sequencing in personalize medicine in chapter 1. How might genome sequencing be useful in infection control?

REFERENCES

Centers for Disease Control and Prevention (2015). Top CDC recommendations to prevent hospital-associated infections. Retrieved from http://www.cdc.gov/HAI/pdfs/hai/top-cdc-recs-factsheet.pdf

Cooper, B. S., Medley, G. F., & Scott, G. M. (1999). Preliminary analysis of the transmission dynamics of nosocomial infections: stochastic and management effects. *Journal of Hospital Infection. 43*(2), 131–147.

Ducel, G., Fabry, J., & Nicolle, L. (2002). *Prevention of hospital-acquired infections: A practical guide* (2nd ed.). Geneva, Switzerland: World Health Organization.

Gawande, A. (2007, December 10). The checklist: If something so simple can transform intensive care, what else can it do? *The New Yorker*, p. 86.

Pronovost, P., Needham, D., Berenholtz, S., Sinopoli, D., Chu, H., Cosgrove, S., … Goeschel, C. (2006). An intervention to decrease catheter-related bloodstream infections in the ICU. *New England Journal of Medicine, 355*(26), 2725–2732.

Riley, E. C., Murphy, G., & Riley, R. L. (1978). Airborne spread of measles in a suburban elementary school. *American Journal of Epidemiology*, *107*(5), 421–432.

Yan, R. (2014). Cost-effective Interventions—Innovative Strategies for Public Health Care. Pennsylvania State University, Ph.D. Thesis, University Park, PA: Pennsylvania State University.

Index

3Ms of waste, 146
80–20 rule, 181, 182

A

A3, 147, 150, 152–154, 156, 157
Accessibility, 1, 18
Accountable care organization, 97
Acquisition, 114, 251, 253
Activity-based costing (ABC), 99–101, 161
Acute care, 60, 62, 64–66, 69, 71, 74, 86, 139, 222, 383, 385
Administration, 14, 17, 61, 98, 99, 109, 129, 138, 140, 171, 181, 212, 220–222, 279, 337, 383, 385, 387
Admission and discharge, 65, 72
Admissions, 59, 71, 72, 80, 301
Adverse drug event, 127, 220
Adverse patient outcomes, 184, 219, 220, 235
Adverse selection, 92, 93, 95, 97
Aging, 13, 18, 56, 323
Agency for healthcare research and quality (AHRQ), 6, 180, 222
Alzheimer's, 251, 268, 269
Ambulance, 4, 68, 69, 305, 327
Ambulatory care, 56–59, 128, 135
American academy of emergency medicine, 80
American college of emergency physicians, 68
American hospital association, 66, 103
American recovery and reimbursement act, 123, 125, 128
Analysis of variance (ANOVA), 176, 188, 192
Andon, 147
Anesthesia, 59–61, 138, 158, 159, 171, 211
Anesthesiologists, 60, 315
Anomaly detection, 248
Antibiotics, 179, 385, 387
Assessment, 33, 41, 74, 75, 78, 148, 164, 165, 175, 179, 210, 211, 215, 218, 224, 225, 228, 231, 234, 236, 271, 279, 393
Assignable-cause variability, 187

Attribute control charts, 187, 188
Automate, 18, 120, 122, 124, 137–139, 387

B

Balancing loop, 38, 39, 41, 198
Banker-Charnes-Cooper, 280, 282, 283, 285–287
Bed management, 81, 297, 299–301, 386
Billing, 19, 95, 97, 98, 101, 115, 118, 120, 125, 126, 135, 171, 182
Binary logistic regression, 252, 254
Bioinformatics, 111, 112
Birth-death diagram, 306
Block time, 55, 56, 60, 69, 310, 311, 322
Block scheduling, 69, 310–312, 314, 315
Blood pressure, 21, 31, 115, 131, 255, 266, 268–270, 300
Boarding, 59, 65, 67–70, 72, 73, 301, 304, 305
Body mass index (BMI), 21, 268
Bottlenecks, 75, 76, 272
Box-Behnkin, 187
Boxplot, 185, 186
Bullwhip effect, 372–374
Bundled payments, 96, 101

C

C-logit models, 335
Cancellations, 53, 192, 301, 315
Capacity management, 297, 298, 300, 302, 304, 306, 308, 310, 312, 314, 316, 321
Capital budgeting, 103, 105
Capitation, 56, 97
Care coordination, 40, 54, 88, 124, 126, 128, 129, 140
Care quality, 124, 133, 180, 188
Care transition, 53, 63, 71–75, 236
Caregivers, 6, 7, 18, 65, 74, 144, 154, 197
Care management (CM), 41, 42, 45, 46, 48, 212, 213, 243, 322, 349
Cass theory, 30
Catheter-associated urinary tract infection (CAUTI), 134, 382, 383
Causal loop diagram (CLD), 38–45, 48, 49
Causal loop maps, 36
Cause and effect, 25, 27, 28, 35, 121, 175, 176, 180, 192
Cause of death, 218, 385
Cause-and-effect diagram, 176, 180, 192
Cause-and-effect matrix, 180
Cause-specific mortality, 219, 239
Centers for disease control and prevention (CDC), 13, 15, 68, 135, 190, 191, 195, 222–224, 271, 353, 380, 383, 393, 400
Center for integration of medicine and innovative technology (CIMIT), 292, 293
Centers for medicare & medicaid services (CMS), 14, 19, 68, 94, 118, 129, 130, 133–136, 145, 195, 223, 381
Central line–associated bloodstream infections (CLABSI), 96, 134, 381, 383
Central sterile services department (CSSD), 166
Central supply department (CSD), 166
Certificate of need (CON), 299
Certification and survey provider enhanced reporting (CASPER), 135
Changeover reduction, 163
Changeover times, 163
Charges, 97, 99, 101, 135, 136, 366
Charnes-cooper-rhodes (CCR), 280, 283, 284
Charter, 175
Checklist, 18, 61, 66, 185, 186, 190, 236, 379, 383, 384, 386, 387
Choices of care, 14
Choices of delivery, 366

Chronic care, 59, 126
Chronic conditions, 3, 11, 13–16, 101, 267
Chronic kidney disease (CKD), 31–34, 40–43, 45–48, 50
Claims data, 131, 135, 250
Claims submission, 124, 125
Clarke-Wright savings algorithm (CWSA), 345, 347, 348
Class/output variable, 252
Classification, 247, 248, 252, 254
Clinical and translational science award (CTSA), 34, 205
Clinical and translational science institute (CTSI), 208–210
Clinical data warehouse (CDW), 120
Clinical decision support (CDS), 125, 126, 128, 139, 224, 231
Clinical decision support system (CDSS), 127, 128, 231, 232, 243
Clinical trial, 14, 21, 22, 187, 202, 212, 250
Clinical workflow, 53, 57, 59, 61, 63, 65, 127
Clinical workflow processes, 53, 65
Closed-loop thinking, 35, 37
Cluster analysis, 248
Clustering algorithm, 207, 248
Clusters, 248, 276, 333
Co-morbidities, 135
Codes, 5, 18, 118, 119, 133, 135, 163, 173
Coevolution, 30
Cognitive biases, 217, 230, 231
Cognitive control, 218
Cognitive function model (CFM), 230
Cognitive reliability and error analysis method (CREAM), 218
Cognitive task analysis (CTA), 230, 231, 237, 238
Collaboration, 6, 29, 30, 134, 201, 205, 207–209, 272, 277, 372
Collaborative planning, forecasting, and replenishment (CPFR), 372
Color coding, 163, 173, 225, 226, 268
Command-and-control, 150
Commonwealth fund, 8–10, 22, 84
Community-acquired infection, 382, 390
Comorbidities, 41, 96, 223
Comparative effectiveness research (CER), 7, 145, 168
Comparative experiments, 176, 185
Compartment model, 388–390
Complex adaptive system, 25, 28–31, 33, 35, 48, 198
Compliance, 41, 128, 138, 190, 271, 383, 386
Computed tomography (CT), 58, 251, 316
Computer-based modeling, 25
Computerized medical record (CMR), 122
Computerized patient record (CPR), 122
Computerized physician order entry (CPOE), 67, 124, 126–128, 137, 140, 215, 221, 263
Conceptual model of situation awareness, 228
Continuity of care, 63, 65, 131, 234
Contract carrier, 364, 366, 367
Control charts, 171, 176, 187–190, 192, 194, 197
Control plan, 190
Convexity constraint, 283, 284
Coordinated care, 10, 11, 19, 20
Cost shifting, 102, 103
Cost-of-care, 97
Cost-shift theory, 96, 109
Crisis, 111, 217
Critical cardiac arrest risk triage, 17
Critical care settings, 231, 385
Critical decision method (CDM), 230, 231
Critical to quality (CTQ), 171, 175–178, 190, 192

Cross-institutional collaboration network, 205, 207
Cross industry standard process for data mining (CRISP-DM), 246
CTQ flowdown, 176–178, 192
Cycle stock, 357, 370
Cycle time, 149, 182

D

Data, 111–122, 124–138, 245–251, 253, 255–263, 265, 267, 269, 271, 280, 282, 283, 285, 286, 293
Data envelopment analysis (DEA), 280
Data mining, 117, 126, 245–247, 249–254, 258, 293
Data visualization (DV), 255–257, 259, 293
Dea, 280, 283, 284, 286, 293
Decision-making units (DMU), 280–283, 287
Defects per million opportunities (DPMO), 170–173, 176, 211
Define, measure, analyze, design, verify (DMADV), 175
Define, measure, analyze, improve, control (DMAIC), 169, 175–177, 185, 188, 192, 193, 195, 197, 199, 210, 216
Departure time, 342, 347
Design for six sigma (DFSS), 175
Design of experiments (DOE), 176, 180, 187, 200
Diabetes, 3, 11, 13, 15, 31–33, 40, 49
Diagnosis-related group (DRG), 96, 101, 135, 136
Discharge process, 65, 72, 154, 171, 178, 179, 193, 194
Discharge time, 66, 193, 194, 305
Display, 114, 115, 157, 255, 262, 263, 264
Diversion, 68, 69, 300, 301, 305, 306, 307

E

Economic order quantity (EOQ), 360, 361
Ehealth, 132
Electronic clinical information system (ECIS), 122
Electronic health record (EHR), 18, 19, 22, 112, 120–130, 249
Electronic medical record (EMR), 18, 57, 66, 67, 112, 115, 118, 120, 121, 128, 136–138, 157, 250, 263
Emergence, 28, 30, 257
Emergency care, 59
Emergency department (ED), 20, 59, 62, 65–69, 72–74, 80–82, 134, 143, 150, 152, 177, 178, 183, 184, 188, 189, 298, 301, 305, 352, 387
Equity modeling, 335

F

Failure mode and effects analysis (FEMA), 176, 182, 200, 234
Fall-prone, 223–225
Falls, 18, 217, 222–226
Fault tree analysis (FTA), 234
Federally qualified health center (FQHC), 336–341
Fee-for-service (FFS), 95, 97, 101, 102
Feedback perspectives, 37
Fishbone diagram, 180, 181
Five S(5S), 157
Flowchart, 33, 76, 77, 148, 150
Fmea, 176, 182–184, 192, 207, 208
Follow-up care, 71, 74, 138
Food and drug administration (FDA), 14, 21
Forecasting, 247, 255, 351–361, 364, 372
Forrester's view, 36
Free on board (FOB), 367, 368
Free-standing centers, 59

Full costing, 99
Full truckload (FTL), 364, 366, 367
Full-time equivalents (FTE), 283, 285

G

General care units (GCU), 299, 308–310
Global burden of disease study, 218, 219
Global health, 12, 13, 23, 118
Gloves, 190, 384
Government-provided insurance, 91, 94
Goal programming (GP), 208–210, 288–292, 335
Grand challenges in global health, 12
Graph theory, 272
Gross domestic product (GDP), 8, 9, 96
Gross national product (GNP), 16
Group model building, 36, 37
Graphical user interface (GUI), 139

H

Health informatics, 111, 113–116, 121, 136–139
Health information organization (HIO), 126, 131, 132
Health information service provider (HISP), 131–133
Health information technology (HIT), 121, 122, 231, 278
Health insurance, 7, 9, 22, 92, 94, 135, 332
Health organization transformation, 197–199
Health policies, 1, 2, 18
Health professional shortage area (HPSA), 337
Health system, 1, 2, 58, 59, 75, 123, 125, 138, 197
Healthcare delivery system, 1–6, 14, 16, 18, 20, 22, 26, 43, 45, 116, 125, 227, 232, 250, 351
Healthcare Information and Management Systems Society (HIMSS), 19, 110
Healthcare information exchange (HIE), 118, 129–131, 140
Health information technology for economic and clinical health Act (HITECH Act), 18
Hendrich, 224
High-efficiency particulate air (HEPA), 393, 395, 398, 400
Hinshitsu kanri, 194, 195
Hitech, 18
Home healthcare (HHC), 323, 341, 342, 344, 345, 347
Hospital acquired infections (HAI), 18, 134, 379, 381, 382, 390
Hospital-acquired conditions (HAC), 223
Hospital-associated infections, 145, 379,
Human error, 217, 218
Human reliability analysis (HRA), 218, 236
Hygiene, 190, 191, 380, 383

I

Icon array, 260
IHI triple aim, 116
Imaging, 58, 104, 105, 123, 126, 134, 247, 298, 316, 387
Infection control, 388, 392, 393
Informatics, 111–116, 118, 120, 122, 124, 126, 128, 130, 132, 136–139
Information sharing, 73, 140, 305, 369, 371, 372
Information technology, 19–21, 73, 75, 112, 235, 236
Inpatient care, 62, 63, 68, 96, 128, 135, 137

Inpatient prospective payment system (IPPS), 135, 136
Institute of medicine (IOM), 17, 18, 25, 27, 123, 124, 145, 218
Intensive care unit (ICU), 17, 20, 62, 63, 69, 72, 138, 237, 277, 297–305, 307–310, 383–388
Intensivist, 62, 63
International classification of diseases (ICD), 118, 119, 133
International classification of primary care (ICPC), 2, 5
Interoperability, 19, 118–120, 122, 130
Inventory control, 147, 163, 351, 357–363, 368
Ishikawa diagram, 180

K

Kaizen kanri, 194, 195
Kanban, 147, 160–164
Keystone initiative, 383
Kidney disease outcomes quality initiative (KDOQI), 32, 33, 40, 41
Knowledge discovery in databases (KDD), 246, 247, 249

L

Law of the vital few, 181
Lean, 141–148, 150, 152, 154, 156, 158, 160, 162–169, 176, 177, 185, 191, 193, 194, 212, 213, 215
Lean six sigma (LSS), 169, 176, 177
Length-of-stay (LOS), 192, 301, 303, 304
Linear perspective, 37
Long-term acute care hospitals (LTACH), 64
Long-term care, 5, 63, 64, 222
Long-term care facility, 26, 62, 64, 73, 226

M

Manipulation, 113, 115
Maximal covering location problem (MCLP), 328
Meaningful use (MU), 19, 118, 128–130, 138, 139
Medicaid, 94, 95, 102, 103, 129, 130, 337
Medical errors, 19, 70, 71, 73, 74, 125, 127, 145, 146, 217–223, 227, 230, 236, 385
Medical informatics, 111, 112
Medical knowledge, 110, 259
Medical specialties and subspecialties, 111
Medicare, 14, 94, 95, 97, 102, 103, 129, 130, 134–136
Medicare severity diagnosis-related group (MS-DRG), 136
Medication error, 17, 180, 181, 220–222, 386
Medication reconciliation, 138, 220, 221
Military health system (MHS), 285, 286
Moral hazard, 92, 93, 95
Morse fall scale (MFS), 224
Mortality rate, 9, 17, 48, 69, 301, 307, 385
Muda, 146, 147
Multiple-criteria mathematical programming (MCMP), 288, 289, 293
Multiple-criteria decision making (MCDM), 232, 288, 289, 291
Multiple-criteria selection problems (MCSP), 288, 293
Mura, 146, 147
Muri, 146, 147

N

National health and nutrition examination survey (NHANES), 32, 271, 338
National healthcare safety network (NHSN), 135
Nationwide health information network (NHIN), 131–133
Neonatal intensive care unit (NICU), 231, 300
Net present value (NPV), 103–106
No care management (NCM), 45–48
Non-dominated (efficient) solution, 289

O

Observational, 121, 236, 242
Observational clinical human reliability assessment (OCHRA), 236
Ohno, Taichii, 141–143, 146, 160, 167
Operating room care, 59, 60
Or scheduling, 65, 69, 70
Ordering, 125, 127, 128, 162, 221, 345, 351, 358, 360, 364, 368, 372
Ordinary least squares (OLS), 283, 284, 287
Organization for economic cooperation and development (OECD), 6, 8–10
Outcome-dependent type, 219
Outpatient surgical care, 59

P

Pareto, 176, 181, 182, 192
Pareto-koopman's efficient, 282
Patient flow, 45, 53, 58, 60, 62, 64–76, 78, 80–82
Patient transfer, 72, 385, 387
Patient-centered care, 2, 10, 65, 235, 250
Patient-centered outcomes research institute (PCORI), 34
Pay-for-performance, 96, 286
People-centered care, 2
Personal health record (PHR), 131
Personal medical records (PMR), 122
Personalized medicine, 21, 22
Physician-directed queuing (PDQ), 67, 82
Plan-do-check-act (PDCA), 147, 150
Poisson model, 301, 306, 308, 320, 394
Poka-yoke, 147
Policy resistance, 38
Patient, Population, Team, organizational, Network, Environment (PPTONE), 26, 198
Practice management system, 19
Precision medicine, 21
Preprocessing, 247, 249, 251, 253
Preventable adverse events, 220
Primary care physician (PCP), 3, 4, 33, 40–42, 57–59, 63, 66, 74, 113, 126, 179, 270, 332
Principle of factor scarcity, 181
Probabilistic risk assessment (PRA), 218, 234
Problem oriented medical information system (POMIS), 123
Process capability, 171
Process mapping, 53, 60, 75–78, 176
Process-dependent, 219, 220
Production kanbans, 161, 162

Q

Quality function deployment (QFD), 201, 202

Quality improvement (QI), 147, 148, 169, 177, 200, 205, 277
Queueing, 301, 302

R

Readmission, 71, 74, 134, 185, 186, 194, 235, 307
Reciprocal allocation method, 98, 100
Regenstreif medical record system (RMRS), 123
Regression, 247, 248
Reimbursement, 40, 95, 96, 102, 103, 107, 123, 223, 338
Reinforcing loop, 38, 39, 198
Roemer's law, 299
Root cause analysis, 147, 200, 207, 208

S

Safety, 1, 17, 25, 53, 61, 65, 72, 74, 115, 121, 125, 127–129, 137–139, 141, 144, 157, 164, 217, 218, 220–222, 224, 226–236, 370, 371, 387
Situation awareness global assessment technique (SAGAT), 229
Sakichi, Toyoda, 142
Sbar, 236
Secondary data, 121
Seiketsu:: standardize, 157
Seiri :: sort, 157
Seiso :: shine, 157
Seiton :: systematic arrangement :: set in order, 157
Sense-making framework, 27
Sepsis, 17, 179, 180, 231, 383–387
Shared savings plans (SSP), 97
Shitsuke :: sustain, 157
Sigma level, 171, 211
Situation awareness rating technique (SART), 229, 242
Six sigma, 169–183, 185–202, 205, 207–210
Social network analysis (SNA), 205, 207, 208, 271, 272, 274, 277–279
Strategic management, 194, 195
Supplier, inputs, process, outputs, and customer (SIPOC), 176, 178
Supply chain, 162, 327, 351, 352, 354, 356–358, 360, 362–364, 366, 368–372, 374, 375
Support vector machines (SVM), 48, 73, 252, 254
Surgical errors, 234
Systems dynamics, 25, 36, 43
Systems thinking, 25, 27, 28, 34–37, 39–41, 43, 45, 47, 48, 198

T

Technique for human error rate prediction (THERP), 218
Telehealth, 20
Telemetry units, 300
Time series models, 352, 353
Toyota production system (TPS), 141, 147, 150, 152

U

Universal healthcare, 94, 95

V

Vaccines, 12, 363, 369, 376, 385
Value, 104, 141, 143, 144, 146, 147, 164
Value stream map (VSM), 147–152, 154
Variable control charts, 187, 188
Vendor-managed inventory (VMI), 368, 374

Visual factory, 157–159
Voice of the customer (VOC), 143, 175, 177, 202, 210

W

Waikato environment for knowledge analysis (WEKA), 253
Waste, 30, 31, 76, 141–148, 151, 157, 162, 166, 176, 177, 185
Withdrawal kanban, 161–163
World health organization (WHO), 1, 2, 22, 61, 118, 381